Breeding and Crop Production

Breeding and Crop Production

Dr. H. Padamalatha

RANDOM

RANDOM PUBLICATIONS

NEW DELHI (INDIA)

Breeding and Crop Production

ISBN 978-93-5111-324-9

Published in 2014 in India by

RANDOM PUBLICATIONS

4376-A/4B, Gali Murari Lal, Ansari Road
New Delhi-110 002
Phone : +91-11-43580356, +91-11-23289044
e-mail: randomexports@gmail.com, sales@randompublications.com, info@randompublications.com

Reprint 2021

Type Setting by : Keystoneprintads, Delhi-110051
Printed at : Replika Press Pvt. Ltd.

Preface

Plant Breeding encompasses a broad array of activities. It can be viewed as a pipeline beginning with initial steps such as pre-breeding, etc. and continuing to development of improved germplasm to be used either for further breeding at other venues, for screening in national or other programmes, or for direct use.

"Plant Bredding" has been used to describe many types of laboratory activities. Some involve primarily tissue culture techniques such as virus elimination, micropropagation, regeneration from explants, protoplast fusion, etc. Others involve molecular biology. Two major areas of plant molecular breeding are analysis of the nucleic acids or proteins in plants/pathogens and production of transgenic plants. Management and retrieval of the data obtained from nucleic acid/protein analysis (bioinformatics) is an increasingly important additional area.

OILSEED crops have been the backbone of agricultural economy of India from time immemorial. Today these crops are cultivated on about 16.5 million hectares, with total production of 10 million tonnes. This area constitutes approximately one-tenth of the total cultivated area in India. On the oilseed map of the world, India occupies a prominent position, both in regard to acreage and production.

The important oilseed crops grown in this country in order of importance are groundnut, rapeseed and mustard, sesame, linseed, safflower, castor, sunflower and niger. The crop is grown both in subtropical and tropical countries. In Asia, it is chiefly grown in China, India and Pakistan. It is also grown in Europe, Canada and the USSR, but the forms of rapeseed and mustard grown there are different from those grown in India.

India occupies the first position, both with regard to acreage and production of rapeseed and mustard in the world. In India the Braddica crops occupy the second largest position after groundnut, with 3-5 million hectares, producing about 2 million tonnes of seed annually.

The chief states producing them are Uttar Pradesh, Punjab, Haryana, Assam, Bihar, Madhya Pradesh, Rajasthan, West Bengal and Orissa. This book explores innovative methods for breeding new varieties of major crops with resistance to environmental stresses that limit crop production worldwide.

I thank all associates of my group who have helped in the preparation of the book. My individual thanks go to "Random Publications" who have in print the book.

— *Dr. H. Padamalatha*

Contents

1

Genetically Modified Food Crops

INTRODUCTION

Genetic modification, otherwise referred to as recombinant DNA (rDNA) technology or gene-splicing, has proven to be a more precise, predictable and better understood method for the manipulation of genetic material than previously attained through conventional plant breeding. To date, agricultural applications of the technology have involved the insertion of genes for desirable agronomic traits (*e.g.* herbicide tolerance, insect resistance) into a variety of crop plants, and from a variety of biological sources.

Examples include soybeans modified with gene sequences from a Streptomyces species encoding enzymes that confer herbicide tolerance, and corn plants modified to express the insecticidal protein of an indigenous soil microorganism, Bacillus thuringiensis (Bt). A growing body of evidence suggests that the technology may be used to make enhancements to not only the agronomic properties, but the food, nutritional, industrial and medicinal attributes of genetically modified (GM) crops.

Regulatory supervision of rDNA technology and its products has been in place for a longer period of time in the United States than in most other parts of the world. The methods and approaches established to evaluate the safety of products developed using rDNA technology continue to evolve in response to the increasing availability of new scientific information.

As our understanding of the potential applications of the technology is broadened, the safety of products developed using rDNA technology and the potential effects of introduced gene sequences on human health or the environment will be more closely scrutinized. In fact, much of the knowledge acquired during the commercialization of the products of rDNA technology in agriculture is now finding application in evaluating the safety of products developed through more conventional means. The objective of this chapter is to provide the reader with an overview of the significant events leading up to the present, science-based, regulatory framework that exists for the safety evaluation of GM food crops within the United States. An attempt has been made to discuss concerns over the sufficiency of existing regulations, as well as to highlight recent initiatives taken by federal regulatory agencies to address them.

Through better communication of how the regulatory process functions within the United States, it is anticipated that current and future applications of rDNA technology in agriculture will be met with a greater level of understanding and acceptance.

HISTORICAL PERSPECTIVE

rDNA technology was first developed in the 1970s. The initial response of the scientific community, including members of the National Academy of Science (NAS), to the prospects of rDNA technology, was to postpone any further research involving the technology until the potential risks to human health and the environment could be evaluated.

Researchers attending the International Conference on Recombinant DNA Molecules in 1975, otherwise known as the Asilomar Conference, tried to establish a scientific consensus on how best to self-regulate emerging applications of the technology. The conditions and restrictions that were proposed at this conference have formed the basis by which federal guidelines and policies for rDNA technology research were drafted within the United States.

National Institutes of Health (NIH)

The National Institutes of Health (NIH) was the first federal regulatory agency to publish their interests in evaluating the safety of rDNA technology in 1976, in the form of guidelines for the conduct of research. Because of the uncertainties that existed at the time, all research into the potential applications of rDNA technology was limited to the confines of federally funded laboratories under NIH control. After continued research, and a more careful assessment and monitoring of the risks, a set of less restrictive guidelines was published in 1978.

However, the environmental release of organisms developed using rDNA technology outside the confines of controlled laboratory conditions was prohibited unless otherwise approved by the NIH director. In the early 1980s, the NIH established an rDNA Advisory Committee (RAC) to review all data and experience gained with applications of the technology under its control. Based on recommendations of the RAC, a more relaxed set of research guidelines was published by the NIH in 1983.

The NIH approved the first environmental release of an organism developed using rDNA technology (ice-minus strain of Pseudomonas) in 1983. In response, they were criticized for failing to prepare a statement or assessment of the environmental impact of their regulatory decision as required under the National Environmental Policy Act (NEPA). Once the legal controversy had subsided, all responsibility that the NIH had for regulating the environmental introduction of GM organisms was relinquished. Nevertheless, NIH guidelines continue to be referenced in assessing the safety of rDNA research performed within industry, federal and other state laboratories. However, it was unclear which federal regulatory agencies would be responsible for ensuring the safety of the products developed using rDNA technology.

FOOD AND DRUG ADMINISTRATION (US FDA)

Authority

The US FDA is responsible for ensuring the safety and wholesomeness of all food and food components, including the products of rDNA technology, under the FFDCA. The US FDA has the authority for the immediate removal of any product from the market that poses potential risk to public health or that is being sold without all necessary regulatory approvals. As a result, a legal burden is placed on developers and food manufacturers to ensure the commodities utilized and foods available to consumers are safe and in compliance with all legal requirements of the FFDCA.

In order to understand the regulatory approach followed by the US FDA in the safety evaluation of GM crops, it is useful to consider food and food safety from a historical context. People had been consuming foods derived from agricultural crops for many years prior to the existence of any food laws or regulations within the United States. Based on this experience, agricultural crops have been accepted as being safe for consumption as food, without additional testing to demonstrate their safety. As long as the new crop variety has exhibited similar agronomic properties, and an appropriate taste and appearance, it has been considered safe to consume. As a result, most foods consumed today, in particular whole foods (*i.e.* fruits, grains and vegetables) and conventional foods, have not been subject to any kind of premarket review or approval by the US FDA. Nevertheless, food scientists have a good understanding that many of the commonly consumed agricultural crops contain natural toxicants (*e.g.* tomatine in tomatoes, solanine in potatoes, cucurbiticin in cucumber, psoralens in celery, etc.). As a result, new plant varieties may be subject to routine chemical analyses to ensure that none of these substances is present at potentially harmful levels. This type of general approach has been used in assessing the safety of thousands of new plant varieties that have been developed over a number of decades of crop breeding without compromising the safety of whole foods.

Role

Consistent with recommendations within the Coordinated Framework, the US FDA considered existing provisions of the FFDCA to be sufficient for the regulation of foods and food components developed using rDNA technology. It was concluded that the scientific and regulatory issues posed by the products of rDNA technology were not significantly different from those posed by conventional products.

As a result, GM foods and food components have been subject to the same standards of safety as already exist for the regulation of other foods and food components under the FFDCA. In order to better communicate interpretations of existing provisions of the FFDCA as they relate to the safety evaluation of foods derived from new plant varieties, including the products of rDNA technology, the US FDA released a policy statement in 1992 entitled 'Statement of policy: foods derived from new plant varieties'.

The US FDA has considered the use of genetic modification (*i.e.* rDNA technology) in the development of new plant varieties to represent a continuum of conventional plant breeding practices (*e.g.* mutagenesis, hybridization, protoplast fusion, etc.), and as a result, the safety evaluation of all new plant varieties, not just those developed using rDNA technology, have been evaluated based on an objective analysis of the characteristics of a food or its components, and not on its method of production.

OFFICE OF SCIENCE AND TECHNOLOGY POLICY (OSTP)

In response to a need for clarification, the Office of Science and Technology Policy (OSTP) began work on the development of a policy to establish a federal regulatory framework for evaluating the safety of products developed using rDNA technology. Following an opportunity for public comment, OSTP published a final version of the 'Coordinated Framework for Regulation of Biotechnology' (Coordinated Framework) in 1986.

The policy provided the basis by which federal regulatory agencies got involved in evaluating the safety of products at later stages of commercial development at that time. According to the Coordinated Framework, the products of rDNA technology should be regulated on the basis of the unique characteristics and features that they exhibit, not their method of production. The products of rDNA technology were considered to pose risks to human health and the environment similar to those posed by conventional products already regulated within the United States.

As a result, no new federal regulatory agencies or regulations were required. The Coordinated Framework did not, however, rule out the possibility of the development of new guidelines, procedures, criteria or even regulations to supplement or alter the scope of existing statutes for the products of rDNA technology. The Coordinated Framework identified three federal regulatory agencies within the United States: the US Food and Drug Administration (US FDA), the US Department of Agriculture (USDA) and the US Environmental Protection Agency (US EPA), as having primary responsibilities for evaluating the products of rDNA technology under development at that time. In 1992, the OSTP released another document entitled, 'Exercise of Federal Oversight within the Scope of Statutory Authority: Planned Introductions of Biotechnology Products Into the Environment', outlining the proper basis by which federal regulatory agencies were expected to exercise their regulatory authority. As with conventional products, dependent upon the intended use and function, more than one federal regulatory agency may share an interest in evaluating the safety of a product developed using rDNA technology.

If more than one federal regulatory agency has an interest, lead agencies are identified as being responsible for coordinating activities to limit any potential duplication of efforts. Although federal regulatory agencies worked independent of one another, it was realised that close working relationships would need to be established in order to evaluate effectively the safety of products developed using rDNA technology.

Recently, the OSTP teamed up with the White House Council on Environmental Quality (CEQ) to perform a six-month inter-agency evaluation of the federal regulatory agency responsibilities in evaluating the environmental safety of products developed using rDNA technology. A case-study approach for a variety of different classes of products developed using rDNA technology was used to evaluate the level of federal regulatory agency involvement, to identify strengths, weaknesses and areas of potential improvement. The review concluded that none of the previously approved products of rDNA technology has had any significant negative impact on the environment.

Although all the case studies were published, OSTP/CEQ failed to reach a consensus on issues relating to the relevant strengths and weaknesses of the existing regulatory structure within the time allotted for the completion of its review. A review of the case studies published provides a comprehensive interpretation of the responsibilities of each federal regulatory agency in ensuring the safety of the products developed using rDNA technology.

NATIONAL ACADEMY OF SCIENCES (NAS)

The National Academy of Sciences (NAS), and its operating arm, the National Research Council (NRC), have served as a primary source of scientific, technological, human health and environmental policy advice during the development of regulatory approaches for the safety evaluation of products developed using rDNA technology within the United States. In 1987, the NRC published a report concerning the potential human health and environmental hazards associated with the commercial introduction of GM organisms, entitled Introduction of Recombinant DNA-Engineered Organisms into the Environment: Key Issues. The risks associated with the introduction of GM organisms were considered to be essentially the same in kind as those associated with unmodified organisms. In other words, rDNA technology did not appear to introduce any unique risks as compared to the products that had been developed using more conventional methods of genetic modification. In reaching these conclusions, the NRC performed an evaluation of the similarities and differences in the properties exhibited by products developed using a variety of different techniques. To this day, the conclusions of this report continue to be referenced by the regulators and developers of

GM crop varieties worldwide. A subsequent NRC report, entitled Field Testing Genetically Modified Organisms: Framework for Decisions, also reached similar conclusions, but provided additional guidance as to how regulatory decisions concerning the introduction of GM organisms should be made. The NRC recommended that regulatory decisions concerning the introduction of GM organisms should be made on a case-by-case basis.

Consistent with the Coordinated Framework, the NRC did not consider the nature of the process used for the genetic modification of an organism to be a useful criterion for determining whether a product requires less or more regulatory oversight. As a result, no valid reason existed to regulate organisms genetically modified via modern techniques (*e.g.* rDNA technology) any differently from organisms genetically modified via more conventional means.

Similar conclusions have been published in the reports of international standards-setting organizations. In retrospect, within both reports, the NRC acknowledged that modern and conventional methods of genetic modification are not without risks to human health or the environment, and as a result, neither could be considered inherently more risky. As a result, regulatory decisions concerning the safety of the products of rDNA technology need only to take into consideration the specific characteristics exhibited by a particular GM organism and the environment in which it is to be introduced, and not the method by which it has been produced.

The NRC has further articulated their conclusions into what is now commonly referred to as the 'Concept of Familiarity'. Although familiarity with the characteristics of a particular organism or the environment to which it will be introduced would not necessarily mean it was safe, it can be expected to provide a sufficient amount of information to allow for a judgement to be made of the risks.

For example, familiarity with a new GM plant variety could be established based on comparisons between characteristics of the parent line or other crop species exhibiting similar traits, as well as through the results of actual field tests involving the GM plant. These principles were further elaborated upon by the Organisation for Economic Co-operation and Development (OECD), and as a result, have been referenced in the development of regulatory policies for evaluating the safety of GM crops on a global basis.

More recently, a committee established by the NRC published a report entitled Genetically Modified Pest-Protected Plants: Science and Regulation, based on a review of all scientific and regulatory data collected during the regulatory approval process for GM crops within the United States. The primary objective was to assess independently the effectiveness of existing and proposed regulations for the safety evaluation of GM crops expressing plant pesticides (*i.e.* plant-incorporated protectants).

No new evidence was identified to suggest plants expressing plant pesticides posed any greater risk to human health or the environment as a result of their genetic modification. In fact, the NRC concluded, 'with careful planning and appropriate regulatory oversight, commercial cultivation of GM plants is not expected to pose higher risks and may pose less risk than other commonly used chemical and biological pest-management techniques'.

However, the NRC report included requests for federal regulatory agencies to further strengthen the current regulatory approval process through better coordination and communication between agencies, on-going investment in the research and monitoring of potential human health (*e.g.* allergenicity) and environmental impacts (*e.g.* insect resistance), and by providing greater access to information evaluated in support of regulatory decisions.

US FEDERAL REGULATIONS FOR AGRICULTURAL BIOTECHNOLOGY

Regulatory systems for the products of agricultural biotechnology have been in existence since the mid- to late 1980s within the United States. The regulatory approach to the safety evaluation of plants developed using rDNA technology has evolved in the best interests of research scientists, industry and the general public. The agricultural products of rDNA technology, such as GM foods and crops, may require approvals from up to three regulatory agencies; the US FDA, the USDA and the US EPA, depending upon the characteristics exhibited by the GM plant, its proposed use and introduced traits. The same standards of safety are applied to all products regardless of the technology used in their development. The US FDA is responsible for ensuring the human safety of all new foods and food components, including products developed using rDNA technology, under the Federal Food, Drug, and Cosmetic Act (FFDCA).

The USDA evaluates the potential of a GM plant to become a plant pest following its environmental introduction under the Federal Plant Pest Act (FPPA). The US EPA evaluates pesticides, including plant systems modified to express pesticides (*e.g.* insect-protected or virus resistance), under the Federal Insecticide, Fungicide, and Rodenticide Act (FIFRA). As a result, the expression of an insecticidal protein in a food crop would undergo review by the USDA, US EPA and US FDA; a GM food crop exhibiting a modified oil content would be evaluated by the USDA and US FDA; and a non-food horticultural plant developed using rDNA technology for any other purpose (*e.g.* flower colour) would be subject to review by the USDA alone. In most instances, obtaining all necessary approvals for the commercialization of an agricultural crop developed using rDNA technology takes a decade or more. However, the exact amount of time required will depend on the need to confirm performance, to evaluate characteristics of the food, environmental effects, and to produce the required amount of seed before the product can be distributed and commercially grown by farmers. Up to five years of field trials (5-10 generations of plants) are required for the developer of a new plant variety to collect sufficient data to meet the reporting requirements of the USDA. An additional five months to two years may be required for the

US FDA, USDA and/or US EPA to complete all necessary product consultations, reviews and approvals. Approval for the first commercial planting of a GM food crop was not issued until 1995. Since 1995, more than 40 new agricultural crops developed using rDNA technology have received approval for commercial planting within the United States.

In 1999, approximately 72% of the total 39.9 million hectares (more than 98 million acres) of GM crops grown worldwide were planted in the United States. Herbicide-tolerant soybeans (54%), Bt corn (19%) and herbicide tolerant canola (9%) accounted for approximately 82% of the GM plants cultivated. With an increasing number of agricultural biotechnology products reaching later stages of commercial development, it is anticipated that the overall area planted with GM crops will continue to rise. The regulatory approach to evaluating the human health and environmental safety of GM crops within the United States is best described as a science-based, case-by-case assessment of hazards and risks. This approach has provided the flexibility required to reduce the regulatory burden placed on products that have been determined to be of low risk or concern. All agencies involved in the regulation of plants developed using rDNA continue to implement and develop policies based on recommendations made within the Coordinated Framework.

In the future, the US FDA, USDA and the US EPA will be dedicating additional resources towards communicating how GM food and food components are regulated within the United States, and how these regulations function to be protective of both human health and the environment.

MAJOR FOOD CROPS

RICE

It is grown over as much as 29 percent (43,70 million hectare) of the cropped area. India stands next only to China in the production of rice contributing 22.3% of the world production. West Bengal is the largest rice producing state contributing to more than 15.81% of the rice production of India. Next is Andhra Pradesh with 12.75% and Uttar Pradesh with 12.13%. However, yield per hectare is highest in Punjab and Karnataka (42.97% and 31.87% respectively).

WHEAT

Wheat is the second important food grain of India, 13.79% of the gross cropped area is devoted to wheat. Area under wheat increased from 9.7 million hectare in 1950-51 to 26.48 million hectares in 2006-07 and wheat production recorded eleven fold increase. The countries producing more wheat than India are the Soviet Union, The United States and China. Over 35.5 percent of the country's total wheat output comes from Uttar Pradesh, Punjab (20%), Haryana (12.68%). The crop shares about 21 percent for the area under food grains and accounts for about 34 percent of the total food grains production in the country.

JOWAR

Jowar ranks third in area among food grains. The leading jowar producing states are Madhya Pradesh, Maharashtra, Karnataka and Andhra Pradesh. It is both Kharif and Rabi crop and about 8.51 million hectares (5%) are devoted to this crop in India. Half of jowar grow in Maharashtra (51.11%)/

MILLETS

Millets like Jowar, bajra, maize and ragi are important cereal crops of India, They are primarily crops of the dry parts of the Deccan. Gujarat, and Rajasthan, These Kharif crops are grown where there is not enough rain to grow rice. They are produced in the order ragi, jowar and bajra, as the rainfall goes on decreasing from the semi-humid to semi-arid regions.

BAJRA

Bajra is generally grown on poor light sandy soils of western Rajasthan (35.94%), Northern Gujarat (13.93%), Uttar Pradesh (16.28%), Haryana (8.85%), and Maharashtra (13.41%). In the south, it is cultivated on shallow black and red and upland gravelly soils. It occupies 9.48 million hectares (about 5.0% of cropped area of the country).

MAIZE

Maize was introduced in India from America in the beginning of the 17th century. It is an important food crop now in the Great Plains and in the hilly and submontane tracts of the North. Green maize provides sweet and succulent fodder. Maize is an important millet crop and has an area of production next only to jowar and bajra. It occupies 7.7% of the cropped area of the country. Its highest concentration occurs in Andhra Pradesh (21.1%, Karnataka (18.56%), Bihar (9.25%), Rajasthan (7.48%) and Madhya Pradesh (8.5%) of all India production. Ragi is an important millet, particularly in south Karnataka where millions use it as staple food. It is raise on the red, light black and sandy loams of Karnataka and Tamil Nadu and the well-drained alluvial loams of Uttar Pradesh, Bihar and Gujarat. Barley is an important cereal crop in many parts of northern India. It is also used for melting in the manufacture of beer and whisky. It is rabi crop in the Great Plains and valleys of the western Himalyas. Uttar Pradesh and Rajashtan together lead in the total area and total production of barley in the country.

PULSES

In our predominantly starchy vegetarian diet pulses form a very important part as they provide us vegetable protein. Being leguminous crops, pulses fix atmospheric nitrogen in the soil and hence are usually rotated with other crops to maintain or restore soil fertility. Pulses are seasonal crops in India. India tops the world in pulses production.

GENETICALLY MODIFIED CROPS

A very promising strategy to evaluate the equivalence of genetically modified crops is gene expression profiling. Conventional methods for the analysis of differential gene expression include Northern blotting, S1 nuclease protection, comparative expressed sequence tag (EST) sequencing, differential display and serial analysis of gene expression. More recently, Kok *et al.* have designed a method of detecting altered gene expression by means of mRNA fingerprinting or reverse transcription polymerase chain reaction (RT-PCR). Based on the original concept of Liang and Pardee specific subsets of the mRNA population of a tomato plant were amplified. This enables the expression of different genes in the GM crop and genes in the parental line to be compared and any empirical information about significant alterations to be established.

The method is still under development and requires further validation, but has been shown to be reproducible and able to detect important differences in gene expression. Some of the most significant remaining problems are that the method largely depends on individual skills based on experience and is time-consuming, even in its most simplified form.

CDNA MICROARRAYS

Over the past few years, the DNA microarray technology has emerged as a powerful high-throughput method for the analysis of gene expression. At present, microarrays for gene expression studies generally consist of cDNAs, which are physically deposited onto small glass surfaces. The major advantage of DNA microarray technology over conventional gene-profiling techniques is that it allows small-scale analysis of expression of a large number of genes in a sensitive and quantitative manner.

Furthermore, it allows comparison of gene expression profiles under a large number of different conditions. The cDNA microarray technology has already been successfully introduced in disciplines ranging from cell and developmental biology to drug development and pharmacogenomics. In the following a number of technical aspects of the technology as well as its potential value for the safety assessment of genetic modification of food plants.

The cDNA microarrays are produced by deposition onto a solid surface of purified PCR products corresponding to specific genes. Typically a few nanolitres of DNA solution (100-500 μg/ml) are spotted using a micro-dispensing robot. Essential information on robot construction and protocols can be found on the Internet. In general, with respect to the spotting devices, a higher density of spots is accomplished when using a passive dispenser (> 2500 DNA spots/cm2).

Thereto, glass slides covered with a positively charged layer (*e.g.* poly-l-lysine) or carrying reactive groups are most commonly used. To analyse the expression of genes in a plant cell or tissue sample, polyA+ RNA (typically

0.5-2.5 μg is required for each reaction) is purified and labelled by reverse transcription using an oligo-dT primer incorporating a fluorescently labelled nucleotide. The labelled cDNA pool is then hybridized to the microarray. The required hybridization conditions (*i.e.* sample concentration, stringency of hybridization) and parameters, such as detection level and the correlation between transcript concentration and hybridization have been determined for each experimental setting.

The intensity of a hybridization signal, which is a measure of the relative expression level of each gene, can be read by charge-coupled device (CCD) cameras or confocal/non-confocal laser scanners. RNAs extracted from two different samples can also be simultaneously analysed with a microarray by labelling each with a different fluorescent label using either Cy-3 or Cy-5. At RIKILT-DLO we are currently testing whether the analysis of differential gene expression using DNA microarrays is an informative strategy for the safety evaluation of GM plants.

To address this question, DNA libraries are constructed and enriched by a subtractive cloning approach for cDNAs preferentially expressed in either green or red tomato fruit. These subtracted cDNAs, as well as control cDNAs representing known tomato genes (*i.e.* sequence information derived from public sequence databases), have been spotted on arrays and are used for comparison of gene expression profiling of control and genetically modified tomatoes.

In the near future we will construct more defined arrays containing genes of toxicologically relevant pathways, such as those involved in the synthesis of natural plant toxins (*e.g.* the glycoalkaloids α-tomatine and α-solanine).

CHEMICAL FINGERPRINTING

The principles for establishing a chemical fingerprint were directed towards the detection of alterations in compositions in 1H-NMR spectra arriving from two populations, for example, non-modified counterpart(s) and an engineered one. In this way the unprocessed parts of the plant used for human consumption were fractionated and analysed, *i.e.* in tomato comparison on the fruit level.

Genetically modified tomato varieties, such as the antisense RNA exogalactanase fruit, have been studied using this innovative technology. The spread of measurements of the individual profiles was comparable for the GMO and its controls. By subtraction of the 1H-NMR spectra the differences of the means were calculated at 99% confidence intervals between the GM crop average (n = 8/batch per line) and its isogenic control average (n = 8/batch per line). The difference of the means was defined here as all differences of the means of the amplitudes of individual NMR signals obtained from spectra of various populations. The differences of the means were used to quantify how the transgenic crop plant differed from its parental line. The normalized 99% confidence intervals of the normalized means of fractions A-E after eightfold independent replicates demonstrated that differences in amplitudes exceeding at least 20% could normally be detected. This is an adequate sensitivity to allow statistical evaluation, and is in line with recommendations of the Nordic Council. They proposed that if the average value of a parameter differs by more than 20% an explanation should be sought.

With the purpose of establishing those differences that may arise from metabolic effects linked to genetic modification the default statistical analysis used a mixed effects randomized block model. Random block effects such as variations in the stage of cultivation, location, logistics or climate had a significant impact on the overall chemical fingerprints of cultivars. For example, chemical fingerprints varied much more between field sites and seasons than between replicates in one plot.

A tomato background mean (*i.e.* crop mean value) was established by constructing an extended range of non-modified tomato references for comparison with natural variability, and to assess the impact of external factors (*i.e.* variations in season, climate, etc.). The crop mean value was used to quantify how the differences of the means between the transgenic crop plant and its control differed from the natural diversity of commercial counterparts.

To differentiate between compositional changes either due to genetic modification, genetic variability or environmental variations the chemical fingerprints were studied in a hierarchical approach.

Hence, the fingerprints of the GMO are compared to those of:

- The isogenic control line grown side-by-side under identical conditions in one plot;
- The isogenic control line grown side-by-side under identical conditions at multiple sites;
- An extended range of commercial varieties of that crop;
- The influence of downstream processing.

It was recognized that 45% of all compounds of the control line harvested in different years varied considerably in concentration. Most differences were found in fractions A, C and D. A similar degree of impact on composition was observed in the case of processing temperatures (*e.g.* 100°C versus 70°C) on tomato juice. These findings were compatible with the protease activity surviving the 70°C treatment. But in fruit harvested from various individual plants, divided into two populations, approximately 95% of all constituents were practically identical in concentration. Also the determination of the differences of the means between transformant tEG to its isogenic control bred side-by-side in the same plot showed that 249 out of 3000 amplitudes (8%) varied in concentration. Assignments by NMR indicated that, for example, citric acid was present in fraction A of the control at 1.4 times the levels found in the transformant. However, in fraction A of the transformant glutamic acid and/or glutamine showed an increase of 1.3 times the levels found in the isogenic control.

On the other hand, one single aromatic compound, a-lycopene, appeared to be present in fraction E of transformant tEG at 2.5 times the concentrations found in the parental line. No alterations were observed in ß-carotene levels. These effects may be the result of the antisense downregulation of the tEG1A gene in tomato. In our view the natural diversity must be taken into consideration, however, when interpreting the biological relevance of any significant difference found in the differences of the means between the GMO and its parental line. The parental line was extended by entering non-isogenic commercial references (*i.e.* crop mean value).

Although limited in the range of lines, the comparison of the crop mean value to the differences of the means showed that the contribution and magnitude of differences (*e.g.* citric acid, glutamic acid/glutamine, a-lycopene) diluted out. It was recognized that in the agricultural practice the conditions of culturing, processing and storage of appropriate references will be mostly unknown and they may have a non-isogenic genotype. Notwithstanding these complicating random block effects we recommend that, for example, the chemical fingerprint should not be limited to the transgene and its isogenic parent bred under identical circumstances only, but should also include a sampling through multiple sites and a comparison with an extended range of controls of that crop.

Otherwise it would be a too limited approach towards the establishment of the claim of substantial equivalence and it would neglect the ranges of natural variations in that crop. In future, if many more lines have to be compared with each other this approach may benefit from multivariate statistical methods.

PROTEOME PROFILING

The introduction of Bt genes of bacterial origin may theoretically influence the nature of post-translation modification (*e.g.* glycosylation) in the transgenic tomato fruit. This is because as well as the impact of the protein moiety, for example, the type of glycosylation pattern may alter due to inactivation of glycosidases and glycosyltransferases in the host cell. In addition, glycosylation of the inserted Bt gene product in the plant might result in altered protein activity and stability.

The importance of structurally elucidating these oligosaccharides, especially the N-glycans, can be attributed to their widespread functions in the cell. These functions include correct folding, biological activity and stability of proteins as well as involvement in plant development. Moreover, N-glycans represent antigenic epitopes by themselves. Immunological as well as biochemical studies revealed the allergenic potency for the a1,3-fucosylation of the proximal N-acetylglucosamine residue. Additionally, also the b1,2-xylosylation of the b core mannose is suspected to be an allergenic epitope.

Recently, there is increasing evidence that these N-linked oligosaccharides may have a substantial impact on the immunogenicity of glycoproteins, as N-glycans are suspected to play a certain role in the adverse food reactions of hypersensitized patients. At the Institute for Agrobiotechnology, Austria, this issue was tackled by developing a generic technology for isolating, purifying and characterizing N-linked oligosaccharides in plant tissues. Our low-temperature acetone powder method appeared to be most suitable for the isolation of (glyco-)proteins from plant tissue (*e.g.* tomato fruit). The resulting (glyco-)protein fraction was proteolytically digested and corresponding glycopeptides purified by cation-exchange chromatography.

The oligosaccharides (N-glycans including a1,3-fucosylated oligosaccharides) were released from the peptide backbones by N-glycosidase A (PNGase A) and fluorescently labelled with 2-aminopyridine. Subsequently, pyridylaminated glycans were separated and analysed structurally by repeated two-dimensional high-performance liquid chromatography (2D HPLC) using a size-fractionation and a reverse-phase column in combination with an exoglycosidase treatment.

MALDI-TOF (matrix-assisted laser desorption ionization time-of-light) mass spectrometry was applied for further confirmation of the observed structures. Sixteen different N-glycosidic structures could be detected in tomato fruit. The two most abundant glycans in all the tomato samples showed identical properties to those of the major N-linked oligosaccharides of horseradish peroxidase (MMXF: Man(3)Xyl-FucGlcNAc(2)) and pineapple stem bromelain (MOXF: Man(2)Xyl-FucGlcNAc(2)), respectively, and accounted for about 65-78% of the total glycan content.

Oligomannosidic glycans occurred in only small quantities (3-9%). The majority of the N-glycans found was b 1,2-xylosylated and carried an a1,3-fucose residue linked to terminal N-acetylglucosamine. A structural element was identified and shown to be an IgE-reactive determinant, which contributes to cross-reactivities among non-related glycoproteins, and is present in a wide variety of plants extracts.

It is proposed that this structural element might also be of importance in some cases of food and pollen allergy. The carbohydrate profiling of, for example, populations of green and red-ripe fruit of non-modified and modified Bt tomato plants (expressing the cryIA5 gene) revealed no differences in the nature of N-glycans, nor were there significant differences in the relative amounts of glycan. Interestingly, the relative amounts found were quite similar across seasons and comparable to the commercial variety Notoro used as an extended control. As far as the N-linked oligosaccharides were involved it was not possible to detect significant alterations in the a1,3-fucose content of Bt tomatoes due to genetic modification.

PROFILING OF THE METABOLOME

Among other factors, identity of a complex GM crop plant with its traditionally bred parent entails metabolism and the level of (un-)desirable substances contained therein. Therefore, we would like to build up an overall picture of components associated with different metabolic capabilities in the (non-)modified plant tissue. A multi-compositional analysis of the so-called metabolome may overcome the test limitations of, for example, single critical nutrient or key toxicant analyses and problems related to their limits of natural diversity.

An attractive alternative is the utility of high-resolution proton (1H) nuclear magnetic resonance (NMR) spectroscopy (1H-NMR) in combination with separation techniques (liquid chromatography (LC)). Recently, we have shown that the approach of chemical fingerprinting using off-line LC-NMR can obtain information on possible changes in complex plant matrices due to environmental effects as long as GLP-like conditions for sample handling, data acquisition and automation of data handling are maintained. In principle, it will be possible to screen all the different low molecular weight components (MW <10 kDa) present in the metabolome of the plant tissue.

TOXICOLOGICAL PROFILING

The significance of detected compositional alterations may need to be further explored by in vitro toxicity assays in order to screen for potential adverse effects and for mechanistic studies. It may even be that the latter approach leads to a redesign, refinement or even replacement of (sub-)chronic rodent feeding trials. In vitro systems derived from organs and tissues from animals and humans, and various types of cultured recombinant cell lines, are successfully used for screening for the toxic potential of single compounds.

However, their ability to screen for the potential toxicity of whole foods or extracts thereof has been insufficiently explored up till now, and has been examined by us within the framework of a tiered safety evaluation of GM plants. The validity of using in vitro test assays for comparative testing of whole food products, food ingredients and extracts thereof depends on the sensitivity and selectivity of the toxic response induced upon exposure. However, since these models are not yet validated, they should be considered as early warning or alert-systems to identify potential changes in toxicity of the engineered product only.

ASSESSING TRANSCRIPTION FACTORS

There is a demand for in vitro models that not only determine the (geno-)toxic potency of plant components or complex mixtures, but also provide information on the mechanism by which that compound exerts its toxic effect. As an in vitro profiling system the eukaryotic stress gene assay (*i.e.* CAT-Tox(L)) was assessed. This utilizes human HepG2 cells stably transfected with chloramphenicol acetyltransferase (CAT) reporter constructs. The aim of the CAT-Tox(L) assay is to collect and quantify molecular responses to cellular stress and toxicity that have a transcriptional component in their regulation. The assay employs an ELISA method to measure the transcriptional activity of the promoter or response element constructs. The advantage is that many of the stress responses occur before any measurable cytotoxicity, thus allowing monitoring of stress pathways at sublethal levels.

Since 14 unique transcription responses can be measured simultaneously, the CAT-Tox(L) provides comprehensive cellular stress profiles. The assay generates specific stress gene induction profiles for a variety of stressful and toxic plant components such as genistein, diadzein and sinigrin.

Aqueous extracts of red-ripe, non-and antisense RNA exogalactanase tomato fruit dosed up to 0.05 g per assay did not cause any cytotoxicity to the HepG2 transfectants nor were there molecular responses to cellular stress or toxicity. Extracts of non-modified and modified green tomato fruit showed cytotoxicity and induced the construct containing the xenobiotic response element (XRE).

This gene-induction profile was not comparable to that of, for example, the reference molecules dioxin, 3-methylcholanthrene, benzo[a]pyrene or b - napthoflavone. These polycyclic aromatic hydrocarbons were characterized with a co-induction of cytochrome P450 1A1 and glutathione-S-transferase Ya subunit- CAT fusion constructs. There was substantial evidence that the systemic fungicides fenarimol and bupirimate did not induce any of the genes in this assay. There was no promotor in the assay that could be linked to universal cellular stress.

A-tomatine caused severe cytotoxicity at high concentrations and yet did not induce any of the genes (*i.e.* causing membrane damage). However, it was observed that a-tomatine diluted in a matrix of red-ripe tomato tissue (final extract concentration 5% w/v) elicited an increased cytotoxicity concomitant with a dose-dependent response profile involving XHF (collagenase involved in inflammatory reactions sensitive to mitogens and cytokine IL-1). So far, the approach of the CAT-Tox(L) model shows potential but there is a need for validation based on molecular mechanisms of plant constituents.

Although the induction of these stress genes represents cellular perturbations, the mechanistic validations need to be extended before any conclusions can be drawn on the value of these findings. Certainly, the development of a database to explore relationships between a wide variety of compounds stressful to plants and meaningful stress gene profiles may serve as molecular fingerprints unique for a particular component.

GENOTOXICITY ENDPOINTS

The mucosa of the gastrointestinal tract is clearly a primary, potential site of action for the effects of novel foods. Therefore effort has been concentrated on the development of a battery of intestinal cell systems (*e.g.* IEC-6, IEC-18, Caco-2 and INT407 cell lines) with different endpoints for initial cyto- and genotoxicity testing.

The choice of a suitable method of sample preparation was considered of utmost importance in assaying vegetables.

Freeze-dried tomato fruit was extracted (10% suspensions, w/v) in water, 0.9% saline or chloroform:methanol (2:1, v/v) immediately before incorporation into the culture medium. It was encouraging that aqueous and chloroform/methanol extracts of, for example, red-ripe tomatoes up to a maximum of 10% suspensions (w/v), had little or no general toxicity once the pH (> 6.0) and osmotic pressure had been taken into account.

No toxicity to the intestinal cells Caco-2 and IEC-18 under short-term conditions was detected in red-ripe, non- and antisense RNA exogalactanase tomato using alpha-glutathione S-transferase (a-GST) lactate dehydroyrenase (LDH) leakage, reduction of 3-(4,5-dimethylthiazol-2-yl)-2,5 diphenyltetrazolium bromide (MTT conversion), neutral red (NR) uptake and total cellular protein as in vitro endpoints.

On the other hand, green tomato fruit displayed a pronounced toxic and cell-damaging effect. No statistical significant differences were found between transformants and isogenic control lines in terms of their activity in the assays. Treatment of tomato tissue with pepsin (ratio pepsin to tomato protein 1:250, w/w) did not result in an increase of or loss of inherent cytotoxicity. Supplementary investigations with naturally occurring plant components, for example, quercetin, b -carotene, a-lycopene, genistein showed that only steroid glycoalkaloids, such as a-tomatine, exhibited a rapid toxic effect, whereas a-tomatidine was ineffective. It was noted that the in vitro models did show general matrix-related toxic effects.

For example, a-tomatine diluted in a matrix of extracted red-ripe tomato tissue (final concentration of 5-10% w/v) showed an increase of dose-response-dependent cytotoxicity in human Caco-2 cells (*e.g.* IC50 7.9 ± 2.1 mg/L versus IC50 33.9 ± 5.3 mg/L of a-tomatine only). Using the COMET assay, suspensions of freeze-dried tomato fruit up to a maximum of 10% (w/v) did not exhibit DNA-damaging effects in a range of gastrointestinal cells derived from rats (IEC-6 and IEC-18) and humans (Caco-2, INT407).

Genetic modification did not result in increased genotoxicity, but fruit extracts were able to suppress the DNA-damaging effects of the known genotoxins H2 O2 and N-methyl-N2 -nitro-N-nitrosoguanidine (MNNG). Although these endpoints are important toxicological measures they do not indicate the molecular mechanisms of toxic damage. So far, pronounced adverse effects could only be observed with glycoalkaloids. The screening for subtle differences in toxic responses of other plant constituents appeared to be limited. However, therefore, the use of critical endpoints for such studies other than cytotoxicity should be clearly defined.

CROP UNIFORMITY

As genetic diversity and possible uniformity is an evident and possible important aspect of clonal forestry. For evaluation of possible ecological impact diversity seems central. I will thus start to discuss disadvantages with a low genetic diversity in a stand. In other respects this chapter focuses on the advantages of clonal forestry.

DISADVANTAGES OF A UNIFORM CROP

There are reasons to believe that the biological production can be higher in a genetically diverse crop than in a uniform:

- A single genotype demands the same things at the same time, and thus utilises the site worse than a mixture of genotypes.
- In a mix, another genotype may take over the ecological space left by a failed genotype.
- A disease spreads faster in a uniform crop.

These statements are supported by the average of a very large number of agricultural mixing experiments and a few experiments and experiences with forestry. On average, mixes performed some per cent better and were less susceptible and more stable than the average of its components, but usually the production of the mix was not better than its best component grown pure.

Even if mixes seem more productive on an average, there are many examples and experiments, when no advantage was found and some experiments where diversity actually seems negative for production. I conclude that from a production point of view an advantage with some degree of diversity is expected, but that does not imply that production is higher the higher the diversity is.

But the superiority in biological production is not expected to be large and it is not certain for individual cases, but just a rough generalisation. This general statement ought to apply to birch also, thus biological production is assumed to be lower and more uncertain in a uniform crop, but the effect is probably small and practically unimportant (a few per cent of biological production).

ADVANTAGES OF A UNIFORM CROP

There must be considerable advantages in uniform monoclonal blocks as they are used frequently. Although the advantages with diversity are repeated and reiterated and very politically correct, it is seldom applied in intensive agriculture or husbandry.

The main reason is that maximizing biological production and maximizing economical production is not the same thing. It makes it easier to manage the crop if it behaves uniformly. For a forester it may not matter very much if individual birches drop their leaves on different days, but a farmer wants the crop to become ripe the same day, as the whole crop is harvested the same day. Diversity is a usually a problem for the customer, who wants a uniform raw material for further processing. For administrative processes, like describing the crop, it is an advantage if it is uniform and does not change from specification over time and space. These problems are evident for many types of agricultural crops, and therefore these crops are usually uniform even when biological production would be higher with a diverse crop.

The dream of many foresters is to find the best genotype and use only that. This has been successfully done for other domesticated species. The uniformity itself in a clonal plantation has a considerable advantage from a management point of view. It is easier in the nursery to raise plants of a single genotype. It is easier to manage and harvest a uniform forest. It has advantages to handle a single type of material. A customer or processor of a product knows more about a lot if it is a specified clone than if just the species is known. A batch of uniform logs is worth more and easier to handle than the same amount of very variable logs.

Even if interest is only in quantitative production and that is lower in a uniform crop, still the higher genetic performance of the best clones compared to the best seedlings available usually much more than compensates for the production loss by a uniform crop. Thus uniform clonal crops produce more than diverse seedling crops. My opinion is that genetic diversity often is overemphasised in comparison with the higher genetic gain achievable at the cost of a reduction in genetic diversity for future forest production plantations in many countries (like Sweden) both for propagation with seedlings and cuttings. To get genetic diversity many clones are used in seed orchards and clonal mixtures and there constraints of relatedness of clones may be seen as desirable. This was not a disadvantage at an early stage of forest tree improvement, as it does not influence gain much if many phenotypically selected plus trees are placed in seed orchards, and in the early stages of cutting propagation there were physical constraints on the number of ramets which can be produced by each clone, so it is necessary to have many clones. But now this constraint often means selection is not intensive and the best genotypes are not exploited to the extent, which would be possible.

However, clonal forestry is unlikely to deal with a single clone only. That does not seem a likely scenario for birch. Many clones will be tested and test results as well as experiences will accumulate of the best clones, so the opinion of what is the best clone will change over time. It is also dependent on the locality, the desires of the user and the availability. Even if uniform plantations are planted, there will usually be a mosaic of clones on the landscape level. From the risk management point of view a clonal deployment philosophy offers advantages. If a clone is sensitive to a pest or pathogen the problem can be identified and managed. If silvicultural means to deal with a problem are insufficient, the stand can be salvage harvested and replanted. The clone which turns out inferior can quickly be taken out of production.

It is an advantage for an operator which uses monoclone blocks to use several clones in the same time that it will become more evident what the strong and weak point of the clones are. A disadvantage with mosaics is that the uniform lots of wood from a stand are often too small to be practical as a specific variety at the end user, and actually the end user may get trouble with larger between lot differences if clonal forestry is used. Alternatively the clones can be grown in intimate mixtures, in that they most of the possible advantages of the diversity of seedlings may be gained by a mix of rather few clones.

When we are talking quality birch - and not quantity birch - the reasons for monoclonal plantings become stronger. When high quality of the product is concerned, it often becomes more important to have uniform quality also, while in a low quality product - as firewood - the quantity is usually relatively more important and a uniform and predictable product is less important. Attention on quality instead of quantity also generally coincides with a higher level of intensity, and at a higher level of intensity, clonal forestry often becomes more interesting. Thus considering the aim the project arranging this symposium has in mind; uniform clonal plantings becomes relatively more important than for forestry in general.

IMPROVED CLONES

There are a number of arguments for forestry with clones. It takes less time to multiply a genotype as a clone than to build a seed production unit with that clone as one of the parents and wait till that works. The time lag is actually much longer than till the first seed production, because a seed production unit is usually designed to last for many years. The risk for getting the seed storage empty means that forestry often want to have five to ten years foreseen seed need in storage. That is expensive, many seeds may never be used and the seeds actually used are less improved than possible.

A seed production unit is designed to meet a certain need of plants with certain characteristics many years ahead. If too many seeds are produced it means that the investment was unnecessary large. If too few seeds are produced it may mean that foresters must use genetically inferior unimproved seeds. If the propagation is done fast by vegetative propagation by *e.g.* micro propagated clones or somatic embryogenesis, it takes only a year to multiply a desirable genotype in the desired number of copies. Thus it is fairly easy to adapt to the latest news from breeders and customers. There are large annual variations in seed production both in the forest and in seed orchards. In a seed orchard the reproductive output is variable among genotypes and the reproductive phenology may make clones incompatible, so the actual genetic set up of the progeny is a bit uncertain. These uncertainties can be avoided by clonal forestry. Many of our forest trees - like birch - can both spread pollen and produce seeds. That means that selfing can occur and the occurrence of selfing in a plant material will decrease the forest production. With the use of clonal forestry selfing can be completely eliminated.

In a wind pollinated seed orchard wild uncontrolled and unimproved pollen may contribute to pollination, clonal forestry is a way to eliminate this problem. However, for birch, seed orchards are small enough to be managed in plastic green houses, where the inflow of unimproved pollen is low, so for birch it is not fair to see this pollen contamination problem as an argument against open pollinated seed orchards.

In a seed orchard it is important to avoid related clones, as relatives get inbred progeny when they mate, which means that forest production drops because of inbreeding depression. Mating between relatives may be a worse problem than selfing in future seed orchards, as self-zygotes usually die; self-seedlings are often culled before planting; or early out-competed in forest, while milder inbreeding is more directly transferred to production loss. If clones are related in a clonal forest, that does not cause inbreeding depression, and thus there is little harm if related clones occur in a stand (actually this is very natural, trees in a natural stands are often related). The clones in the breeding population will be increasingly more related as breeding goes on, it will be easier to handle the consequences of this for the production forest with clonal forestry.

A seed orchard is a big investment intended to meet the seed need for a long period for a specific species used in a specified area, thus a considerable local market is needed to justify the investment. Conifer seed orchards are usually designed for a need of many millions plants per year over some decades. Multiplication of a clone can be seen as a smaller investment during a shorter period, thus the local market is allowed to be much smaller with clonal forestry, *e.g.* micro propagation for curly birch in Finland serves a market of 200 000 seedlings a year. A seed orchard has a stiff structure, its composition can be only slightly modified once it is established and it is not easy to justify several orchards heading for different characteristics for the same area. Clonal propagation is more flexible; an individual customer can choose the clones, which fits the particular needs for the customer and occasion. Sometimes a combination of rather rare characters is desired, a clone can be found which combines several desirable characters in a much more efficient way than dealing with seed orchards.

It is often stated that clonal forestry reduces genetic diversity. This is a generalisation, which need not always be true. Actually, clonal forestry provides a tool to choose genetic diversity at will, while when a seedling material is used, the genetic diversity is essentially left to chance. Probably those using intensively managed birch plantation for producing high quality timber with special characteristics would choose the option to reduce or eliminate genetic diversity on the stand level, because they found it economically sound to do that. It seems likely this is how the tool will be used. But clones in a clonal mix can be chosen so that the genetic diversity is larger than in a seedling material.

Clones can be mixed with seedlings, that may be a good strategy not only for diversity, but economically in cases where cloned plants are much better, but also much more expensive. If the clones succeed well, their share at the final harvest will be larger than at establishment. Over time and space different clones will be used even in a system with monoclone cultures. In such a system the genetic diversity among stands is likely to be much than in nature. Sometimes clones have been used for natural conservation in situation when good seeds could not be obtained. Clonal forestry may strengthen the trend to produce the needed wood at a limited acreage of intensively managed production forests, allowing larger areas to be set aside for other purposes. Thus the question how clones may affect diversity has many aspects and it is an oversimplification to state that it must be severely reduced.

A common situation is that some good seeds exists or can be obtained, but quantitatively insufficient for the plant need. If such a situation occurs unexpected and unforeseen, vegetative propagation may be the solution. A more common situation is that a few superior seeds can be produced by controlled crosses, but too few to meet the plant demand. Clonal propagation is often a way to make forestry with controlled crosses practical feasible. Clonal propagation may be seen as a way to amplify the impact of a few good parents or a tested cross by mating the parents and when multiply the seeds.

The genetic value of a genotype depends on the breeding value of its parents, but also how well the genes of the parents fit together. This may be called dominance. The later reason for genetic differences cannot be exploited in an ordinary seed orchard, but it can be exploited with clonal forestry.

Hybrids sometimes result in good seeds. It may be difficult to make many of those seeds by controlled crosses, but hybrid seeds (or trees) can be multiplied by vegetative propagation. This is used in the Eucalypts and Aspen programs mentioned above. Personally I am not that convinced that hybrid superiority is so common as it is often stated. When hybrids are better than the parents it may be because the parents were not well adapted for the conditions where they were tested or that the hybrid combines characters of the parents for the environment and purpose of the forester in a better way than tested alternatives. Hybrid maize is a best seller, but the reason is more that farmers are unable to produce hybrid seeds them self, but has to buy them, and when it is good economy of a seed company to invest in hybrids, so they can market the seeds.

To get a genetically superior forest, it is desirable to use the genes of the very best clones. In a seed orchard it has to be rather many parents to avoid selfing and problems with lack of overlap in flowering time. This means a drop in breeding value. For clonal forestry the very best clone can be used. This is true even if in the complete absence of dominance. As dominance is seldom the major cause of genetic variation for the most important characters in forest tree breeding, this more effective exploitation of breeding value is usually a more important reason for clonal forestry than the exploitation of dominance variation, at least as long as we do not talk about hybrids of different species or races. Quantitatively the low importance for the superiority of cloning of a limited amount of dominance variation is demonstrated by *e.g.* Lstiburek.

We have also special features for clonal selection. We can intentionally select clones, which do not waist resources on sexual activities but spend resources on stem wood production. Such clones are desirable for forestry. In a seed orchard it is not desirable with clones, which do not produce seed and pollen, so such clones would not be chosen for a seed orchard, and if they were, they would be unable to transmit their genes to the seed crop.

Cloning methods have different characteristics. They are often very depending on the state of the material which is cloned, cuttings usually do not work well on physiologically old material, for somatic embryogenesis other tissue than immature seeds are often a difficult starting material. With micro propagation it is usually possible to multiply old trees or clones, and this is often an advantage.

Cloning and the cloning method can have effects on the physiological status of the tree. *E.g.* conifer grafts have not thick bark at the bottom of the trunk. The protocol for making clones can have effects on the plants. These effects sometimes look desirable and can be exploited. For example Norway spruce cuttings seems to suffer a lower mortality from pine weevil damage than seedlings. Clones usually origin from older and more mature tissues than seedlings and this affects the characters of the plants and may have effects also on the mature trees. I guess these differences to seedlings are not dangerous and nursery raised seedlings can also be claimed to be unnatural, and they my be different because of different growing regimes. I suggest that we can intentionally interfere with the characters of the plants by the production system and that clonal forestry gives additional options to use this in a positive way.

Many of the factors I have mentioned above will contribute to that clonal forestry gives a genetically more improved forest than comparable seedlings. An example of the development of gain over time for the clonal option versus the seed orchard option is shown in Figure. This is based on a simulation of the Swedish Norway spruce (Picea abies K.) long term breeding programme with inputs chosen to correspond to the real programme as close as possible. Clonal forestry and seed orchards are different ways of harvesting gain from the breeding population and clonal forestry means that the genetic improvement reaches a certain level three decades before the same level can be achieved by seed orchards.

Genetic progress is primarily created in the breeding population. The production population is created by creaming the best available material from the breeding population either by clonal forestry or seed orchards. The gain unit can roughly be interpreted as per cent gain. The details of the figure depend on exact specifications, but it gives an idea of magnitudes based on best possible estimates. Clonal forestry gives a higher gain than seed orchards; the difference at a given time is in the magnitude of 15%.

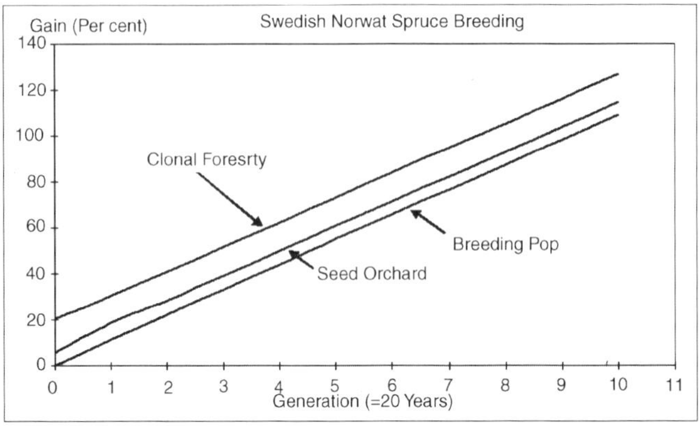

Fig. A comparison of the Genetic value of Three Populations as a Function of Time Based on a Simulation of the Swedish Norway Spruce Breeding Programme.

2

Crop Production

CROP COEFFICIENT APPROACH

In the crop coefficient approach the crop evapotrans-piration, ET_c, is calculated by multiplying the reference crop evapotranspiration, ET_o, by a crop coefficient, K_c: $ET_c = K_c\ ET_o$ where ET_c crop evapotranspiration [mm d^{-1}], K_c crop coefficient [dimensionless], ET_o reference crop evapotranspiration [mm d^{-1}]. Most of the effects of the various weather conditions are incorporated into the ET_o estimate. Therefore, as ET_o represents an index of climatic demand, K_c varies predominately with the specific crop characteristics and only to a limited extent with climate. This enables the transfer of standard values for K_c between locations and between climates. This has been a primary reason for the global acceptance and usefulness of the crop coefficient approach and the K_c factors developed in past studies.

These characteristics are:

- *Crop height:* The crop height influences the aerodynamic resistance term, r_a, of the FAO Penman-Monteith equation and the turbulent transfer of vapour from the crop into the atmosphere. The r_a tenu appears twice in me full form of the FAO Penman-Monteith equation.
- *Albedo (Reflectance) of The Crop-Soil Surface:* The albedo is affected by the fraction of ground covered by vegetation and by the soil surface wetness. The albedo of the crop-soil surface influences the net radiation of the surface, R_n, which is the primary source of the energy exchange for the evaporation process.
- *Canopy Resistance:* The resistance of me crop to vapour transfer is affected by leaf area (number of stomata), leaf age and condition, and the degree of stomatal control. The canopy resistance influences the surface resistance, r_s .
- Evaporation from soil, especially exposed soil.

The soil surface wetness and the fraction of ground covered by vegetation influence the surface resistance, Following soil wetting, the vapour transfer rate from the soil is high, especially for crops having incomplete ground cover. The combined surface resistance of the canopy and of the soil determines the (bulk) surface resistance, r_s The surface resistance term in the Penman–Monteith equation represents the resistance to vapour flow from within plant leaves and from beneath the soil surface.

The K_c in Equation predicts ET_c under standard conditions. This represents the upper envelope of crop evapotranspiration and represents conditions where no limitations are placed on crop growth or evapotranspiration due to water shortage, crop density, or disease, weed, insect or salinity pressures. The ET_c predicted by K_c is adjusted if necessary to non-standard conditions, $ET_{c\ adj}$, where any environmental condition or characteristic is known to have an impact on or to limit ET_c.

FACTORS DETERMINING THE CROP COEFFICIENT

The crop coefficient integrates the effect of characteristics that distinguish a typical field crop from the grass reference, which has a constant appearance and a complete ground cover. Consequently, different crops will have different $K_c$coefficients. The changing characteristics of the crop over the growing season also affect the K_c coefficient. Finally, as evaporation is an integrated part of crop evapotranspiration, conditions affecting soil evaporation will also have an effect on K_c.

Crop Type

Due to differences in albedo, crop height, aerodynamic properties, and leaf and stomata properties, the evapotranspiration from full grown, well-watered crops differs from ET_0. The close spacings of plants and taller canopy height and roughness of many full grown agricultural crops cause these crops to have K_c factors that are larger than 1. The K_c factor is often 5–10% higher than the reference (where K_c = 1.0), and even 15–20% greater for some tall crops such as maize, sorghum or sugar cane. Crops such as pineapples, that close their stomata during the day, have very small crop coefficients. In most species, however, the stomata open as irradiance increases. In addition to the stomatal response to environment, the position and number of the stomata and the resistance of the cuticula to vapour transfer determine the water loss from the crop. Species with stomata on only the lower side of the leaf and/or large leaf resistances will have relatively smaller K_c values. This is the case for citrus and most deciduous fruit trees. Transpiration control and spacing of the trees, providing only 70% ground cover for mature trees, may cause the K_c of those trees, if cultivated without a ground cover crop, to be smaller than one.

Climate

The K_c values are typical values expected for average K_c under a standard climatic condition, which is defined as a sub-humid climate with average daytime minimum relative humidity (RH_{min}) ≈ 45% and having calm to moderate wind speeds averaging 2 m/s.

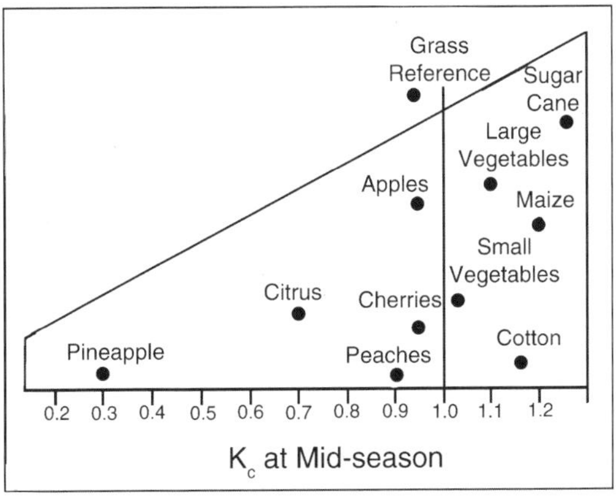

Fig. Typical K_c for Different Types of Full Grown Crops

Variations in wind alter the aerodynamic resistance of the crops and hence their crop coefficients, especially for those crops that are substantially taller than the hypothetical grass reference. The effect of the difference in aerodynamic properties between the grass reference surface and agricultural crops is not only crop specific. It also varies with the climatic conditions and crop height. Because aerodynamic properties are greater for many agricultural crops as compared to the grass reference, the ratio of ET_c to ET_o (*i.e.*, K_c) for many crops increases as wind speed increases and as relative humidity decreases. More arid climates and conditions of greater wind speed will have higher values for K_c. More humid climates and conditions of lower wind speed will have lower values for K_c.

The relative impact of the climate on K_c for full grown crops is illustrated in Figure. The upper bounds represent extremely arid and windy conditions, while the lower bounds are valid under very humid and calm weather conditions. The ranges expected in K_c as climate and weather conditions change are quite small for short crops but are large for tall crops. Under humid and calm wind conditions, K_c becomes less dependent on the differences between the aerodynamic components of ET_c and ET_o and the K_c values for 'full-cover' agricultural crops do not exceed 1.0 by more than about 0.05. This is because full-cover agricultural crops and the reference crop of clipped grass both provide for nearly maximum absorption of shortwave radiation, which is the primary energy source for evaporation under humid and calm conditions. Generally, the albedos, a, are similar over a wide range of full-cover agricultural crops, including the reference crop. Because the vapour pressure deficit (e_s–e_a) is small under humid conditions, differences in ET caused by differences in aerodynamic resistance, r_a, between the agricultural crop and the reference crop are also small, especially with low to moderate wind speed.

Under arid conditions, the effect of differences in r_a between the agricultural crop and the grass reference crop on ET_c become more pronounced because the (e_s–e_a) term may be relatively large. The larger magnitudes of (e_s–e_a) amplify differences in the aerodynamic term in the numerator of the Penman-Monteith equation for both the crop and the reference crop. Hence, K_c will be larger under arid conditions when the agricultural crop has a leaf area and roughness height that are greater than that of the grass reference.

Because the $1/r_a$ term in the numerator of the Penman-Monteith equation is multiplied by the vapour pressure deficit (e_s–e_a), the ET from tall crops increases proportionately more relative to ET_0 than does ET from short crops when relative humidity is low. The K_c for tall crops, such as those 2-3 m in height, can be as much as 30% higher in a windy, arid climate as compared with a calm, humid climate. The increase in K_c is due to the influence of the larger aerodynamic roughness of the tall crop relative to grass on the transport of water vapour from the surface.

Soil Evaporation

Differences in soil evaporation and crop transpiration between field crops and the reference surface are integrated within the crop coefficient. The K_c coefficient for full-cover crops primarily reflects differences in transpiration as the contribution of soil evaporation is relatively small. After rainfall or irrigation, the effect of evaporation is predominant when the crop is small and scarcely shades the ground. For such low-cover conditions, the K_c coefficient is determined largely by the frequency with which the soil surface is wetted. Where the soil is wet for most of the time from irrigation or rain, the evaporation from the soil surface will be considerable and K_c may exceed 1. On the other hand, where the soil surface is dry, evaporation is restricted and K_c will be small and might even drop to as low as 0.1. Differences in soil evaporation between the field crop and the reference surface can be forecast more precisely by using a dual crop coefficient.

EXPOSED AND WETTED SOIL FRACTION (FEW)

Few: Calculation Procedure

In crops with incomplete ground cover, evaporation from the soil often does not occur uniformly over the entire surface, but is greater between plants where exposure to sunlight occurs and where more air ventilation is able to transport vapour from the soil surface to above the canopy. This is especially true where only part of the soil surface is wetted by irrigation.

It is recognized that both the location and the fraction of the soil surface exposed to sunlight change to some degree with the time of day and depending on row orientation. The procedure presented here predicts a general averaged fraction of the soil surface from which the majority of evaporation occurs. Diffusive evaporation from the soil beneath the crop canopy is assumed to be largely included in the basal K_{cb} coefficient. Where the complete soil surface is wetted, as by precipitation or sprinkler, then the fraction of soil surface from which most evaporation occurs, f_{ew}, is essentially defined as $(1-f_c)$, where fc is the average fraction of soil surface covered by vegetation and $(1-f_c)$ is the approximate fraction of soil surface that is exposed. However, for irrigation systems where only a fraction of the ground surface is wetted, f_{ew} must be limited to f_w, the fraction of the soil surface wetted by irrigation. Therefore, f_{ew} is calculated as:

$$f_{ew} = \min(1-f_c, f_w)$$

where

$1 - f_c$ average exposed soil fraction not covered (or shaded) by vegetation [0.01–1],

f_w average fraction of soil surface wetted by irrigation or precipitation [0.01–1].

The 'min()' function selects the lowest value of the '$1-f_c$' and 'f_w' values.

The limitation imposed by Equation assumes that the fraction of soil wetted by irrigation occurs within the fraction of soil exposed to sunlight and ventilation. This is generally the case, except perhaps with drip irrigation. In the case of drip irrigation, where the majority of soil wetted by irrigation may be beneath the canopy and may therefore be shaded, more complex models of the soil surface and wetting patterns may be required to accurately estimate total evaporation from the soil. In this case, the value for f_w may need to be reduced to about one-half to one-third of that given in Table to account for the effects of shading of emitters by the plant canopy on the evaporation rate from wetted soil. A general approach could be to multiply f_w by $[1-(2/3)f_c]$ for drip irrigation.

fw: Fraction of Soil Surface Wetted by Irrigation or Precipitation

Table presents typical values for f_w. Where a mixture of irrigation and precipitation occur within the same drying period or on the same day, the value for f_w should be based on a weighted average of the f_w for precipitation ($f_w = 1$) and the f_w for the irrigation system. The weighting should be approximately proportional to the infiltration depths from each water source.

Table. Common Values of Fraction F_w of Soil Surface Wetted by Irrigation or Precipitation

Wetting Event	F_w
Precipitation	1.0
Sprinkler irrigation	1.0
Basin irrigation	1.0
Border irrigation	1.0
Furrow irrigation (every furrow), narrow bed	0.6...1.0
Furrow irrigation (every furrow), wide bed	0.4... 0.6
Furrow irrigation (alternated furrows)	0.3...0.5
Trickle irrigation	0.3... 0.4

Alternatively, on each day of the application, the following rules can be applied to determine f_w for that and subsequent days in a more simplified manner:

- *Surface is wetted by irrigation and rain:* F_w is the f_w for the irrigation system;
- *Surface is wetted by irrigation*: F_w is the f_w for the irrigation system;
- *Surface is wetted by significant rain (i.e., > 3 to 4 mm) with no irrigation:* $F_w = 1$;
- *Where there is neither irrigation nor significant precipitation:* F_w is the f_w of the previous day.

1 – F_c: Exposed Soil Fraction

The fraction of the soil surface that is covered by vegetation is termed f_c. Therefore, $(1–f_c)$ represents the fraction of the soil that is exposed to sunlight and air ventilation and which serves as the site for the majority of evaporation from wet soil. The value for f_c is limited to < 0.99. The user should assume appropriate values for the various growth stages. Typical values for f_c and $(1–f_c)$ are given in Table.

Table. Common Values of Fractions Covered by Vegetation (F_c) and Exposed to Sunlight ($1–f_c$)

Crop Growth Stage	F_c	$1 - F_c$
Initial stage	0.0–0.1	1.0–0.9
Crop development stage	0.1–0.8	0.9–0.2
Mid-season stage	0.8–1.0	0.2–0.0
Late season stage	0.8–0.2	0.2–0.8

Where f_c is not measured, f_c can be estimated using the relationship:

$$f_c = \left(\frac{K_{cb} - K_{c\,min}}{K_{c\,max} - K_{c\,min}}\right)^{(1+0.5h)}$$

where

f_c the effective fraction of soil surface covered by vegetation [0–0.99],

K_{cb} the value for the basal crop coefficient for the particular day or period,

$K_{c\,min}$ the minimum K_c for dry bare soil with no ground cover [0.15–0.20],

$K_{c\,max}$ the maximum K_c immediately following wetting, h mean plant height [m].

This equation should be used with caution and validated from field observations. $K_{c\,min}$ is the minimum crop coefficient for dry bare soil when transpiration and evaporation from the soil are near baseline (diffusive) levels.

$K_{c\,min} \approx 0.15–0.20$ is recommended. The value of $K_{c\,min}$ is an integral part of all K_{cb} coefficients. $K_{c\,min}$ ordinarily has the same value as the K_{cb} ini used for annual crops under nearly bare soil conditions (0.15–0.20).

Equation assumes that the value for K_{cb} is largely affected by the fraction of soil surface covered by vegetation. This is a good assumption for most vegetation and conditions. The '1+0.5h' exponent in the equation represents the effect of plant height on shading the soil surface and in increasing the value for K_{cb} given a specific value for f_c.

The user should limit the difference K_{cb}–$K_{c\ min}$ to 0.01 for numerical stability. The value for f_c will change daily as K_{cb} changes.

Application of Equation predicts that f_c decreases during the late season period in proportion to K_{cb}, even though the ground may remain covered with senescing vegetation.

Calculation of The Crop Coefficient (K_{cb} + K_e) Under Sprinkler Irrigation

A field of cotton has just been sprinkler irrigated. The K_{cb} for the specific day (during the development period) has been computed and then interpolated from the K_{cb} curve as 0.9. The ET_o = 7 mm/day, u_2 = 3 m/s and RH_{min} = 20%. Estimate the crop coefficient (K_{cb} + K_e).

Assuming h = 1 m, from Equation 72, Kc max for this arid climate is:

$$K_{c\max} = \max\left(\left\{1.2+\left[0.04(3-2)-0.004(20-45)\right]\left(\frac{1}{3}\right)^{0.3}\right\},\{0.9+0.05\}\right)=1.30$$

From Equation, where Kc min = 0.15:

$$f_c = [(K_{cb}-K_{c\ min})/(K_{c\ max}-K_{c\ min})]\ (1 + 0.5h) = [(0.9\text{-}0.15)/(1.3–0.15)]\ (1 + 0.5(1)) = 0.53.$$

As the field was sprinkler irrigated, f_w = 1.0 and from Equation 75:

$$f_{ew} = \min(1-f_c,\ f_w) = \min(1- 0.53,\ 1.0) = 0.47.$$

Assuming that the irrigation was sufficient to fill the evaporating layer to field capacity, so that K_r = 1, evaporation would be in stage 1.

From Equation: K_e = 1.00 (1.30 – 0.90) = 0.40

The value is compared against the upper limit f_{ew} $K_{c\ max}$ to ensure that it is less than the upper limit:

$$f_{ew}\ K_{c\ max} = 0.47\ (1.30) = 0.61,$$

which is greater than the value for K_e. Therefore, the value for K_e can be used with no limitation. The total K_c for the field, assuming no moisture stress due to a dry soil profile, is

$$K_c = K_{cb} + K_e = 0.9 + 0.40 = 1.30.$$

This value is large because of the very wet soil surface, the relatively tall rough crop as compared to the grass reference, and the arid climate (u_2 = 3 m/s and RH_{min} = 20%). In this situation, K_c happens to equal $K_{c\ max}$, as the field has just been wetted by sprinkler irrigation.

Calculation of the Crop Coefficient (K_{cb} + K_e) Under Furrow Irrigation

The cotton field in the previous example has been irrigated by furrow irrigation of alternate rows rather than by sprinkler, and the fraction of the field surface wetted by the irrigation is 0.3.

The f_{ew} in this case is calculated from Equation 75 as:

$$f_{ew} = \min(1-f_c, f_w) = \min(1-0.53, 0.3) = 0.3.$$

Assuming that the irrigation was sufficient to fill the f_{ew} portion of the evaporating layer to field capacity, so that $K_r = 1$, evaporation would be in stage 1.

$$\text{From Equation: } K_e = 1.00\ (1.30 - 0.9) = 0.40$$

The value is compared to the upper limit $f_{ew}\ K_{c\ max}$ which is 0.30 (1.30) = 0.39. Because 0.40 > 0.39, K_e from the f_{ew} surface area is constrained to 0.39. The total K_c for the furrow irrigated field, assuming no moisture stress due to dry soil, is $K_c = K_{cb} + K_e = 0.9 + 0.39 = 1.29$.

This value is essentially the same as for the previous example because the procedure assumes that the soil between alternate rows is the portion that is wetted by the irrigation, so that the majority of the field surface has either vegetation cover or wet soil.

Calculation of the Crop Coefficient (K_{cb} + K_e) Under Drip Irrigation

The cotton field in the previous example has been irrigated by drip irrigation, where the emitters are placed beneath the cotton canopy. The fraction of the field surface wetted by the irrigation is 0.3.

The f_{ew} in this case is calculated from Equation as $f_{ew} = \min(1-f_c, f_w)$. Because the emitters are beneath the canopy so that less energy is available for evaporation, the value for f_w is reduced by multiplying by $1 - (2/3)f_c$, so that:

$$f_{ew} = \min[(1-f_c),(1 - 0.67\ f_c)\ f_w)] = \min[(1-0.53), (1-0.67(0.53))(0.3)] = 0.19$$

Assuming that the irrigation was sufficient to fill the fw portion of the evaporating layer to field capacity, so that $K_r = 1$, evaporation would be in stage 1.

$$\text{From Equation: } K_e = 1.00\ (1.30-0.90) = 0.40.$$

The value is compared to the upper limit $f_{ew}\ K_{c\ max} = 0.19\ (1.30) = 0.25$. Because 0.25 < 0.40, K_e from the f_w fraction of the surface area is constrained by the available energy. Therefore $K_e = 0.25$. The total Kc for the drip irrigated field, assuming no moisture stress due to dry soil, is: $K_c = K_{cb} + K_e = 0.9 + 0.25 = 1.15$. This K_c value is less than that for sprinkler and furrow irrigation (Examples 32 and 33).

SIMULATION OF WATER FLUXES AND WATER LIMITATION TO CROP PRODUCTION

Soil water content W is updated daily for each CFT within each grid cell, following the balance between the amount of infiltrated water (precipitation P, snowmelt M, and irrigation water `Irr' minus interception loss from leaves E_I) and that removed from the soil through surface and subsurface run-off R (generated from excess water above field capacity), soil evaporation E_S, and plant transpiration E_T(all in mm day^{-1}):

$$W = (P + M + Irr - E_I) - R - (E_S + E_T).$$

E_S depends on potential evapotranspiration determined by the Priestley-Taylor method (Epot times the Priestley-Taylor coefficient $\alpha = 1.32$) and the relative moisture in the top 20 cm of the soil W_{r20}. It is computed for the fractions of bare soil of cropland as given by the coverage with vegetation fv and the daily status of crop-specific leaf phenology `phen':

$$E_S = E_{pot} \alpha W_{r20}^2 (1 - f_v \text{ phen}).$$

Crop transpiration E_T is determined for each CFT as the minimum of atmospheric demand D and water supply S:

$$E_T = \min(S, D)$$

S is regulated by soil moisture and plant hydraulic traits; when the soil is saturated, E_T reaches a maximum E_{max} (5-8 mm day-1) and declines linearly with W weighted by the CFT-specific fraction of roots in each soil layer. D (mm day-1) is a function of E_{pot} times a maximum Priestley-Taylor coefficient (α_m = 1.391) and the potential, water-unlimited canopy conductance g_{pot} (mm s^{-1}):

$$D = \frac{(1 - f_{wet}) E_{pot} \alpha_m}{1 + g_m / g_{pot}}$$

where f_{wet} is the fraction of the day that the canopy is wet, and gm is a scaling coefficient (3.26 mm s-1). g_{pot} is calculated from the photosynthesis rate and ambient CO_2 concentration. E_I is determined as:

$$E_I = \min(LAi \quad phen, P + Irr, E_{pot}) f_v$$

where LAI is the leaf area index and i a factor accounting for vegetation type and rainfall regime.

Total evapotranspiration E, *i.e.* crop water consumption, is computed as the sum of E_T, E_S and E_I. Net primary production NPP (Gt dry matter) is defined as gross primary production less autotrophic respiration. The former is estimated daily from the fractional coverage of a CFT in a grid cell, leaf phenology and weather conditions.

The latter is the compound of maintenance respiration (calculated based on the size of the living tissue pools, their assigned C:N ratios and weather data) and growth respiration (the cost of producing new tissues, calculated as a fraction of NPP).

The degree of water limitation of crop production was computed as the ratio between NPP and the production that could theoretically be reached if E_T was always at its maximum (NPP*). The CO_2 effect upon E_S and NPP includes both physiological and structural vegetation responses.

WATER MANAGEMENT SCENARIOS

Several simulations were performed to estimate crop NPP under different conditions:

- *OPT*: a model run in which the theoretical NPP* was simulated by assuming an always saturated soil ($S = E_{max}$).
- *BAS*: the baseline run, in which crops on areas equipped for irrigation were assumed to be irrigated, thus representing irrigation as the only water management (which reflects the present situation for most regions of the world).
- *INO*: a simulation in which no irrigation was assumed.
- *VS*: model runs (based on BAS) that emulate a vapour shift from non-productive evaporation to productive transpiration by systematically reducing year-round ES(equation) by, respectively, 10, 25, 50, and 85% (VS10, VS25, VS50, and VS85 simulations). In practice, such a vapour shift-an increase of the ET:E ratio-can be attained, for example, through soil and water conservation strategies like mulching or different tillage systems. Field studies showed that E_S can be reduced by 34-50% in this way; in semiarid regions, a maximum E_T:E ratio of 85% appears to be feasible.
- *RH*: model runs (based on BAS) that emulate rainwater harvesting strategies by concentrating and storing, respectively, 10, 25, 50, and 85% of R on cropland over a year and by redirecting this water to crops in periods of water stress so as to fulfill supplemental irrigation water requirements on rainfed cropland (RH10, RH25, RH50, and RH85 simulations).

Table. Overview of the Different Water Management Simulations.

		Present Climate
(a)	OPT	Maximum NPP* under always saturated soil
(b)	BAS	Baseline run including irrigation
(c)	INO	BAS, but no irrigation
(d)	VS	BAS with reductions in ES of 10, 25, 50, and 85%
(e)	RH	BAS with run-off harvesting of 10, 25, 50, and 85%
(f)	VS + RH	Same-level intensity combinations of VS and RH
		Projected Future Climate 2041-2070
(a)-(f)	CC	Climate change under constant year 2000 CO_2 concentration
(a)-(f)	CC + CO_2	Climate and CO_2 change

We analysed the differences between these model runs to estimate the effect on NPP of irrigation (INO minus BAS), and of the different management options (VS, RH, and VS + RH minus BAS) for the present, under SRES-A2 climate change (CC), and under climate change with CO_2 increase (CC + CO_2), respectively.

Estimates for the future were represented as averages of the results under three climate scenarios. We also used the spatial distribution of the revised high population growth scenario A2r for the year 2050 to relate the effects of the water management options to global food demand.

Since the A2r population scenario is at the upper end of the SRES range (10.2 billion in 2050) and because food and water demand are strongly affected by the population number, we also analysed results for medium (B2) and low fertility (B1) population scenarios that show, respectively, a world population of 9.4 and 8.7 billion in 2050. To reduce the number of simulations, we did not compute the climate change impacts under B1 and B2 emissions.

COMPUTATION OF WATER REQUIREMENTS FOR CROP PRODUCTION

To estimate the effect of the water management options on securing global crop production, we followed the approach of Rockströmet al, assuming that as a global average 1300 m3 cap-1 yr^{-1} of blue and green water are required under present conditions to produce 3000 kcal cap^{-1} day^{-1} for a healthy diet (20% meat share). Actually, however, this water requirement differs among regions. To roughly account for these differences, we distributed the global average among countries according to their water productivity (*i.e.* the water consumed per unit of NPP), such that regions with low water productivity got values higher than 1300 m3 cap^{-1} yr^{-1}(*e.g.* around 2000 m3 cap^{-1} yr^{-1} in many countries of sub-Saharan Africa), and vice versa.

(Note that these values refer to the normative goal of a diet of 3000 kcal; for producing today's actual diets, water requirements are significantly lower in many regions of the South.) Countries in which the water availability per capita falls below the thus computed value are regarded as water-stressed. Analogously, the water stress threshold changes under each of the scenarios computed here, following the individual changes in NPP and E and, thus, water productivity.

MODEL SETUP

LPJmL was driven by an updated version of the CRU TS2.1 climate dataset (0.5° spatial resolution) of monthly air temperatures, precipitation, and cloud cover for the period 1901-2000, preceded by a 1000 yr spinup during which the climate of 1901-1930 was repeated.

CO_2 concentrations were as in Rost *et al.* For the simulations under future climate, starting from the same equilibrium, the model was driven by the output from three different climate models under the A2 emissions trajectory for the period 1901-2070.

AREA, PRODUCTION AND PRODUCTIVITY OF MAJOR PULSES

India is producing 14.76 million tons of pulses from an area of 23.63 million hectare, which is one of the largest pulses producing countries in the world. However, about 2-3 million tons of pulses are imported annually to meet the domestic consumption requirement. Thus, there is need to increase production and productivity of pulses in the country by more intensive interventions.

ONGOING SCHEMES ON PULSES DEVELOPMENT

Presently, pulses development programmes are being implemented through the Centrally Sponsored Schemes of NFSM-Pulses, Integrated Scheme on Oilseeds, Pulses, Oil Palm and Maize (ISOPOM) and Integrated Crop Development Programme (ICDP)-Pulses under Macro Management Mode. While ISOPOM is implemented in all the 433 districts of 14 States, NFSM Pulses is being implemented in mission mode in 171 districts of these states. In addition, 15 districts of Jharkhand and 10 districts of Assam have recently been included under NFSM-Pulses for implementation with effect from 1st April, 2010. ICDP-Pulses is implemented in states other than the 14 NFSM/ISOPOM States. NFSM-Pulses cover about 80% of the total area under pulses in the country.

CROP EVAPOTRANSPIRATION (ET_C)

Crop evapotranspiration is calculated by multiplying ET_0 by K_c, a coefficient expressing the difference in evapotranspiration between the cropped and reference grass surface.

The difference can be combined into one single coefficient, or it can be split into two factors describing separately the differences in evaporation and transpiration between both surfaces. The selection of the approach depends on the purpose of the calculation, the accuracy required, the climatic data available and the time step with which the calculations are executed. Table presents the general selection criteria.

Table. General Selection Criteria for the Single and Dual Crop Coefficient Approaches

Single Crop Kcb + Ke	Coefficient Kc	Dual Crop Coefficient
Purpose of calculation	• Irrigation planning and design • Irrigation management • Basic irrigation schedules • Real time irrigation scheduling for non-frequent water applications (surface and sprinkler irrigation)	• Research • Real time irrigation scheduling • Irrigation scheduling for high frequency water application (microirrigation and automated sprinkler irrigation) • Supplemental irrigation • Detailed soil and hydrologic water balance studies
Time step	Daily, 10-day, monthly (data and calculation)	Daily (data and calculation)
Solution method	Graphical pocket calculator computer	Computer

SINGLE AND DUAL CROP COEFFICIENT APPROACHES

Single Crop Coefficient Approach (K_c)

In the single crop coefficient approach, the effect of crop transpiration and soil evaporation are combined into a single $K_c$coefficient. The coefficient integrates differences in the soil evaporation and crop transpiration rate between the crop and the grass reference surface. As soil evaporation may fluctuate daily as a result of rainfall or irrigation, the single crop coefficient expresses only the time-averaged (multi-day) effects of crop evapotranspiration. As the single K_c coefficient averages soil evaporation and transpiration, the approach is used to compute ET_c for weekly or longer time periods, although calculations may proceed on a daily time step. The time-averaged single K_c is used for planning studies and irrigation system design where the averaged effects of soil wetting are acceptable and relevant. This is the case for surface irrigation and set sprinkler systems where the time interval between successive irrigation is of several days, often ten days or more. For typical irrigation management, the time-averaged single K_c is valid.

DUAL CROP COEFFICIENT APPROACH (K_{cb} + K_e)

In the dual crop coefficient approach, the effects of crop transpiration and soil evaporation are determined separately. Two coefficients are used: the basal crop coefficient (K_{cb}) to describe plant transpiration, and the soil water evaporation coefficient (K_e) to describe evaporation from the soil surface.

The single K_c coefficient is replaced by:

$$K_c = K_{cb} + K_e \quad (57)$$

where,

K_{cb} basal crop coefficient,

K_e soil water evaporation coefficient.

The basal crop coefficient, K_{cb}, is defined as the ratio of ET_c to ET_o when the soil surface layer is dry but where the average soil water content of the root zone is adequate to sustain full plant transpiration. The K_{cb} represents the baseline potential $K_c$in the absence of the additional effects of soil wetting by irrigation or precipitation. The soil evaporation coefficient, K_e, describes the evaporation component from the soil surface. If the soil is wet following rain or irrigation, K_e may be large.

However, the sum of K_{cb} and K_e can never exceed a maximum value, $K_{c\ max}$, determined by the energy available for evapotranspiration at the soil surface. As the soil surface becomes drier, K_e becomes smaller and falls to zero when no water is left for evaporation. The estimation of K_e requires a daily water balance computation for the calculation of the soil water content remaining in the upper topsoil.

The dual coefficient approach requires more numerical calculations than the procedure using the single time-averaged $K_c$coefficient. The dual procedure is best for real time irrigation scheduling, for soil water balance computations, and for research studies where effects of day-to-day variations in soil surface wetness and the resulting impacts on daily ET_c, the soil water profile, and deep percolation fluxes are important. This is the case for high frequency irrigation with microirrigation systems or lateral move systems such as centre pivots and linear move systems.

3

Commercial Crops Production

OILSEED CROPS

Important points for obtaining maximum Yields:

1. Using the improved variety recommended for the tract.
2. Using good and bold seeds for sowing.
3. Treating seed with fungicide, e.g. Captan or Thiriin, at the rate of 4 g 1 kg of seed.

RAPESEED AND MUSTARD

Under the names rapeseed and mustard, several oilseeds belonging to the crucieferae are grown in India:

- Brown mustard, commonly called rai (raya or laha)—Brassica juncea (L.) Czern. & Coss
- Sarson
- Yellow sarson—B. campestris L. var. sarson Prain
- Brown sarson—B. campestris L. var. dichotoma Watt
- Toria (lahi or Maghi Labi)—B. campestris L. var. toria Duth.
- Taramira or tara (eruca sativa Mill.)

In trade, sarson, toria and Taramira are known as rapeseed, and rai as mustard.

Banarsi rai (B. nigra Koch.) which does not fall under any of the four groups is a garden crop used as spice. The cultivation of white mustard (Brassica alba syn. Sinapis alba) is no longer found in India. Rai and Yellow sarson are self-fertile and the rest of the cruciferous oilseeds, viz. brown sarson, toria, taramira, Banarasi rai and white mustard, are self-incompatible.

The acreage under yellow sarson is scantily (mainly in Bihar) and constantly on the decrease. In the recent past, the accreage under brown mustard is steadily on the increase (over 65%) at the expense of other Brassica due to its higher production, greater to pests and diseases and moisture stress.

Origin

Brassica juncea L. (rai) was originally introduced from China into northeastern India. From where it has extended into Afghanistan via the Punjab. Eastern Afghanistan, together with the adjoining north-western India is one of the independent centres of brown sarson (Brassica campestris var. brown sarson). Yellow sarson (B. campestris var. yellow sarson) is commonly grown in the eastern parts of India where it shows much diversity of forms. Taramira is a relatively recent introduction into India. It is believed to be a native of southern Europe and North Africa.

Economic Importance

Rapeseed and mustard yield the most important edible oil content of the seeds of different ranges from 30 to 48 per cent. In the case of white mustard, the oil content ranges from 25 to 33 per cent. The oil obtained is the main cooking medium in Northern India and can not be replaced by any other edible oil. The seed and oil are used as a condiment in the preparation of pickles and for flavouring curries and vegetables. The oilcake is mostly used as cattle feed. The leaves of young plants are used as a green vegetable. The use of mustard oil for industrial purposes is rather limited on account of its high cost.

Botanical Description

Rapseeed and mustard include annual herbs ranging in height from 0.45 to 1.75 m. Roots, in general, are long and tapering. Toria is more or less a surface-feeder and brown sarson has long roots, with a limited lateral spread, enabling its successful cultivation under drier conditions. Yellow sarson has both extensive and lateral spread. The height of the stem varies from 45 cm (in some varieties of toria) to 1.75 m (in yellow sarson). In toria and brown sarson, the branches arise at an angle of 30° to 40°. In yellow sarson, branches arise laterally at an angle about 10° to 20° and give the plant a narrow and pyramidal shape. The inflorence is coroymbose raceme. In the case of sarson the four petals are spread apart, whereas in brown sarson and toria the petals overlap or may be placed apart, depending upon the variety. The flowers bear a hypogynous syncarpous ovary. In brown sarson and toria, the ovary is bicarpellary, whereas in the case of yellow sarson, it may also be tri or tetra-carpellary.

The fruit is a siliqua. The pods are two-valved, three-valved or four-valved, depending upon the number of carpels in the ovary. The flowers begin to open from 8 a.m. and continue up to 12 noon.

Distribution, area and Production

The crop is grown both in subtropical and tropical countries. In Asia, it is chiefly grown in China, India and Pakistan. It is also grown in Europe, Canada and the USSR, but the forms of rapeseed and mustard grown there are different from those grown in India.

India occupies the first position, both with regard to acreage and production of rapeseed and mustard in the world. In India the Braddica crops occupy the second largest position after groundnut, with 3-5 million hectares, producing about 2 million tonnes of seed annually. The chief states producing them are Uttar Pradesh, Punjab, Haryana, Assam, Bihar, Madhya Pradesh, Rajasthan, West Bengal and Orissa.

Climate and Soil

The rapeseed and mustard crops are of the tropical as well as of the temperate zones and require relatively cool temperatures for satisfactory growth. In India, they are grown in the rabi season from September-October to February-March. The rapeseed and mustard crops grow well in areas having 25 to 40 cm of rainfall. Sarson and toria are preferred in low-rainfall areas, whereas raya and toria are grown in medium and high-rainfall areas respectively.

The rapeseed and mustard thrive best in light to heavy loams. Raya may be grown on all types of soil, but toria does best in loam to heavy loam. Sarson is suited to light-loam soil and taramira is mostly grown on very light soils.

Cultivation

A fine seed-bed is required to ensure good germination. In irrigated areas, the first ploughing is done with a medium sized soil-turning plough, followed by two to four ploughing with a desi plough or a cultivator. Sohaga (Planking) is given after every ploughing.

In rain fed areas, one to two ploughings with a desi plough or a cultivator, each followed by planking, may be given. Toria, in particular, requires a fairly moist seed-bed for good germination, but excessive moisture should be avoided. The rate in the case of mixed cropping depends on the proportion of the rapeseed to the main crop.

When sown pure, 5 kg of seed per ha is used for all rapeseeds and mustard. When sown mixed with other crops, the time of sowing rapeseed and mustard is governed by the sowing of the main crop. The first half of September is best for sowing toria (if wheat is to follow, it should be sown by the end of August), 25th September to 15th October for sarson, 30th September to 15th October for raya, and tarantira is sown throughout October. The seed from healthy and desirable plants, grown in isolation in the case of self-sterile form, should be used. Whenever moisture in the field is inadequate, the seed is mixed with moist soil and kept overnight. For distributing evenly, the seed is usually mixed with sand before sowing.

In mixed cropping rapeseed and mustard are sown in rows 1.8 to 2.4 metres apart across the main crop. The pure crop of rapeseed or mustard is sown at a depth of 4 to 5 cm in lines, 30 cm apart, with a drill, or with a (tube) attached to the plough. Thinning is done three weeks after sowing to maintain a plant-to-plant distance of 10 to 15 cm.

One hoeing in the case of toria in the third week after sowing and 1-2 hoeings in the case of sarson and raya are adequate. Forty kg of W per ha is optimum for all rapeseed and mustard crops in rain fed areas. Under irrigated conditions doses of 40 and 80 kg of W per ha are considered optimum for toria, sarson and rapeseed respectively. For tarantira, 20 kg of W per ha is ancient. All the fertilizers should be drilled before sowing. Among the Brassicae, raya is most responsive to irrigation, followed by yellow sarson. Two irrigations, one at flowering and the other at pod formation, result in the maximum yield in the case of toria, sarson and raya. Its flowering is most economical, if there is water scarcity. Tarantira is not irrigated.

Harvesting is done as soon as the crop begins to turn yellow. Tma, which takes 75 to 90 days to mature, is the earliest crop to be harvested. Harvesting is done with hand-sickler. Threshing is done by beating with a wooden stick the seed bearing part of the plants, taken in convenient bundles or by trampling them under the feet. Winnowing is done with by slowly dropping the threshed produce from a basket held shoulder-high. The seed after being dried in the sun is stored in gunny bags or bins.

Pets and Diseases

The most serious pests of rapeseed and mustard are the mustard aphid (Lipaphis erysinbi (Holt) syn. Rhopalosiphun pseudobrwsicae (Davis), mustard sawfly (Athalia proximo King) and cutworm (Agrotis ypsi" Poott.). To control aphidy the spraying of Methyl Demeton 0.02% or limetboate 0.03% two or three times, depending on the intensity is recommended. The mustard sawfly is controlled by dusting 10% BHC at the rate of 25 kg per ha. Cutworms are controlled by drilling into the soil 5% dust of Aldrin or Heptachlor or Chlordane at the rate of 25 kg per ha.

Alternaria blight (Alternaria brassicae Bork.) is the most widespread and destructive disease of rapeseed and mustard, whereas rust (Cystopus candidus) is serious in certain areas. Spraying with Dithane M-45 or Difolaton at the rate of 11 kg per ha is recommended for controlling these diseases. Orobanche, a phanerogamic parasite, has also been reported to be serious in Punjab and in certain parts of Uttar Flradesh. Hand-pulling of the plants of the parasite and a long rotation are recomnended.

Yield

The average yield of rapeseed and mustard is about 500 kg per ha. Under good crop management, toria, which is a short-duration crop, gives a yield of about 450 to 650 kg per ha, sarson 800 to 1,000 kg and rai 1,000 to 1,200 kg per hectare.

Oil and Protein Quality in the Brassicae

The oil of rapeseede and mustard possesses a Sizable amount of crucie acid (38 to 57%), together with linolenic acid up to 4.7 to 13.0 %. The oleic and linolenic acid, which have nutritive value, together constitute only about 27 per cent. It is desirable to increase the quantity of oleic and linolic acid and erucic acid. A lower proportion of erucic acid will make the oil more palatable, nutritive besides reducing metabolic disorders.

The Protein contained in rapeseed and mustard normally ranges between 24 and 30%. On the whole-seed basis and between 35 and 40% on the meal basis. But the presence of toxic glucosinolates in the mustard cake is unsuitable as source of human protein and is at present use as a manual and as a cattle feed.

SESAMUM

Geographic Origin

Sesamum (Sesamum indicicum L.), also known as sesame, til and gingerly, is an important and ancient oil-yielding crop. Africa is considered to be the primary centre of origin of this crop because of the presence of its diverse wild species in that continent. India is the secondary centre of origin and another secondary centre is Japan.

Economic Importance

The sesamum seed is a rich source of edible oil. Its oil content generally varies from 46 to 52 per cent. Its grains may be eaten fried, mixed with sugar or in the form of sweetmeats. Sesamum oil is used as a cooking-oil in southern India. It is also used for anointing the body, for manufacturing perfumed oils and for medicinal purposes. Sesamum-cake is a rich source of protein, carbohydrates and mineral nutrients, such as calcium and phosphorus. The cake is edible and is eaten avidly by working classes. It is also a valuable and nutritious feed for milch cattle.

Botanical Description

The genus Sesamum belongs to the family Pedaliaceae which has a wide distribution, covering tropical Africa, Madagascar, Arabia, India, Sri Lanka, tropical Australia and a few of the eastern islands of the Malayan Archipelago.

Sesamum is an erect and annual herb. The plant is simple or branched. The leaves are opposite below and alternate above. The flowers are zygomorphic, solitary, occasionally two or three together, auxiliary, short- pedicelled, borne on the upper part of the stem or branches geniculate. The tube, bent downwords and is dialated above the oblique base. The calyx is small, and five sectioned, 0.5-0.6 cm long. The corolla is tubular, campanulate, approximately 3 cm long, with an applicate margin and slightly bilabiate, the upper lip is entire and the lower one is tripartite. There are four functional stamens and often one is sterile, didynamous. The antliers are sagittate. The ovary is superior, bilocular, but may be subdivided by false internal walls. The fruit is a capsule, erect and oblong. The capsule contains numerous small ovate seeds. The testa may be smooth or reticulate and may be white, yellow, reddish-brown or black. The seed is dicotyledonous, albuminous and oleaginous.

Distribution and Production

India, China Sudan, Mexico, Turkey, Burma and Pakistan are the important sesamum producing countries. India ranks first, both in the area and production of sesamum in the world. The annual area put under it in India is about 2-5 million hectares (45 per cent of the world hectarage) and the total production is nearly 52 thousand tonnes. Sesamum is grown on 21 lakh hectares in only eight states, viz. Uttar Pradesh (673,000), Rajasthan (562,000), Madhya Pradesh (345,000), Andhra Pradesh (237,000) Maharashtra (139,000), Gujarat (118,000), Tamil Nadu (117,000) and Orissa (103,000. Among other states only Karnataka has a sizable area (68,000) under sesamum. In the remaining states it is grown only on a small area and hence is a very minor crop there.

Climate and Soil.

Sesamum grows in the plains and at elevations up to 1,200 m. It cannot stand frost, continuous heavy rain or prolonged drought. It is grown in sandy-loam to heavy black soils. Most of the crop is confined to lighter soils and its cultivation in heavy soils is limited to certain regions. The pH of sesamum growing soils ranges from 5.5 to 8.2. The sesamum growing soils are shallow to medium in Andhra Pradesh, Orissa, Andhra Pradesh and Tamil Nadu and are deep in Rajasthan, Gujarat and Maharashtra.

Mixed Crpopping

sesamum is grown both as pure and mixed with other crops, whereas the semi-rabi and summer crops are taken as pure. The common component crops are red-gram (arhar) in Andhra Pradesh, bajra and red-gram in Gujarat, red-gram and jowar in Madhya Pradesh. cotton, groundnut and red-gram in Maharashtra, and black-gram, green-gram and horse-gram in Orissa.

Cultivation

In India, seasmum is grown in three seasons, viz. kharif, semi-rabi and summer. The kharif crop occupies over 70 per cent of the area, whereas the semi-rabi and summer crops occupy 20 per cent and 10 per cent area respectively. The kharif sesamum is sown in June-July with the onset of the monsoon and is harvested in December-January. The kharif and semi-rabi crops are entirely rain fed, whereas summer crop is grown under irrigation. The yield of the kharif crop is poor, whereas those of the semi-rabi and summer crops are high, as they are grown in rich soils and under better management. The preparatory cultivation for the kharif crop is usually not thorough. For the semi-rabi and summer crops, the land is ploughed and harrowed repeatedly to secure a clean and fine seedbed. Except in Maharashtra and Gujarat, where line-sowing is practiced, in all other states it is sown broadcast. The recommended spacing's for the line-sown crop in different regions, however, range from 25 to 35 cm between the rows and 10 to 20 cm between the plants in the row. The seed being small, it is often mixed with sand to ensure its even distribution and the drill is operated rather shallow to avoid deep sowing. After sowing, the seed is covered lightly with a brush harrow. The seed-rate varies from 3 to 5 kg per hectare. The seed should be treated with Captan or Thiram at the rate of 3 g per kg of seed to control seed-borne diseases.

The rainfed crop is weeded and hoed once or twice and the irrigated crop is weeded often. Wherever the weed controlwith mechanical means is a problem, chemical weed control with pre-emergence application of Lasso at the rate of 3 litres/ha is recommended. The crop is generally recommended for different regions as 25-30 kg of W, 20-30 kg of P_2O_5 and 0-20 kg of K_2O per hectare. The crop is harvested when the leaves, stems and capsules begin to turn yellow and the lower leaves start shedding. To prevent the shedding of grains, the crop should not be allowed to become a drip in the field. The ripe plants are cut, carried to the threshing-yard, stacked for a week in the sun with the cut-ends downwards and well shaken or beaten to take out the grains from the dry capsules. Winnowing and cleaning completes the process.

PESTS AND DISEASES The leaf-eating caterpillar (Antigastra Catalaulmlis Dup.) and the gall-fly (Asphon- dylia sesalili Felt) are the serious pests of sesamum. Two disulphur (0.5 kg/ha) sprays of Carbaryl 0.2%, or N on the 40th day and the 60th day, however give an elective control of both the pests.

Stem and root-rot (Macrophominia phaseoli Maubl.) Ashby, phyllody (virus, mycoplasina), bacterial leafspot (Pseudomonas sesami Matkoff) and leaf-curl are the important diseases of sesamum. Seed treatment with Ciptan at the rate of 3-9 Per kg of seed for controlling Macfphomila rot, 3 sprays of Agrimyein at the rate 01-100 ppm at 15-day intervals for controlling bacterial leaf spot are recommended. Both phyllody and leaf-curl can be controlled by spraying, the insecticides, such as Phosphomidon 0.05% or Dimethoate 0.05% or Dichlorovas 0.05%, which reduce the population of the insect vectors. The average grain yield of the kharif crop grown unmixed is 200 to 500 kg per hectare and the average yields of the semi-rabi crop and of the irrigated summer crop are 300 to 600 kg per hectare.

Varieties

The varieties differ not only in respect of the season of growth and the time of maturity but also in the number of capsules formed in the leaf axils, the habit of growth (Single-stemmed or branched) and the size, colour and oil content of the seeds. The seeds may be white, dull white, light brown dark brown, red or black and their oil content varies from 48-52%. The Table gives a list of improved varictica recommended for cultivation in different regions.

Important hints for obtaining high yields:

1. A suitable variety must he chosen for the tract and the season.

2. The crop should be sown in lines.
3. The application of fertilizers, inter-cultivation and plant-protection measures must be taken up at the appropriate time.
4. Harvesting should be done after the crop matures but must not be delayed up to the stage when the capsules shed seeds.

SAFFLOWER

The three centers of origin, as suggested by Vavilov for safflower (Carthamus tinctorius L.), are India, Afghanistan and Ethopia. Decandole was of the opinion that Arabia was the most probable centre of origin.

Economic Importance

The crop is now cultivated, primarily for its seeds which yield oil, though at one time it used to be grown for the extraction of a dye also. The seeds are edible and are eaten after roasting. Their oil content varies from 24 to 36 per cent, depending on the variety, soil, climate and other conditions. The cold-pressed oil is golden yellow and is used for culinary purposes, or for making soap. The oil obtained by dry hot distillation is black and sticky and is used only for greasing well ropes and leather goods exposed to water. Safflower oil has also good drying properties and is, therefore, used in the manufacture of paints, varnishes and linoleum. It can be mixed with white paint without any after-yellowing effects. The cake, particularly from decorticated seed, is used as a concentrated cattle feed, and that from undecorticated seed is sometimes used as a manure.

Botanical Description

The safflower plant is highly branched, herbaceous, thistle-like annual, varying in height from 30 to 160 cm. It has many branches each terminating in a flower. The inflorescence is broad, flat or slightly curved and densely bristled owing to the presence of numerous floral tracts. The flowers are yellow to red, containing 20 individual ficrets, each of which produces a seed.

Distribution Area and Production

The important safflower growing countries, besides India, are the USA, Mexico, Ethopia, Spain, the USSR. and Australia. In India, it occupies 590,000 hectares with a production of nearly 130,000 tonnes. Over 98% of the area is concentrated in the states of Maharashtra (04.4%), Karnataka (26.0%) and Andhra Pradesh (8.0%).

Climate and Soil

The crop is grown during rabi primarily as a rain fed crop, but in some areas it is raised under irrigation. At all stages of growth of the crop, excessive rainfall or humidity increases damage from fungal diseases. Water-logging due to poor drainage or prolonged rains even for relatively short periods can cause substantial fall in seed-yield. Being drought-resistant, it is cultivated on all types of soil, including sandy soil, but it thrives best on water-retentive black soils and alluvial loams. The crop is fairly resistant to saline condition.

Cvltivation

Saffiower is grown mostly as a mixture with wheat, barley, gram and rabi jowar, three rows of safflower being planted after every nine, 12 or more rows of the main crop. Sometimes, it is sown as border rows surrounding the crops of wheat, rabi, jowar, etc., because being spiny, it protects the main crop against cattle trespass. As a pure crop, it is rotated with wheat, gram and rabi jowar. Occasionally, it is grown as a second crop after any quick-maturing kharif crop, e.g. green-gram (moong), black-gram (mash), groundnut, coriander, etc.

The number of ploughings should be as few as possible and be aimed at preserving the maximum amount of soil moisture. One or two ploughings with a country plough will be sufficient to remove the weeds and break the clods. Sowing is generally done in September and October. The seed-rate adopted varies from 5 to 12 kg per hectare, depending upon the soil fertility and the nature of the crop (pure or mixed). When grown in strips, a row spacing of 45 cm is adopted. Plants too close tend to have thinner stems or wear a superficial root-system, with a reduced number of flowers. In general, the pure crop is given one or two weedings, combined with hoeing on the 2Oth and the 45th day after sowing. The application of 20-40 kg of N per hectare results in a substantial increase in yield. The mixed crop shares the preparatory tillage, manuring and cultivation given to the main crop. When the plants have developed the central flowering head, they are topped to promote branching, flowering and seed production.

Harvesting and Yield

The crop matures in 4 or 5 months after sowing. The ripe plants are either cut or pulled out, stacked for a few days to dry, threshed by cutting with sticks, and the resulting material is winnowed to obtain clean seeds. The average yield of the pure crop is 400-500 kg of seed per hectare, whereas that of a mixed crop is about 100 kg.

Pests and Diseases

The safflower aphid (Dactynotus carthami), thrips (FranklinieUa sulphurea) and the flower bud fly (Acanihiphilus helianthi Rossi) are the serious pests of the crop. Spraying the crop with 0.1% Lenition, 0.03% Dimethoate and 0.07% Endosulfan is recommended for controlling aphids, trips and bud fly, respectively.

Rust (Puceinia carthami), alter aria leaf-spot (Alternai carthami) and root-rot (Phytophthora drechskri) are the important diseases of safflower.

The growing of resistant varieties is recommended to avoid damage from these diseases.

SUNFLOWER

Origin and Distribution

The cultivated sunflower (Helianthus annus L.) is a native of southern United States and Mexico, from where it was taken to Spain before the, middle of the sixteenth century. In the nineteenth century, the cultivation of sunflower as an oil aced crop began in the Soviet Union and the majority of the present day varieties grown all-over the world trace back their origin to the USSR. The major sunflower-producing countries of the world are the Soviet Union, Argentina, Bulgaria, Pumania, Turkey and South America. Sunflower, as an oilseed crop, was introduced into India in 1969 and, according to the estimates of the trade and the Vanaspati Manufacturers Association, 3,88,000 hectares of sunflower was grown in India during 1975-76.

Economic Importance

Sunflower is mainly grown for its oil. The oil is used for culinary purposes, in preparation of vanaspati and in the manufacture of soaps and cosmetics. It is especially recommended for heart patients. Its cake is rich in protein and is used as a cattle and poultry feed.

Botanical Description

Sunflower belongs to the family Compositac and the genus Helianthus. The cultivated genotypes are characterized by a single stem terminating in a capitulum. Sunflower is protandrous, in which the male and female elements mature at different times. There appears to be a time-lag of 18-24 hours in the maturity of the male and female elements. So it is essentially a cross-pollinated plant, besides showing varying degrees of self-incompatibility.

Climatic and Soil Requirements

Sunflower is a day neutral plant can be grown successfully in different seasons under varying conditions of day length, provided the temperature is favorable. Sunflower, basically a temperature plant, can also be grown commercially in tropical or sub-tropical conditions. It grows best with clear sky and occasionally rain shower during early stages. This corps requires a cool climate during germination and seedling growth and warm non-cloudy weather and high temperature during flowering to maturity. Sunflower can tolerate temperature range of 8 to 30 °C but the temperature below 16 °C and above 40 °C reduces yield and oil content. High temperature affects pollination and fertilization, thus care should be taken to adjust the sowing date in such a way that flowering does not coincide with high temperature. The optimum temperature for best performance under controlled conditions is 27-28 °C.

Sunflower can be planted successfully when soil temperature exceeds 10 °C, though germination begins at 3-5 °C. Young plants(1 or 2 pairs of leaves)can withstand frost to -6 °C. However, plants are frost sensitive from 6-8 leaf stage until flowering finishes. Late frost does not destroy the whole plant but destroys the fruiting head, leading to increased branching, resulting in small heads with many empty fruits. During flowering temperature should be between 18-20 °C. Goods yields are obtained where the ruling temperature is 18-22 °C showing that generally speaking the crop needs a warm climate. Average minimum temperature below 17 °C between flowering and maturity is necessary to ensure best oil quality. The crop is grown from 45 °S to 55 °N but greater production is between 20-50 °S. The crop can be cultivated upto an altitude of 2500 m but highest yield of oil is obtained below 1500 m. Sunflower cannot tolerate frost and requires a frost free period of 120 days. It is resistant to drought but requires continuous availability of soil moisture for the best performance.

Sunflower can thrive on a wide range of soils such as sandy loam, black soil and alluvial soil, but it performs better on sandy loam soil than on clay soil under similar management. However, good drainage is more important than soil type. Thus, fertile, well drained, medium textured soils having heavy water holding capacity are best suited for the cultivation of sunflower. Sunflower is highly susceptible to water logging. Avoid cultivation of sunflower on soils affected by salts and with poor drainage. the optimun pH requirementis 6.0 to 8.5.

Preparatory Tillage

Sunflower requires a fine seedbed, free of weeds and clods, for proper germination and good crop stand. to bury the residues or previous crop, ploughing with disc harrow followed by 2-3 cultivations and plankings may be required. fields vacated by potato may not require much seedbed preparation. However, adequate soil moisture is essential forgetting good germination.

Variety

MSFH8, PSFH-67,GKSFH-2002, Jwalamukhi, Mega-363 and NSFH-592 have been recommended. These hybrids are medium tall with dark green leaves. their seeds are black in colour and mature in about 110-125 days mega-363 is a long duration hybrid and hence should not be selected for cotton areas. These hybrids have an average yield about 20 quintals per hectare with an oil content of 40.5 to 43.0 percent. In hybrids fresh seed is to be sown every time which adds to its cost of cultivation.

Sowing Time

Spring season is the best for getting higher and stable seed yield sunflower. Sunflower being a cross pollinated crop depends on honey bees for pollination and seed setting. Availability of honey bees is abundant during spring season due to favourable. Planting during spring should be made after the soil temperature exceeds 10 °C at 10 cm depth at 8 a.m. Earlier planting at lower temperature increases time of flowering without much earliness. Best results are obtained when crop is sown in the first fortnight of January which produces higher seed yield than later planting. Sowing of short duration hybrids like GKSFH-2002,PSFH-67 and nsfh-592, however, can be extendedupto second fortnight of January. Besides low yields, the late sowing may also face early mansoon rain at harvesting ans threshing causing difficulty in its management.

Seed Rate and Seed Treatment

Use 5 kg seed per hectare for hybrids to achieve the required plant stand. Low seed rate recommendation for hybrids is mainly because of smaller seed size and wider spacing for hybrids. Seed should be treated with Brassicol or Thiram @ 2 g per kg of seed to control seed and soil borne fungal disease.

Method of Sowing

A spacing of 60 x 30 cm for hybrids has beenfound to be an optimum. These spacings result in better development of head, more seed filling and less stem breakage.

Crust Breaking

Rain after sowing, results in crust formation, which hinders the proper emergence of seedlings. Thus it may be necessary to break the crust, which can be done with harrowing. Harrow as soon as possible after the crust has performed and if possible, before seedlings have emerged.

Fertilizer Management

Fields studies have shown that hybrid sunflower responds well to nitrogen application upto 60 kg/ha. further increase in nitrogen results in lower seed and oil yield. Application phosphatic ferticizers @ 30 kg P_2O_5/ha increase sunflower yield. sunflower responds to K application only on soils deficient in available potassium and it is recommended to be applied @ 30 kg K_2O/ha on such oils. Regarding application method, drilling the fertilizers besides the seed row has been found to be more beneficial for getting higher seed yield. Application of fertilizers below the seed was found to be injurious to germinating seeds and reduced seed emergence count and seed yield. Entire dose of N, P and K should be applied at the same time of sowing. Prefer single superphosphate for P application as it also contains sulphur to which sunflower responds favourably. If single superphosphate is not available apply gypsum @ 25-30 kg/ha at sowing time. A suppllementary dose of nitrogen @ 25-30 kg/ha may be top dressed a month after sowing sunflower if the crop exhibits symptons of nitrogen deficiency. Nitrogen deficiency appears as yellowing of the older leaves, which die prematurely. Leaves are also fewer and smaller. Plants with a severe deficiency are yellow and stuned, while young plants are of purplish colour. Plants are slender with longer internodes. The flower head is small with many unfilled seeds and the oil content is reduced.

Phosphorous deficiency results is delayed flowering, stuned growth and small leaves which prematurelydie back from the leaf tip. Phosphorous deficiency is shown in small dark green leaves later turning pale with necrosis. Growth is restricted with short stiff stems, flowers are small, stuned malformed. Ripening is delayed, oil content and yield are reduced.

Potassium deficient plants are squat with a proportionately greater reduction in stem length then in leaf size. Initially leaves are dark green but turn yellow at the edges and die except the lower leaves. Potassium deficient plants have restricted growth with a rosette arrangement of the younger leaves. later the leaves develop dark brown necrotic patches.

Water Management

Sunflower can withstand moderate drought because of its deep root system. However, it responds favourably to irragation and its seed yield can be increased subtantially by sustantially by sustaining adequate moisture insoil by timely irrigations besides improvements in oil content. sunflower is highly moisture sensitive plant and its broad leaves exhibit wilting with slight moisture deficiency. First irrigation to sunflower should be applied a month after sowing and interval during the subsequent months be reduced to 2-3 weeks during March and 8-10 days during the hots months of Apriland May. Irrigation should be stopped about 12-14 days before harvest. Sunflower is a tall growing plant with heavy head, therefore, irrigation at flowering and thereafter should be given on clam days to avoid lodging. The incidence of Sclerotinia stem-rot is more under irrigated conditions especially where the soil drainage is poor. Excessive and frequent irrigations on fine textured soils should be avoided.

Weed Management

Mechanical methods such as hoeing at 3-4 weeks after sowing followed by earthing up before flowering has been found to be effective in checking weeds. Use of tractor drawn implements before the crop attains 60-70 cm height has also been found effective. for better efficiency of manual weeding an improved 3 tine wheel hand-hoe could also be used. Removal of weeds through mechnical/manual means means quite effective but expensive. Moreover, often timely labour is not available. The studies initiated to find effective herbicides for weed cintrol in sunflower, revealed that pendimethalin (STOMP @ 2.51/ha) as pre-emergence application gave effective weed kill.

Harvesting and Thershing

The crop is ready for harvesting when the back of flower heads turn lemon yellow or yellowish brown near the stalk and the discs start drying. At this stage the seeds give a blakish look and are fooly ripe. the shadding of disc-flower residues from undamaged heads also indicate psysiological maturity. Promptly harvesting seed after psysiological maturity has many advantages; seed is least exposed to bird and rodent attack; the crop cycle is shorter resulting in more time to prepare for the next crop; reduces harvest seed losses; provides high quality seed and reduces exposures to bad weather. The harvested heads are mildly dried and threshed by beating or trampling with tractors or sunflower threshers.The sunflower thershers are more efficient and economical. After threshing the produce should be throughly dried before storing otherwise in storage it gets affected by fungus and decomposes.

Allelopathic Effects of Sunflower

Researches show that sunflower depresses the growth and productivity of succeeding crops. The reason postulated are the allelochemicals being secreted by sunflower crop. Decomposing residues of sunflower could also bring about allelopathic effects. Phytotoxic compounds released into the soil rhizosphere have been observed to cause varying degree of allelopathy/ autotoxicity in crop. Field studies conducted in the department of Agronomy, Punjab Agricultural University, Ludhiana on the allelopathic effect of spring sunflower on succeeding Kharif crops revealed that maize, sorghum, cowpea, pearlmillet, pigeonpea, mungbean and sesamum suffered 41.7, 36.7, 27.6, 28.8, 23.0, 10.8 and 9 percent reduction in yield, respectively, as compared to these when grown after wheat. Studies in allelopathic effects of different plant parts of sunflower indicated that sunflower roots reduced pigeonpea, cotton and mungbean yields the most followed by roots plus shoots treatment. Sunflower shoots did not have any harmful effect on growth of these crops. Bioassay studies involving use of aqucous sunflower extract, decomposing plant residue and soil from area in proximity of sunflower crop revealed inhibition of gemination and growth of cotton, pigeonpea, sunflower and mungbean. These studies further showed that effect of leaf extract/residue was more than roots. The allelopathic effect of sunflower plant residue persisted upto 24 days after incorporation and beyond this period the effect was rather meagre. Chromatograhic and physico-chemical analysis of allelochemicals in sunflower parts showed the presence of chlorogenic and isochlorogenic acids in both stems and leaves besides these some unidentified phytotoxins were also observed. The concentration of these was more in leaves than in stems. Further studies revealed that total phenolics was more in leaves followed by in roots and shoots. Sunflower raised under moisture stress conditions had higher phenolics as compared to the raised without any moisture stress.

Cut Worm (Agrotis spp.)

It is polyphagous pest adn cuts off the tender stalks, near ground, of young crop plants during night. During day time the caterpillars remain hidden in the soil. It is more serious during March_april when sunflower follows potato. When flat sowing is practised, apply 5 liters of Dursban/ Radar 20 EC (Chlorpyriphos) per hectare before sowing. These insecticides should be mixed in 25 kg fine soil and broadcast uniformly after last ploughing but before planking.Sow crop on ridges to avoid cut worm damage in the early stages.

Tobacco Caterpillar (Spodoptera Litura)

The polyphagus pest remain active throughout the year feeding on one or the other crop. The newly hatched caterpillar is pale-greenish with a black head and feeds on the outer epidermis of leaves. It poses a serous threat during April-May and defoliates the plants.

Cabbage Semi-looper (Plusia Orichalcea)

This insects feeds on many crops including the sunflower. The newly hatched caterpillars defoliate the crop and cause severe damage to it during April-May.

Bihar Hairy Caterpillar (Spilosoma Obliqua)

It is also called Bhabbu Kuta. This is highly polyphagous and voraciousfeeder. The nefarious pest causes severe damage and plants looke denuded. Newly hatched larva is bright brown and its body is covered with hair and legs are brown. At the older stage it is much darker and with dark brown hair of almost uniform length. these insects feed gregariously and should be collected along with the plant and buried deep in soil. For the control of above mentioned pests, spray 1.125 L. Folithion/Sumithion 50 EC (fenotrothin) or 1.225 L. Thiodan 35 EC (endosulfan) or 500 ml Nuvan 100 EC (dichlorvos) in 250 litres of water per hectare. Repeat after two weeks, if necessary.

Head Borer (Helicoverpa Armigera)

Head borer or American boll worm has become the noxious polyphagous pest in India adn also causes heavy losses in sunflower. Newly hatched larva is green yellow which changes its colour with every instar. The full grown caterpillar is brownish green having grey longitudinal streaks on it. The adults lay eggs on the flower buds. After hatching the larvae feed on developing seeds and make tunnles through them. Head borer can be controlled by spraying 2.5 Kg of Sevin 50 WP (Carbaryl) or 2.0 litres of Asataf 75 SP (Acephate) or 2.125 litre Folothion 50 EC (Fenitrothion) or 2.5 litres of Dursban 20 EC (Chlorpyriphos) or Thiodan 35 EC (Endosulfan) or 1.25 litre of Nuvacron 30 SI (Monocrotophous) in 375 litres of water per hectare. Repeat after two weeks if necessary. Endosulfan is comparatively safe for honey bees. Spray when honey bee activity is minimum i.e. early morning or late evening.

Precautions for Spraying Insecticides

- Spraying during flowering should be avoided to protect the pollinators.
- Spray in the afternoon when pollinators are less so that there is a time gap till morning or mid day when their activity is more.
- Spray only when pest incidance is observed in abundance.
- Spray on calm day to avoid drift.
- Precautionary measures are advisable if pest outbreak is observed every year.

Diseases and their Control

Steam Rot/ Cottony Rot

It is soil and borne disease. Infestation of basal portion of stem takes place via mycelium that grows with the roots and the stem becomes soft and cotyledon leaves flatten on the ground. At bud formation soft decaying area appears on its stem and base and white mycelia appear later on. The stem becomes weak, leaves dry up and drop off with strong wind. Infestation takes place on heads when fully developed and seed formation has taken place. The back of head becomes soft and turns brown. Fungus like cotton appears and seeds are also infested. The sclerotia of this pathogen, almost of the size of sunflower seed are produced in abundance which are visible in pith of rotton stem and also in the infested head. The affected tissue turns black and shredded. This disease can be effectively controlled by treating seed with thiram @ 2 g/kg seed before sowing.

Root Rot or Charcoal Rot

It is soil borne disease that infects the roots. Roots turn brown and rot. Infection at early stages kills the seedlings rapidaly. This also causes neck rot i.e. stem rotting immediately behind the head where infected stem turns brownish and darker towards the base. Roots turn grey and stunted. At later stages, it causes premature ripening, smaller and poor filled heads and a lower yield. For its control, treat the seed with Thiram @ 2 g/ Kg seed and avoid moisture stress.

Head Rot

High humidity and injury to plants favour infection at lower surface of head where it is attached to the stem. irregular water soaked spots appear, which gradually enlarge and become soft. Shredding of back of heads takes place. Cracking of front side is also common. Seed filling is affected severely and seeds remain small and shrivelled.The yield losses range from 20 to 60 percent and oil reduction to the tune of 25-53 percent and increase in free fatty acids, which is undesirable. To avoid this disease plough the soil with furrow turning ploughto bury the debris of previous crop and residues of sunflower; treat the seed with Thiram @ 2g/kg seed; avoid excessive stress of irrigation; grow recommended hybrid and don't grow sunflower repeatedly in the same field; Spray with Indofil M-45 @ 625 g per hectare in 250 litre of water at 15 days interval starting at about 60 days stage.

Utilization

Sunflower crop has many uses:

- Edible oil of sunflower is of superior quality because of anticholestrol and non-cholestrol contents and is thus preferred over other oils for cooking food.
- Its oil is used for making vanaspati, soap and other allied products.
- Its oil is used in the manufacturing margarine.
- Sunflower meal in the form of cake is used for animal feed or manuring field.
- Hulls and stalks are used for the preparation of furfural.
- Green plants are used as forage or silage.
- Dry stalks are used as fuel.
- Pulp is also extract from plants.
- Sunflower husk finds use in the preparation of ethyl alcohol and yeast.
- Salted and fried kernels of sunflower are eaten as such and its meal is used for confectionery.
- The concentrates and isolates of defatted meals are used for protein additives, meat analogues, meat extenders, emulsifying agents, bakery products, beverages, snak products, etc.
- Seed is included in many commercial bird feeds.

Exploiting Male Sterility

Sunflower being a Cross-pollinated crop can be exploited for increasing the production of edible oil by developing hybrid sunflowers. The development of commercial hybrids is only possible by using male sterility, and fortunately such genetic systems are available in sunflower.

Two types of male sterility occur in sunflower:

1. Genetic male sterility linked with certain marker genes. This sterility is due to a recessive nuclear gene and linked to the green colouring matter in the seedling;
2. The cytoplasmic male sterility was first obtained from the interspecific cross H. petiolaria x H. a.n.nuue by Leetereq in France in 1965. This type is being widely used in developing hybrid sunflowers all

over the world. Work on the development of hybrid sunflower is in progress at the various centres of the A.I.C. O.P., P. 0. (Sunflower).

LINSEED

Geographic Origin

Linseed (Linum uaitdissimum L.) belongs to the family Linaceae and the genus Linum which has 100 species. It has been grown from ancient times for fiber (flax) and for its seed which is rich in oil The oldest regions of cultivated flax are reported to be in Asia and on the Mediterranean coast.

Economic Importance

Linseed occupies a greater importance among oilseeds; owing to its various uses and special qualities. In India, it is grown mainly for seeds, used for extracting oil. The oil content of the seed varies from 33 to 47 per cent. Linseed oil is an excellent drying oil used in manufacturing paint and varnishes, oilcloth, waterproof fabrics and linoleum and as an edible oil in some areas. Linseed-cake is a very good manure and animal feed. Linseed straw produces fibre of good quality. Linseed is used in making paper and plastics.

Area and Distribution

Linseed is extensively grown in the countries of the temperate zone as well as in those of the tropical zone. The major linseed-growing countries are Argentina, the USSR, India, the USA, Canada, Pakistan and Australia. India accounts for about 1.9 million hectares, with a seed production of 4.98 Iakhs of tonnes and occupies the third rank among the linseed-producing countries. Australia and Canada have the highest productivity of about 7 quintals per hectare, whereas India averages 255 kg per ha. This yield is the lowest in the world.

In India, Madhya Pradesh leads in yield and acreage, followed by Uttar Pradesh and Maharashtra, Bihar, Rajasthan, Karnataka and West Bengal also grow linseed in large areas. Madhya Pradesh and Uttar Pradesh together contribute to the national linseed production to the extent of about 70 per cent.

Climate and Soil

Linseed is grown in the range of latitudes between the 10th and 65th parallels, both north and south. Its cultivation is confined to low elevations, but it can he successfully grown up to 770 meters. Areas with the annual rainfall ranging from 45-75 cm are best suited for its cultivation. The seed crop does well under moderate cold, but the fiber crop grows best in cool moist climates. In lndia, the crop is grown in the rabi season from September-October to February-March.

Linseed can be grown on different kinds of soils, except the sandy and badly drained heavy clays or clay loams. It does well on the deep clayey black soils of central and peninsular India and on the alluvium loams of the Indo-Gangetic plains.

Cultivation

Fields of linseed are prepared as for other crops. Frequent ploughings are given to retain moisture and to check weeds in the field during the rainy season and a fine weed-free seedbed is prepared for sowing. BHC 10% or Aldrin or Chlordane 5% dust at the rate of 26-30 kg per ha is mixed within soil during the last ploughing before sowing to protect the seedlings against white-ants and the early damage from cutworms.

The main season for sowing linseed is October to November, depending upon the cessation of the rains. The crop is grown either broadcast or is drilled in lines 20 to 30 cm apart: In the case of broadcasting, which includes uterasowing also, the seed-rate is kept at 40 kg per ha. For line-sowing, the seed-rate varies from 20 to 510 kg per ha, depending upon the seed size. Thin sowing gives a good number of tillers, but for fibre and weed, thick sowing is desirable.

The application of fertilizers and irrigation increases the yield of the crop. Under rain fed conditions, the fertilizer dose including 30 kg of N and 15 kg of P per ha is given. If irrigation is available, two to three irrigations are sufficient. Under irrigated conditions, the linseed crop requires 60 kg of K and 30 kg of P per ha.

The deep placement of fertilizer at sowing in the case of the rain fed crop gives better results. Under irrigated conditions, nitrogen is applied in two splits, half the dose as basal and the other half at the first irrigation, 40 to 50 days after sowing. The utera crop is fertilized at the rate of 10 kg of N per ha applied at the time of sowing. The rain fed crop generally requires no intercultural. When the crop is irrigated, one weeding becomes essential.

Pests and Diseases

Linseed gall midge (Dasyneura lini Barnes) is the most wide spread and serious pest of linseed. The growing of early varieties and mixed cropping are reported to reduce the incidence of the pest.

Linseed rust and wilt are the two most serious diseases. The growing of resistant varieties is recommended to control these diseases.

The linseed crop starts maturing by the mid. of February, depending on winter spread and sowing time. Plants turn golden yellow when the crop is mature and ready for harvesting. Harvesting is done when the crop is dead ripe with a sickle or by uprooting the plants. When the fiber is also desired along with seed, the harvesting of the crop is done at the stage of capsule maturity even when the crop is slight green. Linseed is threshed after it is completely dry. The crop is threshed either by beating it with a stick or by trampling it under the feet of the bullocks. Hand-beating is better to gelien fiber, along with seed. Seed is separated from the chaff by winnowing and is stored in a dry place.

Yield

The average yield of a pure crop varies from 210 to 450 kg per ha of seed under rainfed cultivation. The crop in northern India generally gives higher yield than in central and peninsular India. The irrigated crop may yield 1,200 to 1,500 kg per ha.

The percentage of oil in the seeds varies from 37 to 43, but the commercial extraction in expellers of rotaries gives about 33 per cent from small seeds and 34 to 36 per cent from hold seeds. Crushed in village ghanis, the seeds yield only 25 to 30 per cent of oil.

Varieties

In India, the linseed varieties are of two types:

1. Peninsular types which have deep root-system.
2. Alluvial types which are shallow-rooted and are characterized by profuse tillering. Varieties differ in maturity, growth habit, and seed size and colour.

CASTOR

Geographic Origin

According to the available literature, castor (Ricinus comrnunis L.) is indigenous to Eastern Africa and most probably originated in Ethopia.

Economic Importance

It is generally grown for its,oil-yielding seeds. The oil content of the seeds varies from 35-58 per cent in different varieties, the average being about 47 per cent. Castor-oil is being used widely for various purposes. It is used as a lubricant in high-speed engines and aeroplanes, in the manufacturc of soaps, transparent paper, printing-inks, varnishes, linoleum and plasticizers. It is also uscd for medicinal and lighting purposes. The cake is used as a manure and plant stalks as fuel or as thatching material or for preparing paper-pulp. In the silk-producing areas, leaves are fed to the silkworms.

Botanical Description

Castor belongs to the gellus Ricinus, its a member of the Euphorbiaccae which consists a vast number of plants mostly native to the tropics. The geiut; Ricinus is considered to be momotypic and R. cwnmunis as the only species which includes many polymorphic types. The cultivated types are dwarf annuals

The stem is erect, circular in section, partially, hollow, smooth, glabrous, with good branching. The stem is marked by a number of well-defined nodes, from each of which a leaf arises. The lower internodes are shorter and their length increas with the height. Leaves are alternate, large, palniate with 5-11 lobes, acuminate, margins notchedy serrate or indented. They are carried on long stout petioles. The inflorescences are borne terminally on the main and lateral branches. Flowers are large, in terminal sub-panelled racemes, monoccious, apetalous, the upper portion of the raceme being occupied by the female flowers and the lower by the male flowers. The fruit is a roundish glaucous capsule, with three projecting sides covered with tough spines or smooth, three-loculed and three-seeded. Seeds are albuminous, anatropous, broad, oval, compressed with a marked carnnele and longitudinal raphe. The testa is thin, brittle, varying in colour and mottling. Below the testa is the thin tegmen, covering the whitish oily endospeim containing the embryo.

Distbibutioin Area and Production

The main castor-growing countries are Brazil, India, USSR and Argentina. India ranks first in area (44,8000 hectares) but second in production (140,000 tonnes), Brazil being the, first. Andhra Pradesh (67.2%), Gujarat (12.7%) Karnataka (7.1%) and Orissa (5.8%) account for over 90% of the area and also production.

In lighter soils, the land is repeatedly ploughed in summer, as and when the rains are received, and with the onset of monsoon rains the clods are crushed by working a country plough or harrow. In clay loams, only harrowing is done with a blade-harrow two to three times to bring the soil into proper tilth.

The seed is sown either in the plough furrow or with a seed-drill or by hand-dibbling. The seed-rate used for sowing behind the plough is 12 kg per hectare. For hand-dibbling, 8 to 10 kg of seed will be sufficient per hectare. Before sowing, castor seed should be treated with Thiram 3 g/kg of seed to prevent the attack of root-rot and Alternaria blight. The spacing adopted generally ranges from 90-120 cm between rows and 45-60 cm between plants in the row. One or two weedings and hoeings with bullock-drawn implements are given and sometimes, the plants are earthed up. The fertilizer dose recommended E for castor is 40 N-40 P-20 K kg/ha. The entire dose of P and K and half of W is applied basally and the remaining half of W about 30 days after sowing, as and when the soil moisture is adequate.

The improved varieties cultivated at present are early maturing and take about 150 to 180 days. When one or two capsules in a bunch show signs of drying, the whole cluster is generally removed and stacked, covered and weighed in a corner of the field or in a pit. The harvesting of unripe capsules in this manner has an adverse effect on the oil content of the seed. It is preferable to collect the fruits, as and when they ripen. When the whole crop is gathered, it is dried in the sun for a few days and the threshiug is done either by beating the dried capsules with a stick or by treading them under the feet of bullocks. Winnowing is done in the usual manner.

Pests and Deseases

The castor semi-looper (Achow janaga L.) and the capsule-borer (Dicitocrocis punctiferalis) are the two most serious pests of castor. Dusting BHC 10% in early stages or spraying 0.44% Ekalax or 0.1% Carbaryl or 0.05% 1 arathion on the crop will give an effective control on these pests.

The seedling blight (Phytopiathora colocasic Racib. emend. Thomas & T.S. Ramkar) and Alternaria blight Alternaria ricini (Yoshii) Hansford cause serious losses to this crop. The Sowing of the crop in low-lying and water-logged areas should be avoided to prevent the seedling blight from appearing. For alternaria blight, seed treatment with Thiram at the rate of 3 g/kg of seed is recommended.

Yield

The average yield of rain fed castor varies from 200 to 500 kg per hectare, that of the mixed crop from 100 to 200 kg and of the irrigated crop from 500 to 800 kg. In eleven years (from 1964-65 to 1974-75), the average total yield of castor in India has risen to 350 kg per ha.

Varieties

The varieties differ in the branching habit of the plants, colour of the stem and branches (red and green), the nature of capsules (smooth or shiny), duration (early or late) and the size of seed. The following are the principal improved varieties of castor recommended for cultivation in different states.

NIGER

Geographic Origin

The Niger (Guizotia abyssinica L.F. Cass. crop is said to be indigenous to tropical Africa and more specifically to Ethiopia.

Economic Importance

The crop is grown for its seed, used for extracting oil which is about 37 to 43 per cent of the seed weight. The oil is used for culinary purposes, for anointing the body, for manufacturing paints and soft soaps, for lighting and lubrication and for manufacturing cosmetics. The oilcake is a well-known cattle feed. The crop sheds a large quantity of dry leaves in the field and thereby adds organic matter to the soil.

Botanical Description

Niger belongs to the family Compositae. The plant is a stout, erect, smooth orscabrous leafy annual herb, about one metre tall. The leaves are opposite or the upper ones are alternate. The capitulum is peduncled, axillary and terminal heterogamous. Bay-florets are pistillate, scriate, fertile; ligule 2-3 toothed and yellow. Disc florets are hermaphrodite, fertile; limb campanulate and 5-fid. Achenes are glabrous, dorsally compressed, tip rounded, pappus absent.

Distribution, Area and Production

India is considered to be the chief Niger-producing country in the world, with an area of 4.8 lakh hectares, and with an annual estimated production of 1.2 lakh tonnes of seed. It is mainly grown in the states of Madhya Pradesh, Bihar, Maharashtra, Orissa and Tamil Nadu, of which the largest area of 2.38 lakh hectares is in Madhya Pradesh.

Climate and Soil

Niger is grown as a kharif crop under rainfed conditions. A moderate rainfall of 1,000-1,250 mm suits this crop and, as such, it is not cultivated in regions of heavy rainfall.

Light red soils and brownish loams with sufficient depth and good texture are considered suitable for this crop. It is often grown on poor soils of coarse texture. It can be grown on well-drained heavy soils.

Cultivation

Niger is grown either as a pure crop or is grown mixed with minor millets. Two to three ploughings are enough to prepare the land for sowing. The field is made weed-free and levelled by planking before sowing. The crop is sown from June to August, either behind a country-plough or with a four-coultered drill. A spacing of 30 cm between lines and that of 10-15 cm between plants in the line are generally adopted. A seed-rate of 7-8 kg per hectare is adequate to maintain a good crop-stand and a pure crop. About a fortnight after sowing, thinning is done to regulate spacing between plants. One or two hoeings are required for controlling weeds. In general, the Niger crop is not manured. But the application of 20 kg of N and 20 kg of P_2O_5 per hectare is recommended for obtaining higher yields of the crop. When sown as a mixed crop, Niger shares the manuring given to the main crop. The crop matures in November and December. The mature plants are cut with sickles at the base and heaped in the threshing-yard for a week. They are then spread in the sun to dry for two to three days, threshed by beating with sticks, and winnowed.

Yleld

The average yield of Niger is about 200 kg/ ha, but under favourable conditions and with best management practices it yields up to 600 kg.

BACKBONE OF AGRICULTURAL ECONOMY OF OILSEED CROPS

OILSEED crops have been the backbone of agricultural economy of India from time immemorial. Today these crops are cultivated on about 16.5 million hectares, with total production of 10 million tonnes. This area constitutes approximately one-tenth of the total cultivated area in India. On the oilseed map of the world, India occupies a prominent position, both in regard to acreage and production. The important oilseed crops grown in this country in order of importance are groundnut, rapeseed and mustard, sesame, linseed, safflower, castor, sunflower and niger.

GROUNDNUT

Geographical Origin

Groundnut (Arachis hypogaea L.) is believed to be the native of Brazil to Peru, Argentina and Ghana, from where it was introduced into Jamaica, Cuba and other West Indies islands. The plant was introduced by Portuguese into Africa from where it was introduced into North America. It was introduced into India during the first half of the sixteenth century from one of the Pacific islands of China, where it was introduced earlier from either central America or South America.

Economic Importance

The oil content of the seed varies from 44 to 50 per cent, depending on the varieties and agronomic conditions. Groundnut oil is an edible oil. It finds extensive use as a cooking medium both as refined oil and Vanaspati Ghee. It is also used in soap making, and manufacturing cosmetics and lubricants, olein steering and their salts. Kernels are also eaten raw, roasted or sweetened. They are rich in protein and vitamins A, B and some members of B_2 group. Their calorific value is 349 per 100 grammes. The H.P.S. type of groundnut kernels are exported to foreign countries. The residual oilcake contains 7 to 8 per cent of N, 1.5 per cent of P_2O_5 and 1.2 per cent of K_2O and is used as a fertilizer. It is an important protein supplement in cattle and poultry rations. It is also consumed as confectionary product. The cake can be used for manufacturing artificial fiber. The haulms (plant stalks) are fed (green, dried or silaged) to livestock. Groundnut shell is used as fuel for manufacturing coarse boards, corksubstitutes etc. Groundnut is also of value as rotation crop. Being a legume with root nodules, it can synthesise atmospheric nitrogen and therefore improve soil fertility.

Botanical Description

Groundnut (Archis hypogaea L.) is a member of sub-family, Papilionaceae of the family Leguminosae. Archis hypogaea L. consists of two subspecies each containing two botanical varieties.

Plants of the botanical variety hypogaea are spreading (runner) to upright (erect bunch) in growth habit, have alternate branching, lack inflorescences on the main stem, possess appreciable fresh seed dormancy, flowers are longer and mature later than those of subspecies fasligiata. Variety hirsuta has been used only to a little extent.

Plants of this subspecies fasligiata are upright, have sequential branching and inflorescences in the main-stem leaf axils, possess little fresh seed dormancy, and are of shorter duration than those of the subspecies hypogaea. Subspecies fasligiata includes both the Spanish and Valencia types

Groundnut, in general, has a short-statured plant, with the main axis being upright (15 to 40 cm long) but the major part of the plant consists of the primary branches. Secondary and tertiary branches are found in the semi-spreading and spreading (Virginia) types, giving them a prostrate stature. The leaves are alternate, stipulate and quadri-foliate. The flowers are orange yellow, typically papilionaceous, with a long calyx tube within which is held the style borne on a superior ovary. The calyx and corolla lobes are borne in the axils of the leaves on the fruiting branches. They are bisexual, Zygomorphic, complete and sessile. Petals are five, with one large standard, two wings and two fused keel petals. There are two steering and eight fertile anthers, four of which are globose and four oblong (dimorphic).

Groundnut is predominantly a self-pollinated crop and pollination takes place early in the morning. As soon as the fertilization is complete, the flowers fade. After fertilization, an intercalary meristem becomes active at the base of the ovary above the point of attachment of the hypanthium, producing new tissue below itself and resulting in an elongated stalk in the peg. The pegs are positively geotropic, enter the soil and bend in a horizontal plane. Generally two, and occasionally one, three or four, fertilized ovules are borne at the tip of the peg which later swells to become the pod. The testa is generally pink, but varieties with red, white, purple and blotched testa, with various gradations of colours are available.

Distribution, Area and Production

The major groundnut-producing countries of the world are India, China, Nigeria, Senegal, Sudan, Burma and the USA. Out of the total area of 18.9 million hectares and the total production of 17.8 million tonnes in the world, these countries account for 69% of the area and 70% of the production. India occupies the position, both in regard to the area and the production, in the world. About 7.5 million hectares is put under it annually and the production is about 6 million tonnes. 70% of the area and 75% of the production are concentrated in the four states of Gujarat, Andhra Pradesh, Tamil Nadu and Karnataka. Andhra Pradesh, Tamil Nadu, Karnataka and Orissa have irrigated area forms about 6% of the total groundnut area in India.

Climate and Soil

Groundnut is grown throughout the tropics and its cultivation is extended to the subtropical contries lying between 45 degrees N and 35 degrees S and up to an altitude of 1000 metres. The crop can be grown successfully in places receiving a minimum rainfall of 1,250 mm. The rainfall should be well distributed well during the flowering and pegging of the crop. The total amount required for presowing operations (preparatory cultivation) is 100 mm; for sowing, it is 150 mm and for flowering and pod development an evenly distributed rainfall of 400-500 mm is required. The groundnut crop, however, cannot stand frost, long and severe draught or water stagnation.

Groundnut is grown on wide variety of soil types. However, the crop does best on sandy loam and loamy soils and in the black soils with good drainage. Heavy and stiff clays are unsuitable for groundnut cultivation as the pod development is hampered in these soils.

ROTATION ASND MIXED CROPPING

Generally, as a kharif crop, groundnut is grown year after year. In certain places, it is rotated with wheat, jowar, bajra, garden crops, such as potatoes, onions, chillies, garlic, ginger and turmeric. The yields of the cereal crops following groundnut are usually increased by about 25%.

Pulses, e.g. red gram (arhar), mash and moong are grown mixed with groundnut. In certain places, millets e.g. jowar and bajra, and castor are grown mixed with groundnut.

Season

Groundnut is raised mostly as a rain fed kharif crop, being sown from May to June, Depending on the monsoon rains. In some areas or where the mansoon is delayed, it is sown as late as August or early September. As an irrigated crop it is grown to limited between January and March and between May and July.

Cultivation

For a kharif crop, with one set of rains in May-June, the field is given two ploughings and the soil is pulvarised well to obtain a good tilth. The third ploughing may be given just before sowing. Harrows or tillers can be used for cultivation. When the soils are heavily infested with perennial weeds, e.g., Cynodon or Cyperus, very deep ploughing is needed. If a field is infested with white grubs, chemicals, such as Heptachl or chlordane, are drilled at the rate of 25 kg per ha before final borrowing. For the irrigated crop, beds of convenient size may be made, depending on the tropography of the land, the nature of the irrigation source and the mode of lifting water. Since groundnut is a deep rooted plant it uses up both moisture and nutrients in the deeper layers of the soil. Generally, the following fertilizer schedules are recommended, depending upon the soil fertility, the variety grown, and the quantum and distribution of rainfall.

Farmyard manure		N	P	K	
Rainfed crop	6.25 Tonnes	10-20	20-40	20-40	kg/ha
Irrigated crop	12.5 Tonnes	20-40	40-90	20-40	kg/ha

Usually, the application of only N and P is resorted to, since most of the Indian soils are rich in K. The application of N in two equal split doses, one before sowing and the other 30 days after sowing, is found to increase the yield. In virgin lands, when groundnut is newly introduced, the application of a culture of *Rhizobium* as seed treatment is beneficial in increasing nodulation and nitrogen fixation. The application of gypsom at 500 kg per ha at the pegging stage will enhance pod formation.

Healthy and well developed pods should be hand shelled or shelled with a suitable groundnut sheeer about fornight before sowing. Well defiled kernels should be selected and treatedwith 5 gm of thiram or 3 g of Captanper kg of kernels. The kharif crop is sown with a speed-drill or with a suitable planter at a depth of 8 to 10 cm. specings adopted differ from place to place. For the rainfed bunch groundnut the inter-raw specings vary between the adjuscent rows varies from 20 and 30 and intra-raw specings between 10 and 20 cm. The quantity of well developed seeds required per hectarewith the above specings will be about 110 kg for semi-spreading and spreading varieties and 120 kg for the bunch varities.

Weeds cause considerable reduction in yield. A reduction of 20 to 45% in yield due to weeds has been recorded. For controlling weeds, and also to keep the soil in a friable condition, the crop should generally receive a hand-weeding and one or two hoeing, with bullock-drawn implements, the first about three weeks after sowing and the second and the third about a fortnight and a month later. No intercultural would be done after the pegs have commenced going underground. Earthing up can be done in the case of bunch and semi-spreading types to fascilitate the maximum penetration of the pegs into the soil. Weeding can also be done quickly and efficiantly with a wheel hand-hoe, if the soil is light. It should be run before weeds make excessive growth. Weeds can also be controlled effectively with Lasso or Tok-E-25 weedicides at the rate of 5 litres in 500 litres of water per hectare as a pre-emergence soil spray within two days of sowing groundnut. If the Kharif crop is cought in a long spell of draught, especiallyat the pod formation stage, supplemental irrigation is given, if water is available. For the irrigated groundnut, the frequency of irrigation depends on soil texture.and interval between irrigation ranges fro 8 to 12 days. Studies on the water requirement of indicate that the period between the fifty first and fifty seventh day (i.e. at peg formation stage) the crop depends on timely irrigation to it during this period.

The prominant symptoms of maturity are the yellowing of leaves, the shedding of older leaves, the development of oroper color of the testa and the dark tint inside the shellThe bunch and semispredading varieties are usually harvested by hand pulling when there is adequate moisture in a soil. The spreading types, on the other hand, are harvested by digging with a mammity or by ploughing or working a blade harrow.The pulled out plants are stacked for a few days for drying and stripped afterwards.

The stripped pods are cleaned and dried to a safe moisture content of not more than 5%. Damp nuts, if stored, will ferment and allow the development of poisonous moulds, e.g. aspergillus flavus in the Kernels, leading to contamination with aflatoxin—a health hazard both bor human beings and livestock. The oil express from such produced will be rancid and the cake. When fed to poultry results in a heavy mortality of the birds. It is desirable to store groundnut in gunnt bags are stacked in a storeroom in tires comprising not more in each in such a way that the air keeps circulating planks to aviod damage from dampness, rats, etc. The 4 store rooms should be periodically inspected to ensure that there is no storage pest or to decompose of the stock immediatly after proper cleaning.

Pests and Diseases

Aphids (Aphis craccivora Koch-), Leaf miner (Stomopteryx subseciva Zell) the red hairy cater piller (Amsacta albistriga Walk., and A moorei Bltr.) and the white grub(Lachnosterna consanguiea) are the serious pests of the groundnut crop. Dusting B.H.C 10% on the young larve or spraying Metasystox 0.05% on grown up catrepillers is recomended for controlling the red hairy caterpiller. Spot-treatment with Phorate or Mephospholan granules at the rate of 1.5 kg a.i. per ha into the soil before planting is recomended for controlling the white grub.

The tikka leaf-spot (Cercospora arachidicola and C. personata) and the coller- rot (Aspergillus niger and A. pulverulentum) are important diseases of groundnut. Rust(Puccinia arachidis) has been occuring in a serious form in recent years in certain groundnut growing areas.It is desirable to store groundnut in gunnt bags are stacked in a storeroom in tires comprising not more in each in such a way that the air keeps circulating planks. Dry seed dressing with Thiram or captan @ 4 g/kg of seed is recomended for controlling collar-rot. Yet spraying with Hinosan at the rate of 0.02% is recommended to check the spread of the Pathogen.

Yield

Under rainfed conditions, the average yield of semi-spreading and spreading varieties is 1200-1400 kg of unshelled pods per hectare and that of bunch types is 800-1,000 kg. The crops grown with supplemental irrigations produce 3,000 kg more of pods per hectare. The pods yield 70 to 75 per cent of kernels by weight. The yield of haulms is usually two to two-and-half times that of pods.

Varieties

The varieties under cultivation fall into three groups in respect of the habit of growth, namely bunch (Spanish), semi-spreading (Virginia bunch) and spreading (Virginia runner). In the bunch group, the plants grow erect, possess light-green foliage, produce pods in clusters at the base of the plant and have round, plump non-dormant seeds, with light-rose testa. In the case of the semi-spreading and spreading varieties, the branches trail either partially or completely on the surface of the soil, produce pods all along them, possess dark-green foliage and have oblong, dormant brownish seeds. The semi-spreading and spreading types are usually heavier yielding and later-maturing than the bunch varieties.

AGRONOMY FOR SUGAR-BEET CULTIVATION

Soil. Sugar-beet grows best in loams and clayey loams with a near neutral reaction. It has a high tolerance to soil salinity and alkalinity but does poorly in acid soils. Poorly drained or water-logged soils are unsuitable for sugar-beet cultivation.

Crop Rotation

Sugar-beet is susceptible to a number of soil-borne diseases. Their control through chemicals is difficult, costly and there fore, a long rotation to prevent The build up of diseases is of considerable importance as a Prophylactic measure. A three to Five Year rotation would be the minimum (for fodder) sugar-beet, cotton-Sugarcane (plant)-sugarcane (ratoon); guar-companion cropping of sugarcane and sugarbeet-Jowar for fodder-wheat-maize-sugar-beet.

Seedbed

Sugerbeet is sown in lines, about 50 cm apart, either on flat beds or ridges. The crop is sown before the of October. The crop has poor tolerance to delays in sowing and sowing beyund October are detrimental to Yield and sugar-content.

Varieties

The diploid variety 'Ramonskaya O6' is presently recommended for commercial cultivation. In addition to satisfactory yield. The production of seed for use in this country is established in case of this variety. Seeds of these varieties are also multiplied in the country in Kashmir valley by obtaining foundation seeds from Denmark.

Seed Rate and Seed Treatment

About 10 kg of un- processed seed is sufficient to sow one hectare. Seeds of the varieties currently recommended for cultivation are multigerm Yielding 3 or 4 seedlings per (seed-ball). to facilitate germination, soaked in 0.25% solution of a mercurial Aretan or Agallol overnight. This is done required quantity of seed in a cloth bag into the solution until the seed soaks up and then hanging the bag in the open. The seed can be sown when it is dry.

Sowing

Sowing is done by dibbling manually or by drilling. The seeds are placed rather shallow, that is about 3 cm deep. Cotton-seed drills are found useful for sowing sugar-beet. A push-type drill for sowing sugar-beet has been designed and fabricated at the Indian Institute of Sugarcane Research Lucknow, to sow sugar-beet on ridges as well as on flat land. Dry sugar-beet seed may be sown with a seed-drill, if immediately followed by irrigation. If ridge-sowing is adopted, ridges are made 50 cm apart and 15 cm high with a country plough or a ridger or a double-mould board plough.

Thinning

The multigerm seeds of sugar-beet give rise to a cluster of seedlings and hence it is necessary to thin them to maintain a single robust plant per seed-pocket. A within-row spacing of 20 cm between things should he maintained. Gap-fllling by transplanting is not recommended in the case of this crop, since transplanted seedlings give malformed or 'fangy' roots with poor weight. The dibbling of fresh seed to fill the gap may be done immediately as soon as they are visible.

Infrastructure

Sugar-beet has very poor tolerance to weed competition and hence the crop should be kept weed-free at least for the first two months of its growth. Herbicides such as 'Betanol' and 'Runeat', have given good control of weeds in Bugar-beet. The earthing-up of the crop should be completed by December after the last top-dressing of the fertilizer.

Harvesting

At the time of harvesting, a country plough is run along the sugar-beet rows on both sides. This operation facilitates the lifting of the roots with a slight pull of the KUDALI. The plants after uprooting should be freed from the adhering soil, but the roots should not be washed in water for cleaning, as this practice induces rapid deterioration. The roots must be topped clean. Leaves or leaf bases allowed to remain on the roots seriously affect the recovery of sugar.

Nutrient Requirements

Sugar-beet requires 120 kg of nitrogen per hectare and in soils deficient in phosphorus and potassium, 80 kg of phosphorus as well as 100 kg of potash per hectare, in addition to nitrogen, is applied. The various sources of nitrogen (urea, sulphate of ammonia, calcium-ammonium nitrate, etc.) are equally effective and can therefore, be used as nitrogenous fertilizers for the crop.

The nitrogenous fertilizer is given in three splits at:

1. Sowing,
2. Thinning,
3. Earthing-up in December.

The late application of fertilizers, i.e. beyond December, is contra-indicated, as the practice tends to lower the quality of roots. Top-dressing should invariably be followed by irrigation.

Irrigation Requirements

The crop requires 8 to 10r irrigation, depending upon the seasonal conditions. Sugar-beet is sensitive to the inadequacy of water and, therefore, timely irrigation is very essential to ensure a good yield of the roots. A 2-2-4r irrigation schedule in the formative, leaf-growth and root-development phases, respectively is the minimum irrigation need of this crop. Excessive irrigation to the crop is highly detrimental to root quality.

Plant Protection

Pests

Sugar-beet in young stage is often damaged by cutworms. This pest should be promptly checked at its; first occurrence, otherwise it is likely to spread rapidly in the tender crop and cause serious damage. The pest can be controlled by spraying 5 litres of gamma BHC (20% E.C.) or Heptachlor of the same strength as of gamma BHC, diluted in 3,500 litres of water per hectare along the rows with a water-sprinkler before the first irrigation. The operation is repeated at 4-week intervals in March and April, depending upon the intensity of the pest prevailing at that time.

Diseases

Sugar-beet is susceptible to a number of diseases and disorders caused by fungi, viruses, bacteria and nematodes. The more important diseases are seedling-rots caused by Pythium app., Sckrotiufa rolfsii and Rhizoctonia solani. A partial control of the more important diseases includes the control of the seedling diseases through seed-polishing and seed treatment of root-rot caused by Sckrotiutn rolfeii through 20 kg of FCNB (Pentachloronitrobenzene) per hectare as a soil-drench 'before the appearance of the disease'.

Seed Production in Sugar-Beet

Indigenous seed production is established by the National Seeds Corporation of India and commercial seed is produced in the Kashmir and Kalpa Valleys.

Processing of Sugar-beet for Sugar Manufactures

Suger-beet, unlike sugar-cane cannot be processed by gur or khandsari and can only be used for vaccum-pansugar-production. Sugar-beet roots, deteriorates fast after harvesting and have to be supplied to the sugar-mills within 48 hours of harvesting. Stock piling is not possible even for a few hours and hence well co-ordinated plan of harvesting and supplying roots is of almost importance.

Sugar from sugar-beetroot is extracted by adopting the diffussion process in special diffusers based on the counter current techinique. It is not possible to extract juice from the sugar beet roots with roller mills. The adoption of carbonation process only for clarification and carbonation with disguisers and related accessories are essential to process sugar-beet in the traditional sugar-cane-based sugar-mills in this country. Unlike with sugar-cane extraneous fuel is also required as it does not provide bagasse for use as fuel. However, the beet-pulp and molasses produced as by-products obtained from the manufacturing of beet sugar have a good market and should be expected to more than offset the extra cost on fuel in beet-processing. The establishment of a profitable outlet for the by-products is, therefore, an economic necessity with this crop.

By-products-of Sugar-beet and their Utilization

Beet-pulp, a residue obtained after the extraction of sugar, is a highly valuable cattle feed and can largely replace barley grain in feed concentrates. Beet-pulp can be fed to cattle as fresh or as dried pulp. The mixing of molasses with pulp improves its palatability.

The beet molasses is used as a raw material for several special fermentations and is also a rich source of lactic acid, vitamin B and other pharmaceutical prepa- rations.

Farm Industrial by-products of Sugar-beet

Beet-tops are a highly nutritious cattle feed and are known to improve the milk yield of cows. They, how- ever, contain oxalic acid and, hence the feeding of fresh tops to cattle is contra-indicated. Sun-dried tops with an addition of 60 g of finely ground lime pc 100 kg of the tops is recommended for use as cattle feed

SUGARCANE PRODUCTION

Sugarcane and sugar-beet are two main crops that contribute approximately 56 and 44 per cent respectively of the total sugar production in the world. Sugarcane (Saccharum officinarum L.) is the main source of sugar cane in India and holds a prominent position as a cash crop. It occupies about 1.8 per cent of the total cropped area in the country. The area under sugarcane has been fluctuating from year to year.In the past decade, it has been swinging between 2 and 2.7 million hectares. The maximum crystal sugar produced so far is 4.26 million tonnes in any one year. Uttar Pradesh has the largest acreage under suagrcane, and it is about 57 per cent of the area under this crop in India and accounts for 47 per cent of the annual production in terms of raw sugar. But the production per hectare is highest in Karnataka (Mysore State), followed by Maharashtra and Andhra Pradesh.

Soil

Sugarcane grows best on the medium heavy soils, but can also be raised on the lighter soils and heavy clay, provided there is adequate irrigation available in the former type of soils and drainage is good in the latter type of soils. In northern India, it is cultivated largely on the loams and clay loams of the gangetic and other alluviums, whereas in Peninsular India, it is grown on brown or reddish loams, laterites and black cotton soils. In many places, dark rich clay loams, 120 to 150 cm deep, and lying on a previous substratum of murum (distintegrated traprock) are used for this crop. The provision of proper drainage enables the resumption of cane cultivation on many heavy soils that may have become unfit for this crop owing to water-logging or alkalinity. In some areas in Peninsular India, where the soils are heavy, 70 to 100 cartloads of sand or red earth are added to the soil per hectare to improve its texture. Similarly, in sandy soils, tank silt is added to improve their water-holding capacity.

Climate

Sugarcane is able to grow over a prolonged season. Under warm humid conditions, it can continue its growth, unless terminated by flowering. However, its height is strongly influenced by the age of the crop and seasons. Temperatures above 50° C arrest its growth; those below 20°C slow it down markedly and severe frost proves fatal. The crop does best in the tropical regions receiving a rainfall of 750 to 1,200 mm. It can also be grown in sub-tropical areas, but where the climate is subject to extremes, and dry season is long, the growth period of the crop is restricted to a bare four months in the year and yields, tend to be distincly lower than those in the tropics. In tracts receiving a relatively high rainfall (1,200 to 1,500 mm),e.g. in parts of Uttar Pradesh, Bihar, Orissa, West Bengal and Assam, The crop is grown without irrigation and thrives on the moisture conserved in the soil. For ripening, it needs a cool, dry season; but where rainfall is too heavy and prolonged, the quality of the juice tends to be low, and where the weather remains comparatively warm and moist throughout the year, it does not ripen well. In years of drought, the crop becomes a little fibrous and in the dry areas generally, only the harder and more fibrous varieties can be grown. The climatic conditions of Maharashtra and parts of Karnataka, and the Telengana area of Andhra Pradesh are ideal for sugarcane; in some of these areas, its yield per hectare has shown a steady increase owing ot improved culture. The acreage under sugarcane has also increased.

Rotation

In northern India, sugarcane is grown in rotation with wheat, cotton gram, lahi (Brassica napus), maize, jowar, peas and other crops, and in eastern areas, it is rotated mostly with paddy. Except when ratooned, the crop is not grown on the same land often, than once in three or four years. In southern India, it is usually rotated with paddy, whereas in Maharashtra, the common rotations are sugarcane-ratoon-wheat, or sugarcane-cotton-gram. In many localities, where the crop is grown under irrigation from wells, it is rotated with garden crops, such as potatoes, chillies, sweet-potato, ginger, onion, turmeric, elephant's foot, and even bananas. In some localities, particularly near large towns, onion, radish, coriander, green cucumber and other quick-growing vegetables are grown as companion catch crops on ridges or in interspaces when the cane crop is young.

The practice of taking potato, dwarf wheat, sugar-beet, torai, coriander (dhania), berseem and turnip as companion crops in the interspaces of the autumn planting is gaining popularity in the northern region of the country. In the canal irrigated tracts of Karnataka and Maharashtra, a block system of irrigation is practised and under it a three-year rotation is compulsory, the sequence of crop being rice, groundnut, jowar, ragi, sunhemp and sugarcane, as in Karnataka, or sugarcane, fodder, jowar, groundnut, tobacco, cotton and green manure, as in Maharashtra. In the Telengana area of Andhra Pradesh, the rotation sugarcane, ratoon, monsoon paddy, and sunhemp green manure has been found the best.

Preparatory Tillage

Sugarcane requires a very thorough and clean preparation of land. In Peninsular India, where sugarcane soils are mostly clays and clay loams, one or two deep ploughings and one cross-ploughing with a mould board plough are necessary. These operations should be followed by clod-crushing with a discharrow, or a junior cultivator or a beam clod-crusher or even with wooden mallets and stout sticks. In the alluvial loams of northern India, cultivators usually prepare their fields after winter rains, but when these rains fail, the fields are given a preparatory irrigation. After it, when the soil comes into vattar (right stage of moisture) it is cultivated quickly and thoroughly with a desi plough till it is reduced to a fine tilth. In Bihar, the fields meant for cane are left fallow during the preceding monsoon and are cultivated at intervals up to the time of planting cane. All this is done in order to conserve moisture and keep down weeds.

Cattle manure, compost and other slow-acting bulky organic materials are usually applied to the soil and incorporated into it well in advance of planting. This practice helps to decompose the manure and makes the nutrients available. Where cane is green-manured, the leguminous crop should be ploughed under about a month before planting the cane. In northern India, however, where a clod season intervenes, the green-manure crop is ploughed under in September and cane is planted several months later.

Sugarcane is planted either in furrows or trenches. Depending on the method of planting and the fertility of the soil, furrows or trenches are made half to one-and-a-half metres apart. Comparatively narrow spacings are adopted in nothern India, where yields are usually low, whereas wider spacings are common in Peninsular India where the aim is to obtain high yields. Furrows are made with a padded desi plough in northern India and with a heavy double mould-board ridger in Peninsular India. Furrows made with the desi plough are quite shallow, being about 10 cm deep, whereas those made with a ridger are about 20 cm deep. Where trench-planting is adopted, trenches are dug by hand-labour, using pick-axes and spades, and may be up to 25 cm deep. On large farms, all preparatory cultivation, including ploughing, clod-crushing, harrowing, and ridge-making are mechanized.

The adequacy of moisture is very necessary for the setts. In Peninsular India, arrangements for irrigation are amde well before planting. In northern India and in Bihar, care is taken to conserve as much moisture in the soil as possible, so that the planted cane is able to germinate satisfactorily. All the same, as the planting is necessarily done duirng the comparatively dry season, considerable loss of moisture from the soil and sets takes place, and usually only 30 to 40 per cent of the buds germinate.

Planting Season

In northern India, planting is usually done with the onset of the warm weather and is completed well before the onset of summer. On this basis, the more eastern is the tract, the earlier is the time suitable for planting. Thus broadly speaking, the first fortnight of March is the best time for planting sugarcane in the Punjab and Haryana, February in Uttar Pradesh and January-February in Bihar. In actual practice, however, most of the planting tends to be late, rather than early. On large farms in northern India, where planting work is likely to be prolonged, a part of the planting is also undertaken in October-November.

In Tamil Nadu and Andhra Pradesh, cane-planting is done between December and February. In Maharashtra and parts of Karnataka, it is done in December-February for the 12-month crop which is called eksali, in October-November for the 15 to 16 month crop which is called 'preseasonal', and in July-August for the 18-month crop, which is called adsali. Thus in these areas, for this planted almost throughout a period of about 8 months from July to Feburary.

This is possible because of the more equable climate, low rainfall, the high elevation and the consequent absence of water stagnation, and the easy availability of adequate irrigation. Under such condititions, cane also makes growth and accumulates sugar, more or less, continuously, and hence earlier plantings give higher yields from giving lower yields, are usually badly attacked by the shoot-borer, which causes gaps in the stand of cane and setback to growth. Since crop planted in different months ripens in succession, by suitable adjustment of planting, the necessary areas in different seasons, and a planned supply of cane to the factory, a satisfactory recovery of sugar in factories can be ensured over a reasonably long crushing season.

Seed

Healthy seed material, free from pests and diseases, and having high viability is essential for establishing the crop in the first instance. The top one-third to half portion of a cane, being comparatively immature, has buds of good viability and is best to be used as seed. The bottom portion is usually richer in sugar than the top portion, and it is best to utilize it for sugar or gur-making. In tropical India, setts for sowing are taken from well-manured, erect and healthy canes. In some places in Maharashtra, Andhra Pradesh, Karnataka and Tamil Nadu, crop nurseries are adequately fertilized, and well protected against pests and disease.

Before planting, the leaves of the cane stalks are stripped off by hand in order that buds on them may not be damaged. These stalks are then cut into 3-bud setts each usually 30 to 50 cm long. If seed is required at a distant place, whole canes should be transported without stripping off the leaves or cutting them into setts. If planting is for any reason to be delayed, but the seed canes have been harvested, they should be heaped in shade under a thick cover of straw or cane trash, and periodically sprinkled with water. Under the dry condition of northern India, the soaking of seed in water for four to six hours before sowing improves and hastens germination.

Under irrigated conditions in Peninsular India, the planted setts are liable to rot, and in some areas they are attacked by the pineapple disease. To prevent rotting and infection with the disease, the dipping of the setts into a suitable fungicide, such as 0.5 per cent mercuric chloride or 0.25 per cent Aretan or Agallol solution or 1 kg of Perenox in 400 litres of water is recommended. Aretan, along with gamma BHC, is also recommended in northern India; the former improves the germination and keeps off fungal attack and the latter keeps off termites is serious, gamma BHC at 1 kg actual ingredient per hectare in emulsion form and diluted 300 times is sprinkled with a watering-can on the setts placed in the furrows.

Depending on the distance between the rows and the vigour and soundness of the buds, 25 to 35 thousand 3-bud setts are usually enough to plant a hectare. According as the cane is thick or thin, 1 1/2 to 3 tonnes of cane by weight is needed to provide about 10,000 3-bud setts. In the Punjab, Haryana and western Uttar Pradesh, 2-bud setts planted in rows, about 60 cm (about two feet) apart, have been found to give the best results. In the case of late planting, the seed-rate is often increased slightly, and the setts are planted a little closer in the row in order ot gaurd against gappy germination.

Methods of Planting

There are three methods of planting, viz., flat-planting, furrow-planting and trench-planting:

1. *Flat-planting*: In the case of this method, shallow furrows are opened with a desi plough, 80 to 100 cm apart; the setts are placed in them

end to end and covered with 5 to 7 cm of soil, and the field is levelled with a heavy plank (sohaga).

2. *Furrow-planting*: Furrows are made with a ridger, about 10 to 15 cm deep, in northern India and about 20 cm deep in Peninsular India. Setts are laid end to end or otherwise in the furrows and covered with 5 to 7 cm of soil, leaving the upper portion of the furrow unfilled. Immediately after planting and covering the setts, water is let into the furrow to soak its bed thoroughly and the cane setts are then dropped into the furrow and pressed into the mud by foot. In case of monsoon planting, as a precaution against the stagnant water in the furrow damaging the buds, the setts are placed inclined on the side of the furrow, instead of flat at the bottom of the furrow.
3. *Trench planting*: Trenches, 20 to 25 cm deep and rectangular or trapezoidal in section, are made either by manual labour or with a tractor-drawn rigder; the bottom of the furrow is loosened by digging. Shallow furrows are made in the bed of the trench and the setts are placed end to end or otherwise in the furrows and covered, as in the case of flat-planting. Water is then let into the trenches.

The tractor-drawn sugarcane-planter designed and fabricated at the Indian Institute of Sugarcane Research, Lucknow, is a very suitable device for planting cane in trenches. Trench-making, setts planting applying fertilizers are done simultaneously with this tractor-drawm planter.

In most parts of northern India and Bihar, and in the Malnad tract of Karnataka, cane is planted on flatland. Furrow-planting is practised in parts of eastern Uttar Pradesh and in Peninsular India, and trench-planting is done in some coastal areas where the crop grows very tall and the strong winds are liable to throw it down and damage it badly.

Planting is done dry or wet. In the case of the dry method, irrigation is not applied at planting or immediately after that, and the planted setts germinate in the comparatively dry soil. In wet planting, on the other hand, the field is irrigated just before of after planting or at both times so that setts germinate in wet soil.

Whether the furrows are irrigated just before or just after planting makes little or no difference; indeed, in most cases the planted field gets additional irrigations to keep it wet throughout the period of germination.

Irrigation

In the case of sugarcane, the maintenance of optimum soil moisture during all stages of growth is one of the essential requisites for obtaining high yields. The crop should, therefore, be grown in areas of well-distributed rainfall or under an assured and adequate irrigation. In tropical India, depending on the type of the soil, the seasonal conditions, the variety grown, the method of planting and the rate of manuring, the total water requirement of the crop for optimum growth varies from 200 to 300 cm, inclusive of rainfall. The requirement of an adsali crop is proportionately higher. Where the soil is not retentive of moisture, and where there are no reserves of subsoil moisture, cane requires to be irrigated frequently.

In tropical India, usually one or two waterings are given at intervals of three or four days after planting to help the setts to germinate and the seedlings to establish themselves. Thereafter, in the absence of rains, cane is irrigated every 10 to 12 days during its growing period. In dry areas and in sandy-loams soils, irrigations may be needed at intervals as short as eight days. In deep clay loams, irrigation can be withheld for longer periods, say, up to two or three weeks. Frequent light irrigations, each 40 to 50 mm, adjusted to suit the growing period of the crop and the prevalent weather conditions, are very useful. Towards the time of harvesting, irrigation frequency is reduced, and just before harvest, irrigation is withheld for about a month.

In northern India the summer being drier and hotter the crop needs water more frequently than in southern India, but actually, adequate water for frequent irrigation is not available and irrigations are, therefore, usually given at comparatively long intervals. In canal irrigated areas, the frequency of irrigation depends entirely on the running of the canals. The severity of these conditions is slightly mitigated by the high water-holding capacity of the alluvial loam soil.

In the Punjab State because of the lower rainfall, drier climate, and slightly coarser soils, the crop gets 8 to 10 irrigations during summer. In central and western Uttar Pradesh three to five irrigations are usually given, and they help the crop to tide over the summer. In the eastern Gangetic areas, cane subsists almost entirely on subsoil moisture and rainfall, and receives no irrigation at all. In the post-monsoon season the crop receives only one or two irrigations or none at all; however these post-monsoon irrigations only help to keep the crop in good condition, for cane does not make growth in winter. In areas where frost occurs, irrigations are applied to save the crop from them. Sugarcane responds to irrigation in northern India as it does in the south and it is profitable to apply frequent light irrigations to the crop during the hot weather. Where irrigation facilities are scarce in summer, trash-mulching in the interspaces of the cane rows is done for conserving the soil moisture.

Interculture and After-Care

The first hoeing and weeding should be given to the crop three to four weeks after planting. But in northern India, usually the shoots take a long time to emerge and meanwhile the weeds, particularly the nutgrass come up rapidly in the planted field. This rapid infestation of the sown field calls for at least one blind hoeing which may have to be given within the first one or two weeks. After germination, depending on the field conditions and the frequency of irrigation, two or three more hoeings and weedings may be required during the first three months after planting with the coming up of the crop. Where cane is planted in furrows or trenches these are gradually filled up in the course of interculture.

The use of bullock-drawn implements, like blade harrows for mulching, junior hoes for interculture and weeding and ridgers for earthing up and covering the fertilizer, lessen the costs of these operations. The final earthing-up should be completed before the monsoon rains, and should generally synchronize with the application of the final dose of fertilizer. Good care in weeding and interculture helps the clumps to tiller early and form canes. The crop should be so managed that it grows rapidly to form a canopy over the interspaces. This shading helps to keep down the weeds, the moisture loss from the soil surface.

Where the aim is to get crops of 75 tonnes and over per hectare, it is usual to earth up the standing crop in the rows. But whether the crop is earthed up or not it is very desirable to tie up the canes so that they may not sway during winds, and lodge.

The best way to do the tying is to bring together the stalks from adjacent rows and tie them together with their own trash and old leaves. In many places in southern India, cane is tied at two or three levels with twist of cane trash, the twistline going from one end of the row to the other; sometimes the stalks are further propped up with bamboos. Many times they sow castor or arhar (tur) seeds around a planted field of cane. This acts as a wind-break.

Manuring

Sugarcane is a heavy feeder. A 70-tonne crop removes from the soil 85 to 100 kg of nitrogen, 180 to 330 kg of phosphoric acid, 60 to 190 kg of potash and 70 to 80 kg of calcium. Adequate manuring, therefore, is essential for sustained high yields. At present only 10 per cent of the sugarcane crop is manured, and in most cases, inadequately.

The nitrogen requirement of sugarcane depends upon the soil and climate conditions. It ranges from about 130 kg per hectare in Uttar Pradesh to about 270 kg in Tamil Nadu and 300 to 500 kg in Maharashtra and Karnataka. The general recommendation in most places is to apply half to two-third of the nitrogen in the form of bulky organic manures, such as farmyard manure, compost, green manure or tree leaves, and the remainder in the form of ammonium sulphate or oil-cakes or a combination of the two.

These bulky organics are ploughed into the soil sufficiently in advance of planting. In places, sulphitation press-mud is also used as a manure. Oilcakes are usually given at planting and ammonium sulphate in one or two doses at planting or tillering or at the time of earthing up the crop.

The increase in the yield of cane per kg of nitrogen per hectare also depends largely upon the soil and climate. Provided the weather conditions are not unfavourable, farmyard manure usually gives about one quintal ammonium sulphate and about 2.4 quintals of cane per kg of the nitrogen contained in the manure or fertilzer. Green-manuring increases the yield of cane by about 100 to 150 quintals per hectare.

The response of sugarcane to phosphorus has not been very marked in most areas; however, the general indication is that at high levels of nitrogen application, phosphorus is required at the rate of 40 to 50 kg of P_2O_5 per hectare. The states where phosphorus has given perceptible response are Bihar, parts of Uttar Pradesh, Bengal and Maharashtra. In some states, where cane does not give a good response to superphosphate applied to the preceding summer green-manure crop. This practice improves the benefit from green-manuring.

The response of sugarcane to potash has been obtained only in localized pockets of light soils. Not much research has been carried out on the trace-element fertilization of sugarcane.

Plant Protection

Sugarcane is vulnerable to several pests and diseases and this problem is more serious in the subtropical regions than in the tropical areas of the country.

Pests

The major sugarcane pests are moths, borers, scales, white grubs, termites, white flies, black bugs and mealy bugs. They damage the crop at different stages of its growth. The mechanical (removal of egg masses or adults, spiking and cutting of infested plants), cultural (light earthing-up, avoiding late planting, detrashing of cane stalks) and biological means, and the use of resistant-varieties are some of the means of controlling the pests. Pesticide formulations of chlorinated hydrocarbons, organophosphorus and organomercurials are effectively used to reduce the damage to the crop caused by variety of pests. But the chemical control of moth bores is still not very effective. Some recent formulations of systematic pesticides are however, showing promise in this direction.

RIPENING. The maturity of sugarcane is generally recognized by the lower leaves gradually withering up and leaving progressively fewer green leaves at the top. A ripe cane, cut across with a sharp knife, shows against sunlight a slight sparkling in its flesh in contrast to the more watery cut surface of an unripe cane. A trail boiling of the juice and a satisfactory setting of gur would confirm the proper stage of the maturity of the crop. If the grower can keep and use a hand sugar refractometer, the testing of maturity would be easier. If the samples of juices taken from the middle portions of the stalks show, on the average, the hand refractometer reading of 20, the cane crop may be considered to have reached the stage of maturity. The Fehling's test for glucose provides a better technological assessment of the quality of the canes, the glucose content usually reaching values of less than 0.5 % in the juice at peak maturity.

In nothern India, in a normal season, sugarcane ripens by about early December, but the sugar content of juice continues to rise till about the end of March. After that the temperatures rise very high and the canes standing in the field gradually dry up. In the south the crop continues to grow under favourable climatic conditions and it remains fresh for a longer period. Early varieties everywhere ripen about two to three weeks earlier, and the adsali, and the autumn-planted crop and ratoons also are ready for crushing well before the eksali plant cane; but in the latter part of the crushing season, the differences in juice quality between them are narrowed down. Maturity is delayed by late rains, the warm weather in the late season and high fertilizer doses. Where winters are severe, the quality of the juice gets depressed. Frost may damage the crop seriously, leading to the rotting of buds and the drying of leaves, rendering the cane unfit for gur-or- sugar-making.

HARVESTING. Before cutting the crop, it is customary to give the cane field a good irrigation, although under conditions obtaining in northern India, this irrigation does little or no good. Stalks are cut at the ground level, preferably after digging down the earthed-up ridges. The dried leaves are stripped off from the topmost mature internode where the stalks usually break easily.

The harvested canes should always be processed quickly. In northern India, during the cold season, the harvested canes kept in the may maintain the quality if the juice in a good condition for a day or two, but they lose in weight unless they are kept moist. In the late season and in tropical areas, the quick processing of the harvested cane is essential.

Crushing and Gur-Making

Three-roller iron crushes operated with bullocks, oilengines or electric motors extract 60 to 70 per cent of the juice from the canes and are in common use. Crushing should be done in the cool hours of the morning and the extracted juice should be boiled without delay. Metal vessels are preferable as containers for the juice. Bagasse and trash are almost the universally used fuel. But this fuel is often supplemented with sticks of cotton, tur or castor. For economy in fuel and efficiency in boiling, improved furnaces or bels such as the Lucknow furace and the improved Meerut and Bilari models are recommended in northern India.

The Poona bel, Godavari, Sindewahi and the Pugalur furnaces are popular in various parts of Peninsular India. The principal features of the improved furnaces are heat conservation and fuel economy, a quicker boiling rate, greater facility for working and better control on the striking point.

Much attention is now given to the clarification of gur and its good texture. In northern India, where the juice is usually purer and more nearly neutral, the elimination of acidity with lime is not essential. In southern India, however, the addition of a thin suspension of lime is usually necessary for neutralizing the juice; this operation helps the final product to set satisfactorily. In the north, juice in the course of heating is clarified with the mucilage of bhindi (Abelmoschus esculentus) or with deola (Hibiscus ficulneus) extracted in water.

In Punjab, the bark of suklai (Kydia calycina), and in the eastern areas that of semul (Bombax malabricum) is used for clarification. An emulsion made by crushing castor or groundnut seeds in water is also recommended. The strike stage is reached when the boiling syrup attains a temperature range of 118 °C to 120 °C. At this stage, if the syrup is dripped from a ladle, it forms fine threads which waft in the breeze. The syrup also readily solidifies when a drop is allowed to fall into a mug of water. At this stage, the syrup is quickly transferred to wooden troughs or moulds of various sizes and shapes.

Yield

The average yield of a 11 to 12 month old plant crop under commercial cultivation is about 40 to 45 tonnes of cane per hectare in Assam, Bihar, Uttar Pradesh, Punjab, Rajasthan and northern Madhya Pradesh; 40 to 50 tonnes per hectare in Bengal, Orissa, and Kerala; and 70 to 85 tonnes per hectare in Maharashtra, Karnataka, Andhra Pradesh and Tamil Nadu. With improved inputs and management there is a good scope for substantially improving the average cane yields in all the areas in the country. The 18 month (adsali) crop in southern India usually gives a yield of 120 to 140 tonnes of cane per hectare, and in the heavily manured, canal-irrigated tracts of Bombay-Deccan yields up to 160 tonnes or more are harvested.

Kakavi

Liquid jaggery locally known as kakavi is an intermediate product obtained while concentration of sugarcane juice after purification for jaggery making. It is a semi-liquid syrup like product. A small but significant amount of sugarcaneproduction is utilisedfor jaggery industry to meet the demands for sweetening agent mostly in cooking and alcoholic industries.

The quality of liquid jaggery is largerly depends upon the quality the composition of cane juice, type of clarificants used and striking temperature at which liquid jaggery is collected.The sugarcane juice quality in turn mainly depends upon the parameters such as soil type, fertilizers used, management practices, stage of maturity of cane and the variety. Following table gices the chemical composition of typical kakavi.

Table. The Composition of Typical Kakavi.

Ingradients	> Proportions
Water[%]	30-35
Sucrose[%]	40-60
Invert Sugar[%]	15-25
[Mainly dextrose & levulose]	–
Mineral Matter	
Calcium [%]	0.30
Iron [mg/100g]	8.5-11.0
Phosphorous[mg/100g]	3.00
Nitrogenous Compounds	
Protein [mg/100g]	0.10
Vitamin B [mg/100g]	14.0

Liquid jaggery is collected in semi-liquid form from the boiling pan and packed in suitable containers for marketing. Improper preparation of kakavi during storage results into crysatalization of sucrose which is not preferred by the consumers. Similarly microbial growth causes fermentation and gives unpleasant odour to the product renderingit unfit consumption. Preparation of 'Kakavi' starts with crushing season of a cane from October to February.

Suitable Sugarcane Varieties for Liquid Jaggery

The best varieties for jaggery and its product should possess physiological characters like few nodes, long inter nodes, low proportion of rind tissue, yellow white color,soft stem. At the same time it should have high sugar with low molasses, colouring and colloidal materials. A large number of sugarcane varieties have been screened for liquid jaggery. the following varieties have been found to be suitable for preparation so liquid jaggery.

i]	CoC	671
ii]	Co	8014
iii]	Co	7219
i]	CoM	7215
ii]	Co	740
iii]	CoM	88121
i]	Co	740
ii]	CoM	88121

Varieties wutih brix hydrometer reading more than 190 and purity more than 85% are found to be suitable for liquid jaggery. Better quality liquid jaggery could be prepared from juice with higher ratio fo sucrose: giucose, low N/P_2O_5 ratio, low ash, lime, organic non-sugars,alkaline salts and amide nitrogen contents.

Harvesting of Cane

Harvesting of cane should be cut very close to the ground, since the bottom portion of the cane is rich in sucrose. The cane should be detrashed neatly and soil adhering to it should be removed. Cane should be crushed within 24 hours after cutting to avoid the loss in weight due to evaporation and deterioration due to inversion. In case the cane is required to be kept for a longer time it should be well covered with trash and water should be sprinkled over it at interval to keep down the temperature to avoid dessication and to prevent rapid inversion.

Crushing of Sugarcane

It was observed that horizontal three rollers crushers [Vasant No. 3 and Karajgar] or Shishir with five roller [power driven] crusher gave higher extraction percent [60-70%]. The cane crushing accessories should be cleaned by lime water to to avoid rusting affecting quality. During the crushing the juice should be filtered preferably through nylon net, and collected in storage tanks of 500-600 liters capacity. The juice then is pumped to evaporating pan of suitable capacity by pump using PVC pipes so as to avoid cantact with iron. The juice should not be kept in storage tank for more than 6 hours to avoid fermentation. while transferring the juice from storage tank to the boiling pan it should pass through nylon cloth for filteration. The boiling pans have usually capacity of 725 to 1200 liters. Fire then be started to heat the juice. Usually chimani chullahs are used for boiling.

Addition of Lime

The pH of fresh juice ranges between 5.2 to 5.4. It needs to be raised to 6.5 to 7.00 to make it alkaline to facilitate the coagulation fo suspended impurities of gummy colloidal substances. For this purpose lime water [prepared by dissolving 150 to 200gm lime in 5 liters of water] is added.

Stages in Boiling Process

- Removal of First Scum: When the temperature reaches to 85°C which takes about 25 to 30 minutes, the nitrogenous impurities present the juice start to coagulate anf float on the surface of scum which are removed with the help of long handled steel strainer [Shiba] and collected in scum stainer. The scum is dark and thick and locally called as "Dhormali".
- Addition of Vegetable Clarificants: Bhendi extract is added. This extract is prepared by crushing 1.5 to 2kg bhendi plants [Hibiscus esculantus] added to 15 liters of water. The extract is added to the boiling pan in two to three installments with an interval of 5-10 minutes.
- Addition of Phosphoric Acid: After removal of first scum, the excess of lime is neutalised by addition os phosphoric acid [free from arsenic] to adjust the pH of media to about 5.5.
- Removal of Second Scum: At this stage golden coloured scumappears on the surface which is removed. Locally this scum is known as

'Sonmali'. When the entire scum is removed, the boiled juice looks clear, transparent and brownish yellow in colour. Many farmers at this stage use chemical clarificants like 'hydros powder' [$Na_2S_2O_4$] as a bleaching agent. Excess of hydros powder is injurious to health. The sulphur dioxide [SO_2] content should not exceed 70 PPM in the final product. The hydros 35 g [having purity more than 80 %] per 1000 liters of sugarcane juice is recommended as a permissiable limit for purification.

- Concentration of Juice and Strring: When the temperature raises to about 99 °C-100 °C and the juice begins to forth. To avoid excess of forthing and loss of juice due to overflowing, continuous stirring with wooden laddle is necessary. This helps to avoid charring. Heavy forthing causes the juice to bubble. At this stage the liquid becomes thick. This Concentrated juice now is liquid jaggery [kakavi] and is ready for removal between 104 °C-106 °C temperature. If the liquid jaggery is collected at lower temperature [less 105 °C] its storability is adversely affected since the product is subjected to mocrobial growth during storage. If removed at higher temperature [more than 106 °C] there is a problem os crystallization of sucrose during storage. Thus keeping quality depends upon striking temperature of liquid jaggery - [105 °C-106 °C].

Stages of boiling of sugarcane juice are as under:

Sr. No	Stage	Time [minutes]	Temperature [°C]
1.	Initial temperature of juice	00	25
2.	Ist Scum	25	85
3.	Bubbling	40	97
4.	IInd Scum	70	98
5.	Concentration of Juice	115	99
6.	Vigorous boiling	130	101
7.	Liquid jaggery	160	106

Storage Methods

Liquid jaggery is stored in sterlized glass bottols of various capacities Viz. 650, 375, 250 and 200ml. for consumer acceptance.

a. *Sterilization of containers*:
   - The clean bottols are sterilized by dipping them in boiling water for 15 minutes and dried.
   - Larged sized [20 to 25 liters capacity] cookers are used to sterilize bottles rapidly. The clean dry bottles are kept in cooker and heated upto whistles [7-10 minutes]. After which bottles are removed and dried.

b. *Labelling and storage of liquid jaggery*:

The bottles filled with liquidjaggery should be made air tight and capped with machine. The bottles are labelled with attractive and self informative labels to indicate the details as per specifications of the syrup. These bottles should be stored in clean, cool, dry and dark place to minimise the changes during storages.

Use of Preservatives

a. Use of 0.04% Citric acid has been found to be beneficial in minimising the crystallization and for improving the colour of kakavi.
b. Addition of Citric acid @ 0.1 percent reduced crystalization of kakavi and use of 0.1 percent potassium metabisulphite or 0.5 percent benzoic acid improved the keeping quality of kakavi stored in sterilised bottles

Uses of Liquid Jaggery

- Excellent as table syrup.
- Can be consumed directlywith chapati, bread and bhakari.
- Preparation of sudha ras for consumptionin every day food.
- Recommended as a sweetener in confectionary for various preparations like cake, biscuits etc.
- Best sweetener for ice-cream.
- Advised for jaundice patients in diet as rich source of glucose.
- Recommended asrich source of ironto anaemic patients.
- Better syrup for Pharmaceutical formulations.

Ratooning

The first ratoon only is recommended in most of the states. The second and subsequent ratoons are not advised because of the risk of pests and diseases accumulating and being carried to the next crop, and pernicious weeds establishing themselves in the field. The crop to be ratooned is cut flush with the ground and the above ground stubble should be shaved at the ground level and the trash should be burnt after harvesting the crop. The field is irrigated as soon as possible. At the proper stage, the interspaces are ploughed. There after the end of the cold weather as possible. Ratoons mature earlier than plant crops. They ordinarily give a some what lower cost of cultivation, there being practically no expenditure on preparatory cultivation, seed and planting, they are more remunerative than plant cane. Crops suffering from diseases or sheltering excessive populations of hibernating borers should not be ratooned; such crops should be harvested and disposed of early, and the stubble dug up and used as fuel immediately after harvesting the crop.

SUGAR-BEET

Introduction. Sugar-beet (Beta vulgaris L), another important sugar-producing crop in which sugar is stored in roots, is purely a creation of plant-breeding. It accounts for nearly 44 per cent of the world's sugar production. Unlike sugarcane, which is a crop of the tropics, sugar-beet is essentially a crop of the temperate regions. Genetic and agro-technological improvements have now extended its frontiers to higher latitudes of subtropics as an irrigated winter crop. In countries, such as Syria, Iraq, Iran, Algeria and Israel, sugar-beet is fat assuming and of commercial importance. More recently, its cultivation has been extended to Pakistan. Sugar-beet is now in the experimental and semi commercial testing in subtropical India. In the Kashmir Valley sugar-beet holds out a promise as a rain-fed spring-sown crop.

Winter sugar-beet is a 6-7 month crop, sown in October and harvested in April and May. It is capable of yielding 35-45 tonnes of roots per hectare with a root-sugar content of 15 to 16 per cent. Sugar-beet thus opens up avenues for extending the working season of the sugar-mills of the Indian subtropical regions till the end of May and for generating beet-based sugar-processing capacities at high altitudes in the country.

A beginning has now been made to raise sugar-beet on a commercial scale in the Sriganganagar area in Rajasthan and beet-sugar was made in the Ganganagar sugar-mills. In addition to sugar, the by-products from sugar-beet, viz. beet-pulp and molasses are highly valued as cattle feeds and for special fermentations and pharmaceuticals respectively.

Agro-climatic requirements. The northern and north-western region of subtropical India comprising Punjab, Haryana, Rajasthan and western Uttar Pradesh are potential areas for winter sugar-beet in the country. The Kashmir Valley has a good scope for the generation of a beet-based sugar industry.

4

Fibre Crops Production

SEED SELECTION

Since jute is harvested much before seed formation, special arrangements to raise the seed-jute Crop. (The Cultivators often leave about 4 per cent of the fibre crop for seed production). The seed crop is sown later than the fibre crop for seed production). The seed crop is sown later than the fibre crop. Some interculture is necessary to keep down weeds. Unlike the fibre crop, the seed crop is spaced wider, so that branching is promoted. The induction of branching by clipping the tips at the right time and with right stage of growth, with assured nutrition for the development of branches, has been successfully practised. This practice, however, would not be resorted in the areas with low rainfall and where land is not very fertile. Natural cross-pollination being low genetic mixture is avoided by alternating capsularis and olitorius seed plots. The crop requires much less nitrogen for proper growth. The average yield of seed is 3.8 to 5.0 q per ha in the case of capsularis varieties and 1.7 to 2.5 q per ha in the case of 'JRO 878' and 'JRO 7835' and around 3 q per ha in the case of 'JRO 632'. After harvesting the crop the seed is dried in the sun for about 4 days and stored in gunny bags lined with plastic cloth and kept in dry godowns, avoiding contact with the floor.

Corynespora casseicola and powdery mildew attack pods and capsules. These pathogens are controlled with copper fungicides. Die-back (Diplodia corchori) on the seed crop is rare, but serious.

Defects in the Fibre

The defects in the fibre are its being barky and croppy (incomplete submerging) or incomplete retting; specky (apion damage); leafy (incomplete retting or washing); and sticky (defective extraction). The fibre becomes dirty by using clods of earth and muddy retting water and careless washing. The plants become mossy with adventitious roots emerging from the stems. The fibre becomes shyamla as a result of iron-tanning reaction.

'Roots are, caused by more than one factor but the main one is 'incomplete separation of fibre at the base of the stem during retting'

Quality

The quality of the jute fibre among varieties differs on the basis of anatomical conditions and is genetically controlled. Coarser and light-body fibre is obtained from sandy soils whereas clay-loam soils with silt give fibre of superior quality. Climate and the nutrition pattern also affects the fibre. But the most important single factor is 'retting' which, if faulty, mars the positive contributions of the variety, soil, climate etc. Under-retting gives coarse and over retting dazed and weak fibre.

Marketing

Jute Corporation of India, Calcutta, has the objective of stabilizing the price line, both in the interest of the farmers and the consuming mills.

Grading

Grading system for white jute (capsularis) has 8 classes, viz., WI to W8, on the basis of length, strength, fineness and lustre, and freedom from enlargement and 'roots' (rejectable basal portion). For Tossa jute (olitorius) there are 8 classes, viz. TDI to TDS.

MESTA PRODUCTION

The fibre of mesta is obtained from stem of Hibiscus sabdariffa variety altissima (n = 36) and H. cannaunns (n=18), family Malvaceae. JIS and RC denote varieties of the two species.

Production

The area under mesta has been increa8- ing since 1952 and is around 0.24 million hectares, of which more than 80 per cent is under HS types. The annual average production is around 1.08 million bales (1 bale = 181kg). Within the jutebelt, the yield of mesta is high in West Bengal, Bihar, Assam and Tripura. Outside side the jute belt, the highest acreage is in Andhra Pradesh and the areas assuming importance are Dandakaranya, Orissa, eastern Madhya Pradesh, Maharashtra, Karnataka and Uttar Pradesh. Other states grow these types over small areas, mostly for home consumption. Yield in the drier areas is less than that in humid ones.

At present, the cultivators, in general, grow mesta in poorer soils without much care and profits are marginal. The average yield in different states is between 1.4 and 2 bales. There is great scope of improvement of yield with improved varieties and techniques.

Climate

A warm and humid climate suits both HS and HC, but both grow in drier rainfed areas, the latter being more drought-resistant. In areas with 500-900 mm of rainfall, HC suits better by virtue of its shorter duration and faster growth. Neither of them can stand prolonged water-logging. Both are kharif crops and are sown in April-June with the first showers of the monsoon. Heavy, continuous rains and low temperature are harmful.

Soil

Whereas the rich loams give the highest yield, both grow on a variety of soils, including new and old alluvium; but acid soils are not suitable without amendment. With a high pH (above 7.0) of the soil, chlorosis appears in HS mesta. Both are unfit for low-lying areas, subject to inundation.

Cultivation

Soil is prepared in much the same way as in the case of jute. HS when sown broadcast, requires 12 to 17 kg of seed per hectare. Seedlings, when 10-12 cm tall, are manually thinned in two instalments to a spacing of 15-18 between plants. When raised in rows, seed is drilled 30 cm apart, the seed requirement being 10-12.5 kg per hectare. Seedlings are manually thinned to 12-15 cm apart in the rows. A minor adjustment in spacing is permissible on the basis of the fertility of the soil, rainfall and the nature of the soil. Variety 'HS 4288' (stem with bristles) and 'HS 7910' (stem without bristles) is sown best in mid-April to mid-May, roughly north of 22 °N, early sowing giving better yield. Variety 'AMV 1' is suitable for areas south of 22 °N and can be sown closer to shade the surface of the soil to conserve moisture (22.5 cm x 10 cm) and can be sown till June as in Andhra Pradesh. Dry sowing is possible in anticipation of rain-fall, specially where drizzles do not precede the actual monsoon rains. Variety 'HC 583' requires 17-22 kg of seed per hectare for broadcast sowing and 15-17 kg for line-sowing (30 cm X 12-15 cm). April sowings are best for yield and quality.

Both HS and HC compete well with usual weeds and, unlike jute, can permit delayed weeding. However, two weedings and thinnings are extremely useful, specially when done timely. The use of a wheel-hoe in the row crop, at least twice, within 45 days of sowing leads to better crop.

Nutrition

'HS 4288', yielding 21 quintals of fibre per hectare, lifts 106, 31, 148, 153 and 34 kg per hectare of N,P,K,Ca, and Mg respectively. The corresponding figures for 'HO 867' are 94, 26, 120, 137 and 29. US types respond well to the application of nitrogen from 25 to 60 kg per hectare, depending on the soil type, the method of application and, the rainfall. In drier areas, e.g. Andhra Pradesh, a basal dose of 20 kg of N has been found consistently giving 25 per cent higher yield. In areas having a better rainfall pattern, the quantity of nitrogen can be increased to 40 kg per hectare, and the application may be one-third of the basal dose and the remaining may be, top-dressed in one or two instalments between 45 and 70 days of sowing. Although the application to soil ensures best results, nitrogen as urea can be fed through foliage at the rate of 15 kg of N per hectare. Five kilogrammes of N in 100 litres of water is sprayed with a low-volume power sprayer. A single spray is effective between 30 and 75 days from the time of germination.

A high pH of soil leads to chlorosis in HS varieties because of iron deficiency; such soils are not to be used for growing mesta.

Harvesting

'HS 4288' and 'AMV 1' are harvested when 50 per cent of the plant population is in flower; delayed harvesting gives more fibre, but of coarser quality. 'HC 583' is harvested usually a month or more earlier. In Andhra Pradesh, plants are harvested by uprooting, which practice is not recommended; in the jute belt, they are cut close to the ground, as in the case of jute. All US types are of longer duration (180-210), whereas HC flowers in about 150 days after sowing.

Retting

Retting is done in the same manner as in the case of jute, but the low temperature and the paucity of water or both pose a problem. The harvested stems of both HS and HC can be stored under dry conditions for retting next season during the middle of the monsoon. The resulting HC fibre is better than that of HS. Retting in the current season, if possble, is always preferable.

Pests and Diseases

HC mesta once dominated over the HS type in Andhra Pradesh, but it yielded ground to the latter in recent years, since pest and diseases took a heavy toll of the nondescript local types. 'HC 583' evolved at the Jute Agricultural Research Institute can now defend itself against most of the diseases and Pests.

The important posts are as follows:

- Agrilub acuius Thumb. is a stem-borer, reaching the stem from the soil. Endrin EC 20 (0.04) is effective, but is to be replaced by Methyl Parathion. The fibre from damaged plants is useless.
- Root-knot nematode, causes dwarfing and chlorosis in extreme cases. Its control, with chemicals is costly; therefore rotation is advocated.

The important diseases are:

- Anthraonose caused by Colletotriehum hibisci Poll. (syn. Volutella sp.) has a withering efflect. It is prevented with Copper oxychloride (50% Cu) at the rate of 3 kg per hectare.
- Tip-rot caused by PMma sp. (syn. Trichosphaeria sp.) responds to copper fungicide;.
- Root-rot is caused by Rhizootonia bataticola alone or in combination with Fusarium oxysporum. Seed treatment with organo-merouric compounds or with TWM or Captan is effective.

- Eye-rot of the stem is initiated in the leaves and the lesions develop on the stem by Myrothwium roridum. The disease is prevented and controlled with Copper oxychloride.

Earlier records indicate that old varieties were asuceptible to a virus causing the paling, crinkling and rolling of the leaves and the twisting and elongation of the petiole, with the crowding of leaves at the apex. The pathogen is transmitted by whitefly, Bemisia tabaci. Cotton seems to be collateral host.

The important pest of H. sabdariffa var. altissima is mealy bug, Macollicuus (=Phenaccus) hirsutius Green, causing bunchy top. Methyl demeton (0.05), Parathion (0.04), Monocrotophos O.04) and Fonitrothion (0.04) control the pest effectively.

The important diseases are:

- Root-and stem-rot caused by Phytophthora paracitica F. sriffae. The disease is controlled by drenching the soil and the basal portion of the crop witil Copper oxyechloride and by ensuring good drainage.
- ln areas with high rainfall, leaf-rot caused by Phoma salwa-riffae is severe, but does not affect the fibre directly. The polyphagous hectic, Podagria apicifulva Bry. feeds on leaf-blades and an intense attack leads to damage in the case of HC and HS, more so in the former. HC 10 per cent prevents and controls the pest.

Released Varieties

HS varieties in use are 'R.T 1', 'RT 2', and 'P.T 26'. The new released improved type 'US 4288' has bristled green stem with a flash of released pigment at nodes, slightly proliferating to internodes at maturity; the leaves are deeply palmatised with 5 long narrow lobes; the petiole is pigmented on the upper surface only. The petals are sulphur yellow, with deep-purple throat. The variety flowers in late-November, irrespective of the date of sowing. It attains a height of about 400 cm. The duration of the crop is 180—210 days. The weight of 1,000 seeds is 18 g. The seeds are coffee brown. The yield potential of tile variety is 32 quintals per hectare. With average management, the yield is about 27 quintals. The variety is suitable for areas north of 22 °N.

'AMV 1' has bristled stem, with crim, on-purple pigment, both at the nodes and internode, the upper part of stem being more intensely pigmented. The leaf is deeply palmatisect (as in 'US 4288'); the petiole is red all through; the petals are sulphur yellow with a margin. The height attained is equal to that attained by 'HS 4288'. The variety is suitable for dry rainfed regions, specially those south of 22 °N.

HC varieties in disuse are 'MT 15', 'KT 129',Mt 150' and the NP series. The new improved type 'HC 5M' has, more or less, smooth green stem with an. irregular light flush of red pigment, the pigmentation increasing with maturity. The leaf is entire, cordate and flowers are large, the sepals are green, with red spots and each with a prominent neetary (gland); the petals are lemon yellow, with deep crimson throat. Seeds are greyish black, more angular. A thousands weigh 32 g. The variety is quick-growing and its attainable height is between 400 and 500 cm in warm humid regions, and less in drier areas.

Promising Varieties

The non-bristled stem, equals 'HS 4288' in yield and the cultivators prefer it to the bristled types. 'HC 867' has excelled 'HS 583' in yield at places and possesses better resistance to the spiral borer, A. acutus.

Plant height, base diameter of the stem and the yields of US and HO have the same correlations as in the case of jute.

Quality and Grading

The quality of the mesta fibre is judged on the basis of almost same criteria as in the case of jute and grading is done on the same lines, although ihe I.S.I.sp. are now due.

Natural cross-pollination being high the Varieties of the same species may be grown seed close at the same farm or near the farmers crop.

Ramie

Ramie (boehmeria nivea (L.)gaud) is a shrub belonging to family Urticaceae. It yields from its stem from the strongest vegetable fibre that makes excellent raw material for blending with terylene or rayon and itself can be spun and woven into excellent cloth that takes brilliant colours of dyes. The fibre goes into the making of sophisticated paper, including currency blanks. Though ramie requires special textile machinery, it can be adapted to the existing cotton and silk-waste machinery.

The technology for raising and running ramie plantations has been developed (Ramie in India: JARI). Since a viable plantation requires a large investment and at least 40 hectares, small holdings are thought to be unfit for commercial production. An alternative can be an agglomeration of co operatives in a compact area.

Ramie requires a warm and humid climate for proper growth. It cannot tolerate water-logging and does not thrive at a low temprature. It requires well-distributed rainfall around 1,600 to 2,500 mm. North of Bengal, Assam, the foothills of Meghalaya and the Western Ghats are suitable for its cultivation.

Ramie is unisexual monoecious, and under certain conditions pseudodioecious, and cross pollinated. Its Rhizomes can be used in its propogation. A plantation is raised from rhizome cuttings (two quintals to cover 1 ha) placed 50 cm apart in rows 1 m apart; space variation is permissible. The planting time in north-eastern India is April to June. The new improved varieties suitable for northern India are 'R 1449', 'R 1452', 'R 1411'. Diseases and pests at the moment have not been reported.

At each point of planting, a clump of 8-12 canes (stems) develops. The first years growth being not uniform is cut back close to the ground in April. From the second year, depending on the growth, 4-5 cuttings at the harvest are possible at about 55-day intervals yielding about 12-16 quintals under gummed dry fibre per hectare per year. This yield is commercially attractive. Ramie, being a semi-perennial crop, is replanted in cycles of 7 to 8 years, applying heavy quantities of organic matter. Each cutting, except the last one, is to be followed by an application of ramie-west compost and fertilizers (NPK). Acid soils are to be amended by liming, whereas sandy and clay soils are unsuitable.

Ramie is not retted like Jute or mesta. Extraction is done directly from fresh canes with raspador decorticators indigenous or imported. Extracted fibre retains some gum and markatable as such. The removal of gum is done by consumers by using various techniques, including boiling with alkali using the tri-polyphosphate buffer.

Some Indian textile mills have found in ramie a good source of textile raw material. However, organized plantations are not many. This matter requires the attention of industrialists. There is a great potential for the export of ramie.

TOBACCO

The genus Nicoliana has more than 60 species, of which two are commercially cultivated for the production of tobacco. They are N. tabum and Nrustica. N. is widely cultivated in most countries of the world, and as far north as Sweden, and as far south as New Zealand. The cultivation of N.rustica is restricted to India, Russia and some other Asiatic countries.

Though the tobacco plant is tropical in origin, its production at present is concentrated mostly outside the tropics, except in India. In tobacco production, India ranks third after the U.S.A. and China, with an annual production of 380 million kg of leaf from an area of about 445 thousand hectares. Almost all.states in India grow tobacco, but the important ones are Andhra Pradesh, Gujarat, Tamil Nnadu, Karnataka, Bihar, West Bengal and Uttar Pradesh in that order, and account for about 90 per cent, of the annual production.

The species N. tabacum is grown in almost all the states, whereas the cultivation of N. rustic is confined to the northern and north-eastern states, where the temperatures are considerably lower during the growing season. India ranks fourth in the world in tobacco' exports, after the USA, Turkey and Greece, Pride exports about 80 million kg of tobacco to about 50 countries and earns about 635 million rupees of foreign exchange annually. About 85 per cent of the export consists of flue-cured Virginia tobacco, and India is the world's second largest exporter of this variety after the U.S.A.

For the improvement and development of different types of tobacco grown in the country, the Indian Central Tobacco Committee was established in

1945. In 1947, this Committee established the Central Tobacco Research Institute, Pajahmundry (Andhra Pradesh) for fundamental reference on all tobaccos and applied work on tobaccos grown locally. Subsequently, the committee had taken over the research station at Guatur (Andhra Pradesh) from the then lindustrial Agricultural research Institute, New Delhi, and started the Regional Research Station at Yedasandur (Tamil Nadu), Hunsur (Karnataka), Pusa (Bihar) and Dinhata (West Bengal) for the improve- ment of the types of tobacco grown in these states. Besides the above stations, a number of schemes run by the state governments for the improvement of the crop were financed by the Committee.

With the abolition of the Commodity Committees in 1965, the Central Institute and the regional research stations attached to it came under the control of the Indian Council of Agricultural Research and the development activities were taken over by the Ministry of Agriculture, Government of India.

SOIL AND CLIMATE

India produces a wide range of commercial types of tobacco and each state specialise's in the production of specific types. However, all these types are mostly used for internal consumption except the flue-cured Virginia tobacco which is exported. The flue. cured Virginia tobacco is normally grown in other countries during the rainy season in light soil with a sandy surface, a friable sandy clay subsoil, low in organic matter content and acidic in nature.

In India, on the other hand, it is mostly grown in heavy black- Boils of Guntur, Prakasam, Krishna, Bust and West Godavari, Khammam and Nellore districts of Andhra Pradesh in winter

Cultivation

Usually, 6 to 10 ploughings are given by way of preparatory cultivation. Experiments have shown that deep ploughing of the fields once in two years during summer, followed by 2 to 3 ploughings and 1 to 2 harrowings are beneficial for most of the tobaccos, particularly for the flue-cured virginia tobacco grown in Andhra Pradesh. Digging with a spade, followed by ploughing with a mould-board plough and a country plough and then a harroeing is recommended for Bihar. Farmyard manure is usually applied and the dose varies from 10 to 125 cartloads per hectare for different types of tobacco. In addition, oil-cakes or ammonnium sulphate was found beneficial for some tobaccos. Based on the research findings of the Central Tobacco Research Institute, Rajahmumdry, and its regional stations, the following manurial schedule is recommended:

1. Seven-and-a-half tonnes of farmyard manure, six weeks before planting and 22 kg of N + 50 kg of P_2O_5 + 50 kg of K_2O per hectare, a week before planting for the flue-cured Virginia tobacco on black soils of Andhra Pradesh.

2. Ten to twelve tonnes of farmyard manure, six weeks before planting and 30 to 40 kg of N + 80 to 100 kg of P_2O_5 + 80 kg of K_2O + 12 to 15 kg of MgO, of which the entire quantity of Mg and P_2O_5 and half the quantity of N and K_2O are applied a week before planting and the remaining quantities of N and K_2O are applied 3 to 4 weeks after planting as top-dressing just before irrigation for the flue-cured Virginia tobacco on light soils of Andhra Pradesh.
3. Fifteen tonnes of farmyard manure or 300 kg of groundnut-cake 2 to 3 months before planting and 44 kg of N per hectare a week before planting for natu tobacco in Andhra Pradesh.
4. Forty-five to 90 kg of N per hectare, half as oil- cakes and, half as ammonium sulphate for the bidi tobacco in Karnataka.
5. One hundred and eighty kg of Wilha, half as castor-cake or groundnut cake and Half as ammonium sulphate, 2 weeks before planting for the bidi tobacco in Gujarat.
6. Twentyfive tonnes of farmyard manure and 75 kg of N/ha (213 as ammonium sulphate and 113 as groundnut. cake, half before planting and the remaining half 6-7 weeks after planting for cigar and cheroot tobaccos of Tamil Nadu.
7. Sixty tonnes of farmyard manure and 112 kg per ha, half as mustard-cake and half as ammonium sulphate for the chewing tobacco in Bihar.
8. Thirty tonnes of farmyard manure and 150 kg of N+150 kg of P_2O_5 per hectare for hookah and chewing tobaccos; 17 tonnes of farmyard manure and 125 kg of N+112 kg of P_2O_5+224 kg of KaO per hectare for the wrapper tobacco in West Bengal.

Spacing

The The tobacco seedlings are always planted in rows. The distance between the rows and between the plants within the two varies with the type of tobacco. Based on the research findings, a spacing of 80 cm x 80 cm for natu and the flue-cured Virginia tobacco in black soils and 100 cm X 60 cm for the flue-cured Virginia tobacco in light soils of Andhra Pradesh; 75 cm x 50 cm for cigar and cheroot tobaccos and 75 cm x 75 cm for the chewing tobacco in Tanil Nadu; 100 cm x 70 cm and 105 cm x 45 cm for the bidi tobacco in Gujarat and Karnataka, respectively; 90 cm x 60 cm for the chewing tobacco in Bihar, 60 cm x 45 cm for N. rustica and 90 cm x 90 cm for N. tobwum in West Bengal, are recommended.

Interculturing

When after the plants get established, i.e. about 20 days after planting, the tobacco fields are given frequent inter-culturing and weeding to conserve; the soil moisture and cheek the growth of weeds. Tined harrows, blade harrows and the country plough are normally used for interculturing.

Mulching the crop with paddy straw at 3,600 kg per hectare after the first interculture was found beneficial in the case of the flue-cured Virginia tobacco on black soils of Andhra Pradesh in increasing the yields and improving the quality by understanding soil moisture.

Irrigation

In Andhra Pradesh, flue-cured Virginia tobacco on black soils is not normally irrigated, but the crop on light soils is given up to six irrigations. The irrigation water should not contain more than 50 ppm of chlorides, as otherwise the leaves get burnt and other qualities suffer. In black soils also, in adverse conditions, one irrigation on 40-day-old plants is recommended. The cigar, cheroot and chewing tobaccos in Tamil Wadu and the chewing tobacco in Bihar are normally irrigated. The bidi tobacco in Karnataka, Maharashtra and Gujarat and the hookah and chewing tobaccos in West Bengal are not normally irrigated, but in years of deficient rainfall, one or two irrigations are desirable to maintain the yield.

Topping and Suckering

The removal of the flower head alone or along with some of the top leaves of the plant is known as topping. After topping, the axillary buds grow and their removal is known as suckering. Most of the types of tobacco grown in the country are topped and suckered for improving the size, body and quality of the leaves, except wrapper tobacco where the texture is one of the important criteria. The flue-cured Virginia tobacco grown in the black soils of Andhra Pradesh was also not topped till recently, but of late, topping is recommended to improve the body of the leaf. Even now, if the growth of the crop is rank and the colour of the leaf is dark green, topping is avoided. All other types of tobacco are topped. The level of topping varies from type to type and ranges from high topping in the case of the flue-cured Virginia tobacco, in which case only the flowerhead is removed, leaving all the 20-24 leaves intact. To the very low topping in the case of chewing tobacco of Tamil Nadu where only 7 to 8 leaves are retained on the plant. Based on the research findings, topping at 14-16 leaves for the cigar and cheroot tobaccos in Tamil Nadu, at 13-15 leaves for bidi, tobacco in Karnataka and Gujarat and the chewing tobacco in Tamil Nadu and at 8-9 leaves for N. tabacum and at 6-8 leaves for N. rustica in West Bengal, is recommended.

To get the full benefit of topping, the suckers are to be removed periodically. This is a laborious and time consuming job. Hence, a number of experiments for the suppression of suckers by the application of oils, chemicals, hormones, etc. were conducted and it was reported that the application of coconut-oil to the top six leaf axils soon after topping suppressed the suckers in the flue-cured Virginia tobacco. Two per cent naphthalene acetic acid triethenolemin applied to the topped portion of the-steni suppressed the majority of the suckers in cigar and cheroot tobaccos of Tamil Nadu. In

Bihar, though vegetable oils and hormones suppressed the suckers by 34 and 47 per cent respectively, in the case of chewing tobacco, they did not prove to be of much practical importance to the cultivators. However, piercing the stem after topping improved the yield and quality and is widely practised. Naphthalene acetic acid and ESSO oil did not produce significant effect in controlling suckers in the case of the bidi tobacco at Anand.

HARVESTING, CURING AND FERMENTATION

The signs of maturity and the methods of harvesting differ with the type of tobacco. The flue-cured Virginia tobacco is harvested during December-March in Andhra Pradesh and during July-September in Karnataka. The leaves are considered ready for harvesting when the normal green colour changes to yellowish green or to light yellow. Harvesting starts from the bottom and each time 2 or 3 leaves are harvested. In another week, the next 2 or 3 leaves mature when they are harvested. Thus the 20-24 curable leaves, that become available per plant in the case of the flue-cured Virginia tobacco, are harvested in 6 to 8 primings at weekly intervals. Soon after harvesting, the leaves are strung on bamboo sticks at the rate oF about 100 leaves per stick and loaded in the barn for curing.

The bidi tobacco is harvested in January-February when the majority of the top leaves develop red rusty spots known as spangles. The cigar and cheroot tobaccos are harvested 90 to 100 days after plant when the leaves pucker and become brittle and yellowish-green. The chewing tobacco is harvested 110 to 120 days after plant when the leaves develop Pronounced puckering. The hookah tobacco (rustica) is harvested in May or June, and the tabacum harvested when broad flecks appear on the leaves. The whole plants are harvested in the case of the bidi, cigar and cheroot, chewing and hookah tobaccos. except in West Bengal where harvesting is done by prIming.

Tobacco leaves are cured after harvesting in order to impart the required colour, texture and aroma to the final product. Different methods of curing are adopted for different types of tobacco, depending on its quality requirements and the use to which it is put to. Flue-curing, air-curing (or shade-curing), sun-curing (or rack-curing or ground-curing), smoke-curing and pit- curing are the different methods of curing in vogue in the country. The Virginia tobacco is cured in special chambers, known as barns with artificial heat passing through metal pipes, called flues. Hence this tobacco is known as the flue-cured Virginia tobacco and is exclusively used for manufacturing cigarettes. Curing takes 100-120 hours, after which the leaves are unloaded from the barn, bulked for a few days and then graded, depending on the colour. Since the quality of the leaf depends, not only on colour but also on its position on the plant, the plant-position grading is now recommended.

The wrapper tobacco of West Bengal and the Lanka tobacco of Andhra Pradesh are air-cured. The wrapper tobacco is cured under atmospheric temperature and at 70-80 per cent relative humidity in barns with sides and the roof closed. The lanka tobacco is cured by hanging the strung leaf in thatched sheds for 2 to 21 months after which the curing is completed in pits.

Most of the tobaccos grown in the country are suncured, as in this method the construction of costly strue tares can be avoided and also because the sun is quite-bright and hot during this part of the year. The cigar and chewing tobaccos of Tamil Nadu are cured in the sun for 15 to 20 days by stringing the whole plants on bamboo poles. The natu tobacum of Andhra Pradesh is cured by hanging the Strung leaves on sea golds in the open for 1 to 11 months. In the case of the bidi tobacco, the plants are cut and left upside-down on the same spot for 3 to 7 days. After the leaves are dry, the lamina is stripped off. The chewing tobacco in Bihar is cured in the sun, the cut plants in the field for 4 to 7 days. In Uttar Pradesh, the plants are cut and exposed to sun for 1 or 2 days, after which the leaves are separate and spread on the ground or made into heaps for a further drying. In West Bengal, the leaves are primed and kept on the ground upside-down for a few hours for drying, after which they are tied into bunches of 4 to 5 leaves and cured in sheds for 4 to 5 weeks.

Some of the chewing tobaccos grown in Tamil Nadu are smoke-cured in cylindrical buts with open tops. The cured leaves, except that of the bidi tobacco are then bulked into heaps of convenient size for fermentation which improves the quality and aroma of the leaf. The bulks are broken up and remade occasionally to avoid excess moisture and decay. The leaf is then graded and packed, as required by the trade.

Yield

The average yield of tobacco leaf per hectare is about 750 kg and 950 kg for the flue-cured Virginia and the natu tobacco respectively in Andhra Pradesh; 1,000 kg, 450 kg and 350 kg for the bidi tobacco in Gujarat, Maharashtra and Karnataka states respectively; 1,250 kg and 1,600 kg for the cigar and the root and chewing tobaccos, respectively in Tamil Nadu, and 950 kg, 850 kg and 800 kg for the hookah and chewing tobaccos respectively in Bihar, West Bengal and Uttar Pradesh.

Varieties

There are local or commercial varieties grown in the various tobacco-producing areas under both N. tabacum and N. rustica, Wo numerous W mention. It is with the help of 'the Central Tobacco Research Institute, Rajahmundry, and its regional stations and other organizations connected with the crop.

BIOTECHNOLOGICAL APPLICATIONS FOR THE IMPROVEMENT OF INDIAN TOBACCO

India is one of the top 10 countries cultivation tobacco for the part ten decades both for domestic and international market, India is the third largest producer of tobacco in the world with an output of approximately 520 m kg. Of which FCV tobacco account for 120 m kg. The economic importance of to Indian exchequer is peogressively increasing in the last 10 years. Tobacco generates 10% of total excise revenue and tobacco 4% of the value of total agricultural exports, India has the unique distinction of growing different types of tobacco in different Agroclimatic zones.

Tobacco provides direct and indirect lively-hood to 26 million in India including 6 million farmers and workers. Many small and marginal farmers were benefited by growing tobacco. Research efforts to identify alternative crops to tobacco indicated that no other crops is as remunerative as tobacco. Stability of the worldwide demand of the crop, the crops hardiness and ability to grow in climate and soils non-suitable for other crops, relative ease in transporting tobacco and tobacco's profitability compated to other crops are incentives for farmers growing tobacco thus making tobacco one of the most appealing cash crops to farmers. Tobacco crops is considered as major employer worldwide. Despite technological progress tobacco cultivation remained as one of the most labour intensive crops among all other agricultural crops. Women in particular and skilled labourers, who would otherwise have little chance of employment are able to earn living by working in tobacco cultivation. WHO in its report has acknowledged that in developing countries tobacco can create vast employment. In view of its great role in the economy of the country and farmers, further improvement of tobacco crop through Biotechnological tools will benefits them a great deal.

Biotechnological applications: Scientific system of tobacco cultivation in India was introduced with the establishment of Central Tobacco Research Institute (CTRI) at Rajahmundry in 1947. Tobacco research in India over the past 50 years through CTRI has made significant contributions to the economic upliftment of tobacco farmers by identifying high yielding varieties and suitable package of practices This is reflected in the good reputation for Indian Tobacco in the International market, as leaf of desired quality traits as per the specific requirement of traders is produced. Initially, conventional breeding methods were utilised for improving the yield and quality of tobacco. However, later application of biotechnology tools for supplementing the conventional tobacco improvement programmes or initiated.

Different biotech techniques like anther culture, fertilized ovule culture micropropagation, protoplast culture somatic hybridization and also genetic engineering are being utilised. These techniques are helpgul is strengthening the conventional breeding programmes where ever they are week and failing to serve the purpose.

YIELD IMPROVEMENT

Yield improvement through conventional breeding programmes takes fairly long time. This can be reduced substantially (3-4 years) by culturing anthers (Anther culture). Of elite lines from early breeding populations in tissue culture. Through anther culture several dihaploids are produced and evaluated every year. An improved line D1 was developed, in this way, which is superior in yield. Efforts to incorporate budworm resistance through someclonal variation in this line is in progress. Stable resistant D1 someclones are there in field testing stage.

Resostance to Biotic and Abotic Stresses

Wild species of tobacco possesses resistance to various tobacco pests and disease. It is difficult and in some cases impossible to transfer this resistance from wild species to cultivated tobacco through conventional methods. In such cases, inter-specific hybrids were rescued through fertilized ovule culture and mass multiplied for further testing by micropropagation, In this way, interspecific hybrids possessing resistance to leaf spot diseases, black shank, root knot nematodes leaf eating caterpillar, aphids, budworm etc. were developed and evaluated for their resistance. Thus, stable pest and disease resistant lines were identified through Biotechnological tools. Efforts to transfer desirable characters from wild species through somatic hybridization is in progress.

Somaclonal variation was attempted to fortigy tobacco lines resistance to white fly (Bemisia tabci Genn.) (leaf curl) and budworm (Heliothis armigera Hb) and the results are encouraging. Somaclonal variation is highly effective to create variation when desired variation is impossible to get through conventional breeding methods and also variation is required only for one character. Transgenic Bt tobacco cultivars processing resistance to leaf eating caterpillar (Spodoptera litura F) and budworm (Heliothis armigera Hb) were developed. Production of transgenics will give scope for transferring desirable characters from different species and genera.

Preliminary attempts were made to screen lines for drought tolerance in tissue culture. Such efforts will save expenditure on extensive field evaluation and work can be carried out in the off season also. Transgenic tobacco having wheat desiccation tolerant protein were evaluated water stress tolerance.

Herbicide Resistance

Orabache cernua os a root parasite of tobacco causing heavy loss in yield and quality. Glyphosate at lower does was found to kill Orabanche, but it is phytotoxic to tobacco. Hence research on production of herbicide (glyphosate) resistant tobacco lines through tissue culture was initiated. When farmer grows herbicide resistant lines in the field he can use berbicides without any harmful effect on tobacco, for the effective control of weeds and Orabanche.

Conservation of germplasm: Germplasm is the valuable source for various characters that are useful in future breeding programmes. This can also be conserved under in vitro conditions for long time. Also lines that cannot be maintained under field conditions due to non-flowering and sterility can be effectively maintained under tissue culture conditions. Interspecific F1s, mammoth mutants, in vitro rescued hybrids protoclones, BT transgenics various mutants are routinely being maintained.

Future Thrusts

Biotechnology can help Indian tobacco farmer in many ways, Particularly in tobacco improvement research and seed purity disputes, if offers a large scope. Research on tobacco improvement benefit farmer as he can get high yielding and pest and disease resistant tobacco for cultivation. In research various aspects like in vitro conservation and molecular characterisation of germplans, isolation, characterisation and transfer of desirable genes from across genera, molecular aided selection in breeding programmes, someclonal variation can play a great role. Utilising Biotechnological techques tobacco lines that can yield higher amounts of phytochemicals which are useful in the production various drugs and pesticides can also be developed.

In recent times many disputes about seed purity among farmers are coming up. At present morphological criteria is only being used for varietal identification. As morphology is highly subjective in nature and also varies with the type of environment the plant is grown, it is becoming highly difficult to solve the disputes. Under this situation molecular characterisation of cultivated varieties will help to identify correct varieties beyond doubt.

In future, giving emphasis on the above said lines will go a long way in benefitting the India tobacco crop as well as the farmer. However, as biotechnology got wider applications, any thing that put the farmer and environment at disadvantage need to be discounrages.

JUTE PRODUCTION

Jute, a bast fibre, is obtained from one of the most important cash crops of eastern India Exported as manufactured goods and as raw fibre, it earns foreign exchange around 3,000 million rupees. The crop is grown in West Bengal, Assam, northern Bihar, south-eastern Orissa, Tripura and eastern Uttar Pradesh on 0.80 million hectares annually and the average production of dry fibre is 6.5 million bales.

Currently, about 82 per cent of the total annual production is used by jute-mills in India, 8 per cent is retained by cultivators and the remainder is exported to Europe and the USA.

The fibre is chiefly used for manufacturing hessian, sacking and carpet backing. It is used for storing and transporting grains, pulses, spices sugar, cement, fertilizer, minerals, cotton and wool all the world over. Jute is also used for making mats, tarpaulins, ropes and twines. Woollenized jute is used for manufacturing cheaper rugs, and mixed with cotton it is used for producing decorative cloth, curtains and upholstery.

The jute sticks are largely used as fuel and also for making gunpowder charcoal. Since recently, the paper industry has been using it as a raw material for coarser paper. Resin-bonded pressed jute sticks make durable hard board.

CLIMATE

Jute requires a warm and humid climate, with temperature fluctuating between 24 °C and 37 °C, the optimum being around 34 °C. The permissible diurnal variation in relative humidity favourable to growth is between 57 and 97 per cent. Incessant rain water-logged conditions are difinitely harmful. Whereas the capsularis varieties can stand water logging to some extent. In the seedling stage, water-logging is not tolerated by both the species.

Areas receiving early showers in late February or early march, followed by a relatively short spell, are suitable fore sowing the capsularis varieties. Olitorius varieties, until the introduction of 'JRO 878' and 'JRO 7835' have been sown with showers in April or even in May. The monsoon sustains the subsequent growth. Rain below 1,000 mm is not helpful. Alternate sunshine and rainy days are most conducive to growth. The dry spell preceeding the break of the monsoon enables the essential weeding and thinning operations to be carried out.

Soil

The new grey alluvial soil of good depth, receiving silt from the annual floods, is nature's best gift, but jute is widely grown in sandy loams and clay loams, with varying soil-management practices. Sandy soils and heavy clays are unsuitable. Soils with a low pH give a poor crop, the optimum pH being around 6.4.

Cultivation

Jute requires a clean, clod-free field with fine tilth. The land is, therefore, ploughed, cross-ploughed, and planked several times. All weeds are thoroughly removed.

Sowing

The sowing of capsularis varieties starts in late February in low-lying areas that retain moisture of the previous flood or monsoon.

Sowing in midlands and highlands starts with showers in March or April and continues till early June in the western part of the jute belt. For broadcast sowing, the seed-rates are 10 and 7 kg per hectare for capsularis and olitorius varieties, the germinability of seed being 80 per cent or above. The high seed-rates ensure an even emergence of seedlings, which are thinned out in broadcast plots to a spacing of 10 cm. This operation is preferably carried out in two instalments, once when the seedlings are about 10 cm and subsequently when they are about 15 cm tall. While thinning, care is taken to remove all weeds with a hand-hoe. These operations account for 30-40 per cent of the cost of cultivation.

Row-cropping has proved advantageous and a single row seed-drill is used to sow the capsularis varieties 30 cm apart and the olitorius varicties 20 cm apart. The plant-to-plant spacing is manually adjusted between 5 and 7.5 cm. Sowing is always done shallow, i.e. not exceeding 3-4 cm.

Cropping Pattern Under Irrigation

In jute-paddy potato, jute-paddy-wheat, (jute-moong)-paddy-potato, cowpea-jute-potato, and jute-paddy-berseem rotations, jute may need one pre-sowing irrigation and one sustaining irrigation cropping pattern under rainfed conditions. Jute-paddy-mustard, jute-paddy-pulses or jute-paddy rotations are possible under this cropping pattern.

Nutrition and Manuring

The gross uptake of N, P, K, Ca and Mg by the capsularis variety 'JRC 212' is 84, 16, 147, 84 and 29 kg respectively and by the olitorius variety 'JR0 632' is Ill, 28, 164, 124 and 25 kg per ha, respectively. The resulting fresh crop weighs 40 to 50 tonnes, yielding 20 to 25 quintals of dry fibre, which is around 5 per cent of the fresh weight of the crop.

Acid soil require amendment with 3-7 tonnes of lime in 3 years. Compost or farmyard manure at the rate of 4-7 tonnes per hectare promotes proper growth, specially when fertilizers are not available. P and K are applied as basal nutrients, whereas nitrogen is better top-dressed in two instalments-N (40-80 kg per ha for capsularis varieties, 20-60 kg per ha for olitorius varieties)-P (half of the quantity of N) and K (quantity equal to N).

If the jute crop is preceded by potato, no fertilization is recommended; if it is preceded by paddy or wheat, P and K are not so much required. Nitrogen is the most important nutrient required.

The ash of water-hyacinth is a rich source of potash. The application of magnesium on the North Bank of Brahmaputra and in northern Bengal produced good results.

In case of short supply of nitrogen, foliar feeding with nitrogen at the rate of 15 kg of urea per ha at 10 per cent concentration, with a low-volume power sprayer in 2 instalments between 35 and 60 days' age of the crop proved useful. Capsularis varieties respond better up to 30 kg of urea.

Interculture

Weeding, specially in the early stage, is a must. For a broadcast crop, the pre-sowing application of 2,2,3,3 tetrafluorpropionate of sodium proved to be beneficial. A post-emergence application of mono-sodium methanarsonate with or without 2,2 dichloropropionate, as a directed spray, proved to be beneficial in row-cropped fields.

Yields, with two hand-weedings, however, are significantly superior and constitute the prevailing standard practice. Smothering the weeds between rows with a wheel-hoe saves labour and helps to mulch the soil.

Major Pests

The female of jute apion (Apion corchori Marshall) lays an egg at a node where a knot is formed; it makes one puncture or several on the same plant. The jute semilooper (Anomis sabulifera Guen.) eats tender leaves and is widespread; it usually comes in three waves between late June and August. The vehow mite (Hemitarsonemus Iatus Banks) sucks the sap from young apical leaves which curl, and the growth of the plant is arrested. The hairy caterpillar (Diacrisia obliqua, Walker), when young, is gregarious on a single leaf and later scaterrs out and feeds on the foliage.

The root-knot nematode (Meloidogyne incognita Kofoid and White) attacks the capsularis varieties more than the olitorivg varieties and induces chlorotic Symptoms and s and arrests growth. What rovides foci for root infection bataticola (Taub.) Butler. hytrypes achatinus Stoll is a menace to seedlings which are cut. The pest is mostly confined to Assam.

Minor Pests

The indigo caterpillar (Lapltygma exigua Habn.), the leaf-miner (Trachys dasi Thery.), the red mite (Olygonychus coyfeae Neither), the jute stem- girdler (Nupserha bicolor postbrunnea, and the leaf-eating weevil (Myllourus discolor Boheman) are not of commercial importance.

Control Measures

DDT and Gammexane, with a limited efficacy in the field, are unprofitable. Ethyl parathion and Endrin are highly efficacious against all major pests. But because of their hazardous nature, they have been withdrawn. Endosulfan, Leptophos, Phosalone, Fenitrothion, Chlorfenvinphos are effective to different extents against the major pests; Monocrotophos and Tamaron on semilooper; lime sulphur on yellow mite and Heptachlor and Aldrin dust on crickets and Dicofol on red mite, are effective.

Timely sowing and clean cultivation guard against afion; hand-picking of the leaves with gregarious clusters young hairy caterpillars controls the pests.

Diseases

The important diseases are:

- Seedling blight, stem-rot, collar-rot, and root-rot caused by Macrophomina phaseoli (Maubl.) Ashby, both seed-and soil-borne, secondary spread through air-borne pycnospores. The predisposing factors are low pH of the soil, the lack of adequate potash water-logging, high temperature and excess nitrogen. Seed treatment prevents seedling blight, whereas soil amendment with lime and potash retards its incidence. Jute-paddy and jute-paddy-wheat rotations are useful in reducing the incidence of the disease.

- Hooghly-wilt, caused by a microbial complex, including Macrophomina phaseoli, Fusarium solani. Pseudomonas solanacearum, is confined to the jute potato fields only; control through rotation with paddy and legumes is possible.
- Colletotrichum gloeosporioides causes anthracnose in Assam.
- Leaf mosaic is caused by a pollen-transmitted seed-borne virus, also by white fly Bemisia sp. in Assam, specially in capsularis varieties.
- Physoderma stem gall on the seedlings of olitorious varieties occurs in Uttar Pradesh and Tripura.
- Soft rot of stem Pellicularia rolfsii (Sace.); the disease can be controlled with deep ploughing, and clean cultivation.
- Nematodes provide foci for infection by Macro phomina phaseoli in sick soils.

Harvesting

Jute may be harvested any time between 120 and 150 days. Early harvesting gives finer fibre of good quality, whereas late harvesting gives a larger yield but a coarser fibre.

The compromise between quality and quantity is found in harvesting at the early pod stage or around 135 days of cropping. The other consideration for early harvesting is to accommodate paddy-transplanting in accordance with the cropping pattern chosen.

Harvesting is done by cutting the plants at or close to the ground level. In flooded land, plants are uprooted. The harvested plants are left in field for 2 to 3 days for the leaves to shed. Next, the plants are tied into bundles, 20-25 em in diameter, and the branching tops are lopped off to rot in the field.

Retting

The bundles are kept standing in water, 30 cm deep, and later placed side by side in retting water, usually in 2 to 3 layers and are tied together. They are covered with water-hyacinth or any weed that does not release tannin and iron. The float is then weighed down with seasoned logs or with concrete blocks or are kept submerged (at least 10 cm below the surface of the water) with bamboo-crating. Clods of earth used as a covering material or as weighing agent produces dark (shyamla) fibre of low value.

Gently flowing, fairly deep, clear and soft water is ideal for retting. The optimum temperature is around 34 °C; ditches, tanks and pools are also used for retting. Incomplete submergence produces 'croppy' fibre of extremely low value. Most of the defects in fibre are due to faulty retting. Over-retting results in 'dazed' weak fibre.

Retting is a microbiological process and, therefore, the end-point is determined by inspecting a few plants each day from the tenth day onwards. If fibre slips out easily from the wood on pressure from the thumb and fingers, retting is considered complete.

Extraction

The fibre, when extracted separately from each reed (stem) with fingers; is sleek, clean and free from entanglement. By the beat-break-jerk method, ten or twelve reeds are taken at a time; their stiffer root-ends are beaten with a mallet to loosen the fibre. The bundle is then broken in the middle and the fibre is loosened. By gripping this loosened fibre in the middle, the broken bundle is jerked in water so that the sticks slip off. The fibre is then washed in clean water, wrung and. eventually spread to dry, preferably in shade or mild sun. This second method often leaves the broken sticks and makes fibre somewhat entangled, resulting in 'sticky', fibre.

The extraction of fibre from the green stem with a machine, followed by a short-period retting has proved to be successful. Adaptive research continues at the Jute Agricultural Research Institute.

Yield

The national average is 13 quintals, per hectare. In general, an olitorius variety gives better yield than a capsularis variety, and the former is a more efficient user of nitrogen. Under the National Demonstration Programme, the average yield has been improved to 27 quintals in the case of the olitorius and to 20 quintals in the case of the capsularis varieties, the maximum potential being 40 quintals and 37 quintals, respectively.

A strong positive correlation exists between the plant height (PH) and the yield, between the base diameter of the stem (BD) and the yield, also between PH and BD. Hence, taller and stouter plants give higher yield. However, if the yield from a smaller number of stout plants is equal to that from a greater number of thinner plants, the quality in the case of the latter is always better and, there-fore, preferred. The leaves of both the species are used for culinary purposes, olitorius being sweet and capsularis being bitter, but the latter is considered to be good for the stomach.

Varieties

The capsularis varieties cover 60 per cent of the jute area, whereas the olitorius varieties cover 40 per cent of the area, mostly in the southern West Bengal.

Released Capsularis Varieties

- 'JRC 321' (Sonali) has a light coppery-red stem. It is suitable for low-lying areas for early sowing from late February to mid-March. It comes to 50 per cent flowering in about 125 days (variable); it is fast-growing and has good-quality fibre. It yields 20-25 g per hectare in northern West Bengal and Bihar.'JRC 212' (Sabuj sona). This variety has a full-green stem. The sowing time is from late March to mid-April. It is suitable for medium and highlands where inundation does not occur. It is particularly good for Assam, Bihar, northern West Bengal, Uttar Pradesh and Orissa. With 60 kg N per ha, it yields around 30 q per ha.

- 'J.RC 7447' (Shyamali). This variety has a full-green stem, and is an X-ray derivative of 'JRC 212'. It responds better to higher doses of N above 60 kg per ha; it is suitable for Orissa, Nadia, Murshidabad and 24-Parganas.
- 'D 154'. It is an old standard variety, originally selected in 1918 from 'Kakya Bombai'. It is a hardy variety and can adapt itself to all conditions. It yields 20-25 q per ha.
- 'JBC 1108'. A pre-release variety. It yields better than 'D 154', equalling 'JRC 212' at places, and is fairly resistant to stem-rot and root-rot.
- 'JRO 632' (Baisakhi tossa). This variety has a full-green stem; its sowing time is from 15 April onwards on midlands and highlands. It takes 130-140 days to flower; its seed colour is steel grey; its yield potential is 40 q per ha; its yield averages 32 q per ha with 40:20:40 "NPK plus farmyard manure. Its fibre is of good quality.
- 'JRO 878' (Chaitali tossa). The stem of this variety is red-pigmented. Its sowing time is the second week of March to mid-April on highlands. It is non-lodging and has non-shattering pods. It does not flower prematurely. It is suitable for West Bengal and Orissa. Its Yield Potential is 38 q and the average yield is 32 q per ha,with 40 N: 20 P: 40 K plus farmyard manure.
- 'JRO 7835' (Basudev). The stem of this variety is green. It is sown in early March to early May. It is non-lodging and bears non-shattering pods (It does not flower prematurely. It is suitable for all areas with highlands). It is high-yielding and produces around 34 q per ha with 40: 20. 40 NPK plus farmyard manure.
- 'JBO 620'. The stem of this variety is red-pigmented. It yields fibre of superior quality. It yields 3-4 quintals less than 'JRO 632' and, therefore, has gone out of cultivation.
- 'C.G.' (Chinsurah green) is a Poor Yielder and has been withdrawn. The pre-release variety 'JRO 524' (Navin),yields as high as, or more than 'JRO 7835', with added qualities of fine fibre and a stem that rets quickly.

SISAL PRODUCTION

Sisal (agave sisalana perrine), introduced from Mexico, is a semi-perennial plant and yields cream-white fibre from its leaves. The fibre is used for making ropes, matresses, brushes, doormats and fancy articles of various kinds. Moribund plantations of sisal are found in southern Bihar, Orissa, Andhra Pradesh, Karnataka, Maharashtra, Madhya Pradesh and West Bengal, raised in the early part of this century.

The technology for raising and maintaining a sisal plantation in India has been developed (sisal in India-JARI). As a result old plantations are being renovated and new ones are being planted on scientific lines to cater for commercial needs. The planting material can be obtained from the Sisal Research Station, Bamra, and from other plantations, particularly in north-western Orrisa. The plant popularly known as century plant, has a short thick stem (bole) which is almost concealed in early stages by the surrounding sessile leaves which are thick, long and broad. From slender rhizomes, buds grow and sprout above-ground to form baby-plants or suckers. After 8-10 years the plants flowers. The stout floral axis develop slowly and produces flowers which drop off and bulbils are small baby-plants develop on the seat of flowers. Both bulbils and suckers furnish the propogation material for plantations.

Bulbils and suckers are nurtured with copious nutrients in nutrients for a year in close spacings and later transplanted when 10 to 18 months old. Depending on growth the harvesting of the leaves starts from 2 to 3 years old plants and such annual harvests last for 7 to 8 years comprising a full cycle. The fibre is extracted with power-driven decorticators. Raspador decorticators are being manufactured in India. After a cycle is completed, the plantation is cleared, heavily manured and is preferably put for a year under a green manure crop and then replanted.

The yield of two tonnes per hectare per year may be considered good in the dry, infertile areas of north-western Orissa or Chhota Nagpur. With good management, the yield can be much improved, more so in better types of soil, with a rainfall between 650 and 1300 mm. Good drainage is essential. Gravely or highly acidic soils, poor in nutrients, are unfit for fibre production.

When exposd to rainfall above 1500 mm, sisal leaves become susceptible to a number of fungal pathogens, including colletotrichun, diplodia and alternaria spp. in the wilting of plants has been observed, but not frequently.

Sisal suffers most from imbalance in nutrition. It requires large quantities of calcium, magnesium and potassium, besides nitrogen and phosphorus. In acid soils, with pH below 6, this imbalance begins to be reflected in various kinds of symptons and the remedy lies in the application of soil correctives, organic manures and fertilizers. Phytophthora and pythium spp. have been reported on exotic varieties introduced recently. This is a serious disease and planters must use only indigenous resistant types. With assured internal consumption, the present target is 8000 hectares. There is scope for many more plantations.

The other two species found to have been used in India for plantations are Agave cantala Roxb. and Agave veracruz Mill. The fibre of these two species is often marketed as sisal fibre.

SUNNHEMP PRODUCTION

Sunnhemp (Crotaria junea L.) or Bombay hemp or Banaras hemp, is a crop grown for the bast fibre it yields. It is also a green-manuring crop, but a minor fodder crop of minor importance. The fibre is dull yellow somewhat coarse, strong and durable. Owing to the fibre having a high cellulose content, low lignin and negligible ash, the paper industry has identified it as the most suitable indigenous raw material for manufacturing tissue paper and paper for currency. In rural areas, it is also used for making ropes, twines and nets. It is also used for making canvas and screens (iat-patti). sunnhemp fibre is not used for textile purposes, unlike jute and mesta.

The area under this fibre crop is 1,38,000 hectares and the annual production of fibre is 50,000 tonnes.

CLIMATE

The crop is grown in almost all parts of India. But the states of Uttar Pradesh and Madhya Pradesh are most prominent in this respect. The crop grows best in tropical and subtropical climate. It is grownas kharif in the northern states. In the southern regions, Where the climate is, more or less equable and the winter is not pronounced, it is raised in rabi also. A minimum of 40 cm of rainfall distributed in not less than 50 rainy days during the growing season is a primary requisite.

Soil

Well-drained alluvial soils, having a sandy loam or loamy texture are suitable. For uncultivated fallows or freshly reclaimed soils it is an ideal first crop. Being a leguminous crop, the nodule formation depends on the calcium and phosphate status of the soil, hence, old alluvial soils with low pH are not suitable, unless acidity has been corrected by liming.

Cultivation

The sunnhemp crop sown in the kharif preceeds rabi crops, e.g. wheat and oilseeds. It is sown broadcast, but should better be sown in rows. In broadcast sowing, the seed-rate is about 25 Kg per hectare; in lines (30 line-to-line spacing and 5 to 7 cm plant to plant spacing), the seed requirement is only 15 Kg per ha.

Since it is a crop that fixes its own nitrogen fertilizers are added. Doses of P and K, ammounting to 20 Kg per ha is beneficial. Trases of boron and molybdenum are required, but these elements are not generally added, since most soils already contain them. Calcium is required in large quantites. The traditional sunnhemp belt of Uttar Pradesh, i.e. Banaras, Pratapgarh, Sultanpur, Azamgarh, Deoria; etc. do not require liming. After the application of the fertilizer during the preparation of the land, sowing follows immediately. After sowing, the soil is raked and laddered in order to put the seed 2-3 cm below the surface.

Sunnhemp seeds, not older than one year, generally retain 90 percent germinability. The germination of sunnhemp seeds is epigeal. The seedings grow very fast and in the process, they smother all weeds, except nut sedge. Weeding is not, therefore, necessary. Yet a common weed associated with sunnhemp is Ipomoea sp., whose seeds are disseminated along with the sunnhemp seed itself. The Ipomoea plants are twiners and grow along with the crop neck-to-neck, flower and fruit at the same time as sunnhemp. Unless manually weeded out, they hamper the growth of the crop. Seeds of these weeds can be separated from sunnhemp saads with some care.

Cultivation

The fibre crop gets ready for harvest in 120 to 150 days. Pod formation is considered the proper stage of harvesting. The crop flowers from September onwards, and this period sychronizes with the onset of short days in latitudes around 23 degree N to 24 degree N, where the main sunnhemp belt lies. The flowering time for most of the varieties remains almost the same, irrespective of the date of sowing. The minimum vegetative period needed before the initiation of flowering is 45 days in the case of most of the varieties (e.g.'k 12 Yellow').

But there is an early-maturing strain ('t-6'), which is also day-neutral and flowers in 30 days, irrespective of the sowing time. This type is generally harvested at the dead-ripe pod stage in order to yield both fibre and seed, as is practised by the farmers of Midnapore (West Bengal) and the adjoining districts in Bihar.

It is characteristic of this type that even after the maturity of the pods, the stalk remains green and the fibre is extractable. In this case the total cropping period is 80 days. The plants are harvested with sickles and the bundles are kept in the field for 2 to 3 days for the leaves to be shed.

Retting and Extraction

The process of retting is similar to that of jute. Unlike the case of jute, the source of water for retting in the sunnhemp-growing areas is not abundant. In most cases, retting tanks are shallow muddy ditches adjoining the fields where the run-off water of a shorter spell of the monsoon in collected. The retting time needed by sunnhemp is only 4 to 6 days.

The extraction of the fibre of sunnhemp is more cumbersome than that of the fibre of jute. The beat-and-jerk method is unsuitable in the case of sunnhemp as the tendency of the fibre to fibre to stick to the wood is more and the fibre gets entangled with broken twigs. Hence the extraction is done stalk by stalk from the lower end upwards.

The fibre is washed in water and dried in the sun. Incomplete retting makes the fibre coarse and green, although its strength is not impaired. Inadequate washing leaves behind some gum which renders the fibre coarse. Paper-making concerns, which make a bid to produce a small chunk of their raw-material stock by themselves, resort to mechanical decorticated lump goes directly to the bleaching and digesting plant.

Varieties and Yield

Sunnhemp has a fibre content of 2-4 percent on the basis of the weight of green twigs, or 8 to 14 percent in terms of dry weight. This is a low figure contributing to low yields. The national average is such a low figure as 3 to 4 quintals of fibre per hectare.

The cultivated types acclimatized in the diverse ecogeographical belts exhibit variability to their vegetative growth, branching, leaf size and thickness, and more pronouncedly with respect to their sensitivity to photoperiod.

Sunhemp-A

Varieties and Yield [SSH]

On an sunnhemp has a fibre content of 2-4 percent on the basis of the weight of green twings, or 8 to 14 percent in terms of dry weight. this is a low figure contributing to low yields. The national average is such a low figure as 3 to 4 quintals of fibre per hectare.

The cultivated types acclimatized in the diverse ecogeographical belts exhibit variability with to their vegetative growth, branching, leaf size and thickness, and more pronouncedly with respect to their sensitivity to photoperiod. geographical belt 8 exliibit variability witil respect to their vegetative growth. branching, leaf size and thicknesa, and more pronouncedly with respect to their sensitivity t,o photoperiod.

There were a number of improved varieties;

S.No.	State	Variety	Special features
1	Utter Pradesh	'Kanpur-12' ('k-12')	High yielding, good-quality fibre resistant to wilt
2	Madhya pradesh	'M-18'	Matures early; suitable for areas having light soil, with low rainfall
		'M-19'	same as 'M-18 plus moderate resistance to shoot-border
3	Bihar	'BE-1'	Good Yield and Quality
4	Madras	'Bellary'	High yielding
5	Marashtra	'D-IX'	Wilt resistant
6	West bengal	'ST 55'	Found yielding higher than K-12

Varieties under serial Nos. 1 to 5 were evolved he fore 1948 and No. 6 before 1958. Except the 'K 12' variety, much is not known about other varieties at present. 'K 12' has been the most widely-cultivated variety, which in course Of tinle, an suffered from genetic mixing witi-i poor-yielding local types owing to cross-pollination.

Later this variety has been further improved (1971) and released as 'K 12 yellow'. This is easily di,.tingui- r,hable from other types because of its yellow seeds in place of black ones of every other type. This variety has a high-yield potential up to 14 qlha, and is moderately resistant to the wilt disease. The fibre,. is of good qlnality and lustre-.

Seed and Seed Production

Sunnheinp seed occurs in shades of black in most cases and yellow in one case. The hundred-seed weight ranges from 3.4 to 6 g, depending on the variety. The statistics of seed requirement are lacking, since. the massive quantity used up ill green manuriiig is not accounted for. Sunnhemp seed is a,Aonree of seed protein used for manufacturing adhesives.

The seed production of sunnhernp is yet to be organised. BY tradition, the farmer lows a few lines of sunnhemp on the borders of ba'ra, ragi and jowai. or on the bunds of paddy fields for obtaining seed.

The seed-crop is raised late in the kharif season, i.e. in early August, in the stinnhemp-growing districts of the northern states. But it is ideal to grow a seed-crop in the rabi season in a zone where winter is not severe, and the temperature seldom falls below 10 °C. Under optimum conditions, the seed shield will touch about 20 quintals per hectare.

They ace:

- Xyllatipes Faber.,
- Xylocopa fenistroides Faber. (the bumble-bees)
- Mogachile lanata fibre (111)elonging to the bee family.

Sunhemp for Green Mannuring

A sunnhemp crop Grown thick by using a heavy seed-rate (60 kg per hectare),vith profuse branching and foliage is ready for incorpo- ration into the soil within 2 to 3 months of sowing. At this time, the stalks will be tender with very little fibre formation, and hence, easy for quick disintegration in the soil. The green matter at harvesting weighs kbroun 20 tonnes per hectare.

It contains about 80 kg of N per hectare, besides P and K. The crop is either ploughed in it stands or is cut close to ground and ploughed down and levelled with a wooden plank (sohaga).

In the case of double-cropped paddy, the green-manuring crop is cut and carted to the field and is incorporated into the soil. Green-manuring with sunnhemp is practised invariably in every type of soil. It is particularly found suitable in reclaiming saline alkaline soils: The other greeni-manuring crop used is dhaincha (Sesbania sp.).

Diseases

Important diseases are wilt (Fusarium udum) mosaic (virus) spread by Bemisia tabaci; anthra- enose (Colletotrichum curvature); rust (Uromyres deco- ratus) and powdery mildew (Oidiuin sp.). Good manage- ment, crop rotation and seed-dressing can control these diseases.

Pests

The important pests of sunnhemp are:

- Top-shoot borer (triuntra Meyr) whose larvae burrow.into the growing tips arresting their apical growth add inducing branching;
- The hairy caterpillar (Utdkoim pile" Lian.) whose larvae feed on the foliage and the pods.

Spraying with Diazinon (0.04%) 0411 control the former, whereas 1% B.E.A. or Methy) Parathion (0.04%) can control the latter

FLAX

Flax or linwed (Linum usitatimsutn L.) fibre con from the stems and the oil from the seeds. The fibre obtained from the linmwda is inferior in respect of length and quality. There are some types exclusively for fibre. The flax fibre is valued ort its strength and durability which are superior to those of cotton. It is soft, shining and poeon high water absorbency. It has low elasticity and is stronger wheis wet than when dry. It is resistant to high temperatures, moiature and mudow. ltd not readily reaettomordants and dym, unlike cotton. B'abrim made of flarlaunder well. Max contains 80 to 90 per cent cellulose

Flax in woven into fine fabrics, Much an lawns, cambric and drills, canvas and buckrarns. It is used in linen. stitching, making twines and nets for fishing, ropes, carpet- backing. wrapping cloth and house furnishings. It is a good raw material for tissue paper, firefi htirig hose and water-baga.

Although in the Gangetic flax types introduced Italy, were not found the bre amp; on the cont- rary, they deteriorated into the linaced types. The attempta to produce a dual-pur, i.e. strain, have so far failed. We are bout SW tonnes of flax to meet the demands, chiefly f'rom the Defence Organization. India's own efforts to cultivate flax have not been successful because of the climate.

Climate

Flax is a crop of the temperate zone. A cool humid climate with temperature ranging from IT to 27'0 and a diurnal fluctuation of relative humidity between 30 and 95 per cent, cloudy weather during the growing period is preferred. In the hot and dry climate, flax types tend to branch and grow as linseed types.

The ideal geographical regions envisaged for successful flax production in India are the valleys of the Himalayan ranges where the winter iinulates a European summer; this possibility bm been proved by some trials.

Soils

Heavy clay and light soils are unsuitable. Rich loams or clay loams are considered best. The 'char' lands of the Gange and the Brahmaputra which receive silt from the annual floods are suitable in winter.

Cultivation

For the Gangetic plains, the sowing time is early November. For the high rangu, it will be early October. The seed is drilled in rows 20 em apart. After germination, the seedlings are thinned to keep a plant-to-plant distance of 1 to, 2 em. The wed-rate is 40 kg, the expected germination heirs 90 per cent. One or two irrigations will be needed. The crop matures in 100 days, one month after the initiation of the first flower. Harvesting is done by cutting the stems close to the ground,or by pulling them out.

Retting

Bundles of stalks (straw) steeped in retting water just enough to cover,, It take about 3 days for the retting to be completed under the normal temperature prevailing at the harvest time. Over- retting weakens the fibre; under-retting makes extraction difficult. Betted straw is then dried.

Extraction

The extraction of fibre requires a simple machine. The machine is indigenously designed on the, principle of passing the straw through fluted rollers to break the woody core and separate the fibre. The eeparat ed fibre stranda are then brushed and rolled into bundles. This machine can also he used to salvage fibre from the linseed straw that go or at best can he ntmed as fodder or can be burnt a fuel. The fibre obtained way he inferior in quality, but it can be used for paper-Making.

Varieties and Yield

Flax Mtivation of the put depended on varieties straightaway impot from abroad. The recurring imports of wed in large quantities (owing to the deterioration of the crop) was not economical.

CONGO JUTE

Cotton is one of the most important commercial crops playing a key role in economic oplitical and social attairs of the world. some new selections ('F 895' and 'F 896') have been made and they have become acclimatized.

They attained a heialit up to 1.5 m in Hima-chil Prade:

- Congo Jute: Congo jute (Urena lob L.) selections from the acclimatized type yield 27 q per ha on the average.. Ribre rets easily and is dazzling white and is spinnable on jute machinery.
- Cotton: Cotton is one, of the most important commercial crops playing a key role in economic, political and social affairs of the world. Chiefly as a fibre crop cottt)ll is cultivated in ibout 60 countries of the world but ten countries, viz. the USST, the USA, China, India, Brazil, Pakistan, Turkey, Egypt, Mexico and Sudan account for nearly 85 per cent of the total production.

The organized sector of the Indian. textile industry constitutes the largest single industrial segment in the country in terms of the annual value of output and labour employed, both direct and indirect. There are 688 cotton textile mills in India, as enumerated on 1 January 1973, having an installed capacity of 18.4 million spind- les and 207 thousand looms.

The industry provides direct employment for nearly 900,000 workers and indi, rect employment for several millions. The decentralized sector comprising the power-loomE;, hand-looms, charkha, etc. is reported to provide -employment for over 2.5 million.

With nearly 8 million hectares under the cotton crop, India ranks first in the ii,orld in respect of area and fourth in total production which -reached the level of 63.3 lakhs of bales during 1971-72. The other major cotton- growing countries of the world are the United States of America, the USSR and China.

Speciics Composition

India grows on a commercial scale varieties falling under all the four cultivated species of Gossypium, viz. G. hirauguin, G. barbadense, G. arboi.eum and G. herbacin. The predominant species cultivated is G. hirsulunt, which covers about 50 per oent of the area of 8 million.hectares, followed by G. ambwmm with 29 per cent and G. Aeum with 21 per cent. The area under G. barba nse is negligible and covers only a few thousand hectares. Cotton research and development efforts are directed towards progressively increasing the area under G.-hirmaum and replacing, as far as possible, the area under the old world species.

The cotton crop in the United States is now approximately 99 per cent G. hirsutum and one per cent G. barbadmst In the USSIT, G. hirmauits is grown over 92 per cent of the area and G. barbadenje over the rest of the area.

Area Under Irrigation

Cotton in India is largely cultivated under rain-fed or barani conditions. lit is mainly raised during the tropical monscon season, although in the southern parts of the country, it in cultivated during the late monswn season in winter. In nearly 75 per cent of the area, the cotton crop is entirely dependent on rainfall, whereas supplementary irrigation facilities exist in about 25 per cent of the area. There has been a progressive increase in the area under irrigated cotton in India during the past 25 years.

Year		Area under cotton (in million hectares)	Percentage area irrigated
	Total area	Irrigated area	
1947-48	4.26	0.29	7
1951-52	6.56	0.58	9
1956-57	8.05	0.93	12
1961-62	7.72	1.08	14
1966-67	7.83	1.23	16
1971-72	7.78	1.95	25

Climate and Soil

Cotton is tropical subtropical crop. For the successful germination of its seeds a minimum temperature of 15 degree is required.The op- timum temperature range for vegetative growth is 21-27 °C. It can tolerate temperatures as high as 43 °C, but does not do well if the temperature falls below 21′0.

During the period of fruiting warm days and cool nights with large diurnal variations are conductive to goog boll and fibre development.

Crop Season and Rotation

The sowing season of cotton varies considerably from tract to tractand is early sowings done in the Dharwar- Gadag area in the beginning of SePtember have been found to give better yields. In addition, summer sowings in Tamil Wadu ire done during February-March.

The sowings of cotton in the rice fallows of Andhra Pradesh and Tainii Nadii extend from the second half of December to the, middle of January. The mmonest rotations for cotton in the irrigated.areas of the Punjab, Haryana, northern Rajastban and Uttar Pradesh are cottori-wheat, cotton-jewar and cot- ton-wheat-toria. The growing of′ berseen& and guar has been found to have a beneficial effect on the succeeding cotton crop.

In the central and western India, cotton- jr, cotton-bajra, cotton-wheat, cotton-gram tnd cot- ton-sesamum are the usual rotations followed. The raising of the cotton crop after a legume in rotation and preceded by a grain crop has been found to give a better yield. Cotton following groundnut has been found to give 15-20 per cent higher yield than that follwoing wheat or jowar. Another common practice in cotton cultivation is mixed cropping with maize, jowar, ambadi, seiamum, pulses or vegetables as practised in many parts of central India. Intercropping with ragi, other millets or ground. nut is also quite common in parts of Tamil Wadu, Karna- taka and Andhra I'Yadesh. Intercropping and mixed cropping. under rain-fed conditions serve both as an insurance against crop failures and as a preventive against soil erosion.

Seed-Rate and Spacing

For obtaining mali. yields, proper seed-rate and spacing are essential. Depending on the variety, the level of soil fertility and the cultivation practices, smd-rates and spacing & have bets recommended for different improved varieties.

A rate of 15 to 25 kg per hectare and a of 75 to 90 cm between the rows are generally recommended for cotton grown under irrigated conditions. For the raiu-fed deei cotton, a seed-rate of 12 to 16 kg per hectare and a spacing of 45 to 60 cm between the rows are adopted. For the rain-fed American cotton, the seed-rate is 12 to 16 kg and the apacing is 60 to 75 em between the rows. B,roadeasting the or sowing it in lines either with a nkeehanical ~- drill or by hand, are the two commoii methods. Line-gowing with wod.drills is recommended in order to ensure uniform germination, better stand and easy inter-cultivation.

In the case of the ridge-sowxk irrigated crop, hand-dibbling of the seed at the spacing is commonly followed. Thia method is also followed in dryland aras of parts of Madbya Pradesh and Andbra Pradesh, adopting the square-pocket systeni of cultivation.

The Cotton crop in most of the areas in India is usually not adequately manured. The soils in which cotton is grown are generally sufficient in phoa-phate and potash, but since they are deficient in nitrogen and matter, it is necessary to apply them for better Yields. Any leguminous crop. cithe grecii-manure crop or as a crop in rotation preceding cotton, increased the yield Of cotton.

Farmyard manure is rarely applied to the cotton crop in the states of Punjab, Haryana, Rajasthan and Gujaret. Of late, sonic of the rogreegive farmer, states apiply 6 to 12 tonnes farmyard manure per hectare to the rain-fed crop and 15 to 28 tonnes of it to the irrigated crop.

In Maharashtra, the cotton crop is manured with farmyard manure once in 3 or 4 years at the rate of 12 to 15 tonnes per hectare. In Tamil Nadu, the irrigated American cotton crop is usually manured heavily and auppleinente with fertilizer.

The need for applling chemical fertilizers for increas. ing cotton yield ham been amply proved. The usual dose recommended is 40 kg of nitrogen per hectare in the case of the irrigated crop and 20 kg in the case of the rain. fed crop. In northeria India, a higher dose of 60 to 100 kgN/ha has been found to he economical. Good response to hiah doses of 80 to 100 kg Niba has obtained in parts of Maharashtra, in the fallown of Andhra Pradesh and in the case of winter-sown coton in Tamg nadu.

The application of nitrogen by band placement has been foun to give better yield broadcasting. For the irrigated cotton crop, it has been found to be economical to apply half of the basal dressing at the tigne of sowing and the other half as top-dressing during thin. ning or just before flowering.

In the case of the irrigated and row-gown cotton crop, intercultivation is done fairly regularly,with either a blade harrow or with a three. tined hoe or with a dui plough. In the case of the crop that is gown broadcaat, one or two hand-booings are given to remove weeds. InterciLlitivation not only cheeks the growth of weeds, but also lead to better soil aeration and soil-moisture conservation.

The thinning of the cotton crop is a special feature of the irrigated crop sown on ridges in Peninsular and in parts, of southern Guiarat. Mannuring is desirable for maintaining the optimum population of plants to obtain a high yield. During this operation, the vigorous things are retained and weak and oil-type ones are removed.

In the case o irrigated cotton, the optiown moisture supply during the critical stages of growth and developmuent is the most important aspect of cultivatio A proper irrigation schedule in the case of this crop in relation to soil type, eather conditions,cultivation practices and the variety is the dessisive factor in getting the high yield.

The cotton is grown in variety of soils it require to amenable to good drainage as it does not tolarate water logging.It is grown mainly as a dry crop in the black cotton and and mediam black soils as an irrigated crop in the alluvial soils

In tamil Nadu, the major portion of irrigated and rainfeed crop is planted in August-September. Early sowings done in the Dharwar- Gadag area in the beginning of SePtember have been found to give better yields. In addition, summer sowings in Tamil Wadu ire done during February-Marcii. The sowings of cotton in the rice fallows of Andhra Pradesh and Tamil Nadu extend from the second half of Decemberer to the, middle of January.

The commonest retations for cotton in the irrigated.areas of the Punjab, Haryana, northern Rajastban and Uttar Pradesh are cottori-wheat, cotton-jewar and cot- ton-wheat-toria. The growing of' berseen and guar has been found to have a beneficial effect on the succeeding cotton crop. In the central and western India, cotton, cotton-bajra, cotton-wheat, cotton-gram tnd cot-ton-sesamum are the usual rotations followed. The raising of the cotton crop after a legume in rotation and preceded by a grain crop has been found to give a better yield. Cotton following groundnut has been found to give 15-20 per cent higher yield than that follwoing wheat or jowar.

Another common practice in cotton cultivation is mixed cropping with maize, jowar, ambadi, seasamum, pulses or vegetables as practised in many parts of central India. Intercropping with ragi, other millets or ground. nut is also quite common in parts of Tamil Wadu, Karna- taka and Andhra Pradesh. Intercropping and mixed cropping. under rain-fed conditions serve both as an insurance against crop failures and as a preventive against soil erosion.

Preparation of the Land

For the irrigated cotton crop, following wheat in the northern states of India, the preparation of land is usually hurried through, since the interval between the harvesting of the grain crop in March- April and the sowing of the irrigated cotton in April- May is very short. The soil is first given a heavy irriga- tion and one or two ploughings. Thereafter, a light irri- gation is given. followed by one or two ploughings and then the soil is planked with a wooden plank (sohwa) before cotton is Sown.

The black-soil areas of central and southern. India are prepared for sowing the rainw cotton by harrowing the field 3 or 4 times with the blade harro.. The land is given a deep ploughing only once in 4-5 year to remove,weeds.

In the red and lateritic soils of southern India, which are poor in retaining soil moistures, the field is usually given two or three light ploughings or hamwings for raising cotton as a rabi crop. The irrigated crop in these soils is given one watering, ploughed once and the land is prop~ well for aowiiig the irrigated winter crop.

Seed Rate Spacing and Sowing

For obtaiwng mali. mtim yields, proper seed-rate and spacing are essential. Depending on the variety, the level of soil fertility and the cultivation practices, smd-rates and spacing have bets recommended for different improved varieties.

A seed rate of 15 to 25 kg per hectare and a ng of 75 to 90 cm between the rows are generally recommended for cotton grown under irrigated conditions. For the raiu-fed deei cotton, a seed-rate of 12 to 16 kg per hectare and a spacing of 45 to 60 cm between the rows are adopted. For the rain-fed American cotton, the seed-rate is 12 to 16 kg and the apacing is 60 to 75 cm between the rows.

Broadeasting the seed or sowing it in lines either with a mehanical drill or by hand, are the two common methods. Line-sowing with wod.drills is recommended in order to ensure uniform germination, better stand and easy inter-cultivation. In the case of the ridge-sowxk irrigated crop, hand-dibbling of the seed at the spacing is commonly followed. Thia method is also followed in dryland ar as of parts of Madbya Pradesh and Andbra Pradesh, adopting the square-pocket systeni of cultivation.

Manuring

The Cotton crop in niost of the areas in India is usually not adequately manured. The soilsin which cotton is grown are generally sufficient in phoa-phate and potash, but since they are deficient in nitrogen an matter, it is necessary to apply them for get Yields. Any leguminous crop. either as green -manure crop or as a crop in rotation preceding cotton, increased the yield of cotton.

Farmyard manure is rarely applied to the ootton crop in the states of Punjab, Haryana, Rajasthan, and Gujaret. Of late, sonic of the rogreegive farmexr, ilk tlicee states apiply 6 to 12 tonnes of farmyard manure per hectare to the rain-fed crop and 15 to 28 tonnes of it to the irrigated crop. In Maharaehtra, the cotton crop is manured with farmyard manure once in 3 or 4 years at the rate of 12 to 15 tonnes per hectare. In Tamil Nadu, the irrigated American cotton crop is usually manured heavily and auppleinente(t with ferti- lizer&. The need for appling chemical fertilizers for increasing cotton yield ham been amply proved. The usual dose recommended is 40 kg of nitrogen per hectare in the case of the irrigated crop and 20 kg in the case of the rain. fed crop. In uortheria India, a higher dose of 60 to 100 kgN/ha has been found to he economical. Good response to hiah doses of 80 to 100 kg Niba has Ceii obtained in parts of Maharashtra, in the fallown of Andhra libra. desh and in the case of winter-sown coton in Tamg Wadu. The a lication of nitrogen by band placement ban,PI been foun to give better yield t-an broadcasting. For the irrigated cotton crop, it has been found to be economi.

cal to apply half of the a basal dressing at the time of growing and the other half as top-dressing during thin. ning or just before flowering.

In the case of the irrigated and row-gown cotton crop, intercultivation is done fairly regularly,with either a blade harrow or with a three.tineii hoe or with a dui plough. In the case of the crop that is gown broadcaat, one or two hand-booings are given to remove weeds. Interculitivation not only cheeks the growth of weeds, but also leads to better soil aeration and soil-moisture conservation. The thinning of the cotton crop is a special feature of the irrigated crop sown on ridges in Peninsular lndia and in parts, of southern Guiarat. Thinning is desirable for maintaining the optimum population of plants to obtain a high yield. During this operation, the vigorous lings are retained and weak and oil-type ones are removed. lawonor. In the case o irrigated cotton, the optiown moisture supply during the critical stages of growth and developmuent is the moat important aspect

Well-drilled holls alone are picked, as otherwise the quality will deteriorate. It is better to remove the dry stalks an won as the picking is completed, as they usually harbour insect pests if they are left in the field.

The yields vary widely from tract to tract and from to Redtnding upon the agm-elimatic con- pcdilone. The unit area of cotton in India during 1947-48 was only 90 kg of lint per heotate and it lead been estimated at 151 kg/ha during 1971-72.

Year		Area Under Cotton (in Million Hectares)	Percentage Area Irrigated
	Total area(in Million Hecrare)	Production in Lakh	lint yield
1947-48	4.26	0.29	90
1951-52	6.56	0.58	85
1956-57	8.05	0.93	105
1961-62	7.72	1.08	114
1966-67	7.83	1.23	157
1971-72	7.78	65.3	157

Verietal Improvement

A cotton Research in India is said to have started with establishment of Agricultural department and 1904. The inception of was an loping better varictis through the 20 litre constituted in 1923 ion of research, gave a great mological research on cotton.

First Five-Year Plan in ecterated and many schemes so and extending their cultivation and distribution were of new varieties is a continual process and improved strains evolved in the earlier years were progreealvely replby still superior ones

An-India CO-ordinated Cotton-Improvemacat Project

Main Improved Varieties

The All-India co-ordinated Cotton-Improvement Project was sponsored by the Indian Council of Agricultural Research in 1967 and as a result of interaffication of research 16 new varieties of cotton were released during 1968-72 for cultivation. They are, 'Kriobna'for the rice. fallow tract of Andbra Pradesh, 'MCU-5' -and 'Sujata' for the irrigated Cotton tracts of Tamil Wadu, 'Khandwa-l' and 'Khandwa-2' for the rain-fed Wimar tract of Madhya Pradesh, 'G-27' for the desi cotton tracts of Punjab and Haryana, 'J. 205' for the central sistriete of Punjab, 'Barathi' and 'JK.S' for the rain-fed Karunganni tract Tamil Nadu. 'Mabalaxwi' for the rain-fed northern tract of Andhra Pradesh,'H Ibr' 'ld-l' cotton in Gujarat and other state, 'Sujay' 94i for the southern Gujarat tract, 'IMCU-7' for the medium-duration rice-fallowing of Tamil Nadu, 'M-89'for the Briganganagar tract of Rajasthan, 'Varalaxmi' hybrid for the Turigabbadbra Project area in Karnataka and 'GS-23' for the low-rainfall areas of Karnataka. Out of the above mentioned varieties the release of extra long-staple variety 'MCU-5' and 'Hybrid-4' and 'Varalaxmi' hybrid cotton constitute a sufficient mile- stone in cotton research. Variety 'Sujata' is the first high-spinning cotton released in India. The spread of 'Vyplbrid-4' cotton had a great impact on production and the quality of the Indian cotton crop has been significantly stepped by variety 'MGU-S'. Besides the above varieties have already been released, a number of promising varieties have been built up to meet the requirements of different zones in the various quality groups. Based on further evaluation they are expected to be released during the 1970s.

Pests and Diseases

The cotton crop is subject to damage by a number of pests right from the, time of germination till the final picking. Very intensive work has been carried out on the cotton pests to evolve measures to control them and obtain economic yields. The pests of major significance in most of the cotton-growing tracts include aphids, jamids, thrips, bollworms, lca f-worms and mites, which affect the yield and quality of cotton. Ephylactic control schedules have been developed for checking the damage from pests and they have formed the basis for the crop-protection programmes in cotton cultivation.

Diseases of cotton cause reduction in Yields by affect- ing germination, by killing the seedlings, by lowering the productivity of plants and also by affecting the quality of lint. Root-rot and wilt, caused by several fungi, bacterial blight and anthraenose are the major diseases of cotton in India. Control measures, including seed treatment, soil-drelaching, spraying the foliage with fungicides and the use of resistant varieties have been developed for the major diseases.

Processing and Technological Evaluation

Ginning

The ginning of seed-cotton occupies an important place in the processing of material from the field to the factory. The act of separating the fibres from the seed cotton is known as ginning. This can be done either with roller-gins or with saw-gins. In India, the gins used are mostly roller gins.

Cotton Quality Evaluation

Cotton. First spun into yarn before being used ultimately to manufacture fabrics. Hence, the price paid for a sample of cotton is related to the quality of yarn which çan be spun from it. In India, this spinning performance is usually expressed as the 'Highest Standard Count' (HSC) to which it can be spun. The fineness of yarn is usually expressed in terms of counts. A count is the number of hanks of 840 yards each, which weigh one pound; for instance 20 counts of yarn would indicate that one pound of a particular yarn would contain 20 hanks of 840 yards each. Based on the end uses of the yarn, certain strength or standards have been prescribed for different counts by the Cotton Technological Research Laboratory of the Indian Council of Agricultural Research.

The Highest Standard Count for cotton is that finest count of yarn which can he spun to satisfy the yarn-strength standards.

The important fibre characteristics which determine the quality are the fibre-length parameters, fineness, maturity and strength. Other properties, such as convolutions, rigidity and inter-fibre friction, play only a minor role.

In addition to the above, two other characteristics, viz:

1. The colour and appearance of cotton,
2. The trash content are also taken into account for the fixation of price in the market.

Fibre Length Parameters

Fibre length is one of the most important characteristics and the market price is determined to a large extent by the staple length. Other factors being equal, longer linted cottons give better spinning performance than shorter linted cotton.

Staple-Length Classification

Based on staple length, the cottons produced in various countries of the world are classified into three main groups, viz. short staple (9.5 mm-25.4 mm), medium staple (13 mm-40 mm) and long staple (24 mm-63 mm) which are somewhat overlapping each other. The staple-length classification of cottons adopted in India is not identical with that adopted in the USA or other foreign countries, but is based on indigenous production. The groups now recognized in India are superior long staple (27 mm or 1.1/16 inch and above), long staple (24.5 mm-26 mm or 31/32 inch-33/32 inch), superior medium staple (22 mm-24 mm or 28/32 inch-30/32 inch), medium staple (20 mm to 21.5 mm or 25/32 inch-27/32 inch), and short staple (19 mm or 24/32 inch and below

The Cotton Making in India

The marketing of crops close of harvesting of kapas end after the lint is produced by the mills. Between the set points, it passes through several stages, viz. sale as kapas in the primary and secondary marketing, ginning and processing, storage, transport to terminal markets and the sale of lint to the consuming mills.

In primary markets, kapas is sold by the grower to the village merchant or itinerant merchant without the intervention of any intermediaries. However, in recent years direct sales in the village have dwindled considerably. The bulk of the growers, nosy dispose of kapas in the secondary markets, i.e. important trade centres. Here, the produce is bought by stockists, ready dealers, brokers. or commission agents and by the mills themselves which own ginning factories. In the terminal markets, cotton lining old to the textile mills, exporters and traders dealing with the consuming mills or engaged in inter-state trade. Bombay, Coimbatore, Ahmedabad and Kanpur are some of the important terminal cotton markets of which Bombay is the largest.

Apart from the above system of marketing, farmers co-operative cotton sales societies are functioning in many states. The first co-operative cotton sales society was opened at Gadag in Mysore in 1917. Later, such societies have been established in other parts of the country and at present co-operatives not only undertake the marketing of cotton but also undertake its ginning and pressing.

The functions of these co-operatives are to provide the grower-members with credit, supply agricultural inputs to them, and give it to the mills.

Monopoly Purchase in Maharashtra

The Government enacted a scheme of monopoly state under the Maharashtra Raw Cotton (Procurement, Processing and Marketing) Act, 1971.

The Act has three objectives, viz.:

1. To ensure that the farmers get a fair share in the market price,
2. To supply unadulterated cotton to consumers at reasonable prices,
3. To guarantee the purity of cotton and honest trade practices at the processing centres.

Under the scheme, at the commencement of every cotton season, the state Government shall, in consultation with the Central Government, fix guaranteed prices for different varieties or grades of kapas and purchase them from the growers through collection centres. Trade in kapas is prohibited.

The farmer gets 80 per cent of the guaranteed price as 'advance' on the delivery of, kapas and the balance on the disposal of lint and seed by the Government.

The Cotton Corporation of India has also entered the marketing field of internally produced cotton from 1971-72 onwards.

The operations of this Corporation are, however, mainly confined to limited purchases to solve the problems of marketing.

Seed Multiplication

Specific schemes for the multiplication and distribution of the pure seed of improved strains had been financed by the former Indian central cotton committee in colaboration with the state government since 1929. Further sponsored schemes for the production of nucleus and foundation seed of all the improved varieties of cotton released under the research programme. The subsequent stages of the multiplication of improved seeds are handled by the departments of agriculture through co-operative societies and the National Seeds Corporation, with the assistance of seedgrowers. The Central State farms are also involved in the in the production of pure seed cotton. the

Seed multiplicatton 1950 onwards, the Government of India has sponsored Cotton Extension Schemes in different states under these schemes, the distribution of good seed was one of the items. The work hitherto carried out under the above two schemes was later integrated under the Comprehensive Cotton development Scheme.

The Central Farms and the state seed farms are also involved in the production of the pure seed of cotton varieties.

Cotton Development Programmes

The Government of India has been implementing a number of schemes in different cotton-growing tracts for increasing production. The government is involved mainly and an Intensive Cotton District Programme has been implemented in selected districts, based on the area approach to crop development.

The Directorate of Cotton Development under the Department of Agriculture, Government of India, supervises the cotton-production programmes in the states which implement the cotton development schemes. The Indian Cotton Mills Federation's, Cotton Development and Research Association is financing a limited programme of research and extension activities.

The yield data from the farmers fields in the Government Sponsored Package Programme areas as well as from the Development Project areas supervised by the Indian Cotton Mills Federation have indisputably proved that much higher yield of cotton than what is obtained at present is quite possible through the adoption of improved practices.

Impact on cotton production. Consequent on the release and spread of the better yielding varieties of cotton in the various quality groups and the adoption of improved agronomic practices and crop-protection schedules, there has been a progressive improvement in cotton production in the country both quantitatively and qualitatively during the past 25 years.

Table

Year	Total	Long Staple	Medium Staple	Short Staple
1947-48	21.6	6.09 (28.2%)	11.93 (55.2%)	3.55 (16.4%)
1951-52	31.0	9.80 (31.6%)	12.08 (39.0%)	9.0 (29.3%)
1956-57	31.0	9.80 (31.6%)	12.08 (39.0%)	9.0 (29.3%)
1956-57	46.7	7.68 (16.4%)	19.83 (42.5%)	19.14 (41.0%)
1961-62	44.6	8.07 (18.1%)	17.47 (39.2%)	19.03 (42.2%)
1966-67	49.7	7.48 (15.1%)	30.77 (61.9%)	11.48 (23.1%)
19671-72	65.3	16.01 (24.5%)	39.65 (60.8%)	9.61 (14.7%)

Staple Length Classification

	Up to 1961 62	After 1961-62 Ung
Long	14-mm and above	24 mm and above
Medium	18 mm to 21 mm	20 to 24 mm
Short	17 m- and below.	19 mm and below.

Owing to the interaction of three major factors, viz. the availability and spread of new technology, favourable weather conditions and adequate financial resources with the farmers, because of the higher prices for cotton during the previous season, the country achieved a record production of 65.3 lakh bales of cotton during 1971-72 from an area of 7.78 million hectares.

The highest production previously achieved during the most favourable weather of 1964-65 was 56.6 lakh bales from an area of 8.27 million hectares, thus bringing into focus the role of new varieties and improved techni)- logy in increasing production during 1971-72.

5

Crops: Establishment, Coefficient and Production

PEDIGREED SEED

The genetic characteristics of a seed influence its response to environmental conditions. Plant breeders develop varieties with the desired genetic traits for optimum yield in certain environments and to meet quality specifications demanded by the marketplace. The small amount of seed for each new variety produced by plant breeders must be multiplied rapidly with a minimum of genetic change or contamination before release to commercial canola growers. In Canada, this is achieved with a pedigreed seed production system.

The Canadian Seed Growers' Association (CSGA) is designated by the Canada Seeds Act to be the sole pedigreeing agency for most crops in Canada. The CSGA establishes genetic standards and ensures that these standards are maintained. The CSGA's responsibility covers varieties developed by the public and private sectors.

There are two stages in the production and identification of pedigreed seed. First is the production and inspection of the seed crop. Approved crop inspectors inspect pedigreed seed crops and report to the CSGA on the conditions of these seed fields. The inspectors verify the variety sown, previous land use, isolation, weed infestation, plants of other crop kinds, and the degree of contamination with other varieties or off-types. Crops that meet the standards are issued a crop certificate by the CSGA.

The second stage is the inspection of the seed by a licensed grader at an Authorized Establishment or by an inspector to determine the grade under the Canada Seeds Act. Factors assessed include germination, freedom from weed seeds, sclerotia bodies, other Brassica crop kinds and overall quality.

SEEDING RATE EFFECT ON MATURITY AND QUALITY

The rate of seeding affects crop maturity. Higher seeding rates produce plants with fewer branches that mature earlier than those at low seeding rates, especially under high levels of nitrogen. In the Peace River region of Canada, an Agriculture and Agri-Food Canada study reported that increasing the seeding rate from 5 to 18 kg/ha (5 to 16 lb/ac) reduced the maturity by five to seven days. A reduction of two to three days in maturity has been reported in the Edmonton, AB and Melfort, SK areas with seeding rate increases from 5 to 12 kg/ha (5 to 11 lb/ac). The seeding rate effect on maturity is more pronounced under cool, late summer conditions than under warm conditions. The slight gain in maturity at higher seeding rates must be weighed against possible increases in lodging and sclerotinia.

Higher seeding rates will have variable effects on oil quality. Research has occasionally found highest oil contents and lowest free fatty acid levels at lowest seeding rates. In cases where high seeding rates decrease maturity enough to avoid fall frost, this can result in reduced green seed and chlorophyll contents compared to low seeding rates.

RESEEDING

Each year there are fields that suffer from poor emergence or damage after emergence by frost, hail or insects. The grower often has difficulty deciding whether to replant. To aid in the decision, carefully consider the cause and severity of damage, soil moisture, reseeding costs, potential herbicide residues, and the date. The key question is: will reseeding likely result in greater net income without significantly increasing risk?

The first crucial step in the reseeding decision is to accurately assess the crop injury then estimate the yield potential of the surviving stand. Growers often overrate the injury and thus underestimate the potential of canola seedlings to recover. Canola seedlings injured by hail, wind or frost need several days of recovery before accurate assessments of survival can be made. Severe damage to cotyledon and true leaves that causes yellowing/browning/blackening does not mean the seedling is dead. A seedling can survive as long as the growing point and the hypocotyl (the stem from the seed to the above-ground growing point) remain intact and turgid. In the seedling stage, canola's small growing point is at the plant tip in the centre between the leaves. After several days of recovery, a small leaf will begin to emerge from the growing point, indicating survival. Also, the stem should have remained turgid and can vary in colour from white to purple.

In years with successive nights of frost, the recovery period may be one week or more from the initial injury before new leaves begin to be noticeable. Seedlings that have been mortally injured will have pinching and browning of the stem below the growing point. In some cases, only the top of the stem is pinched causing the top to lean, but in other cases the whole stem withers and the seedling collapses on the ground. In summary, to assess seedling survival, look closely at the stem for pinching/withering, and look for new growth out of the growing point several days after the last damaging event.

During early growth, most canola stands can sustain substantial seedling mortality without significantly affecting yield due to compensatory branching of surviving plants. Make random plant density counts from the damaged field. The larger the area of damage, the more counts made. If the field is poor due to uneven emergence, then make an allowance for later emerging plants.

To justify leaving a poor field, ensure the majority of the field contains the minimum plant number that can produce an acceptable yield. Remember there is no single number that represents a minimum number of plants as growing conditions can have a significant impact on the ability of the plant to compensate.

As a guide, a reasonable threshold is 32 to 43 plants/m^2 (3 to 4 plants/ft^2) for fields with conventional varieties early in the seeding season. However, with herbicide-tolerant systems or late in the seeding season or with dry seedbed conditions, 11 to 22 plants/m^2 (1 to 2 plants/ft^2) can be adequate. Thin stands such as these probably will yield 90 per cent of the normal stand seeded at an early date but will be later in maturity.

Crops replanted late in the season typically yield less than earlier plantings that had poor stands. Although some reseeded crops manage to yield the same as very thin crops not reseeded, they incur more costs due to reseeding and, therefore, result in lower net returns. For an example of reseeding yields. In addition, late seeded crops will mature later and thus have a higher risk of fall frost damage.

Table. Example of Reseeding Yields with B. napusCanola

Date Seeded	Emergence		Yield	
	Plants/m^2	Plants/ft^2	Kg/ha	bu/ac
May 6	35	3	1456	26
May 18	79	7	1232	22
May 31	70	6	1400	25
May 6, reseeded 31	62	6	1512	27
Least significant difference	17	2	168	3

Farmers in central Alberta experienced a record string of killing frosts in late May to early June, 2000. Seedling canola was severely injured by these frosts and significant reseeding occurred. However, many growers overestimated the mortality and probably reseeded unnecessarily. The following is an actual field situation followed over that growing season. A grower requested the local extension crop specialist to look at a frozen canola stand. The field was seeded in early May to a Roundup Ready B. napus canola, had good straw cover and suffered severe visible damage from the late spring frosts (Figure).

The producer was not sure if there were enough survivors to produce a reasonable crop and had already reseeded another damaged field. Close inspection of the damaged seedlings indicated that the surviving density was on average 32 to 43 plants/m^2 (3 to 4 plants/ft^2) and about 11/m^2 (1/ft^2) in the worst areas. He was encouraged to leave the field considering the fact that good weed control could be obtained with the herbicide-tolerant system. Continuing frosts hampered the recovery of the canola seedling but eventually new growth appeared after 10 days. The canola stand began to recover slowly over the next few weeks but initially looked quite poor (Figures).

By flowering, the stand began to fill in but differences in maturity were evident between areas that suffered different amounts of stand thinning (Figure). The stand continued to improve through flowering and pod fill. The crop matured in September with the thin areas maturing later than the better areas by one to two weeks. Plant counts at harvest averaged 43 plants/m^2 (4 plants/ft^2).

The thin crop that was questioned in spring eventually yielded 2,128 kg/ha (38 bu/ac) gross with 2 to 3 per cent green. Although this crop probably would have yielded higher if frost had not occurred, the yield was satisfactory and equivalent or better than reseeded B. rapa canola.

Thus the decision to reseed is often difficult and must be based on thorough agronomic and economic assessments, including accurate evaluation of seedling mortality, surviving stand density, and yield potential of the existing stand versus reseeding.

CROP COEFFICIENTS

Changes in vegetation and ground cover mean that the crop coefficient K_c varies during the growing period. The trends in K_c during the growing period are represented in the crop coefficient curve. Only three values for K_c are required to describe and construct the crop coefficient curve: those during the initial stage ($K_{c\ ini}$), the mid-season stage ($K_{c\ mid}$) and at the end of the late season stage ($K_{c\ end}$).

CROP COEFFICIENT FOR THE INITIAL STAGE ($K_{C\ INI}$)

Calculation Procedure

The values for $K_{c\ ini}$ are only approximations and should only be used for estimating ET_c during preliminary or planning studies. For several group types only one value for $K_{c\ ini}$ is listed and it is considered to be representative of the whole group for a typical irrigation water management. More accurate estimates of $K_{c\ ini}$ can be obtained by considering:

Time interval between wetting events

Evapotranspiration during the initial stage for annual crops is predominately in the form of evaporation. Therefore, accurate estimates for $K_{c\ ini}$ should consider the frequency with which the soil surface is wetted during the initial period. Where the soil is frequently wet from irrigation or rain, the evaporation from the soil surface can be considerable and $K_{c\ ini}$ will be large. On the other hand, where the soil surface is dry, evaporation is restricted and the $K_{c\ ini}$ will be small.

Evaporation Power of the Atmosphere

The value of $K_{c\ ini}$ is affected by the evaporating power of the atmosphere, *i.e.*, ET_o. The higher the evaporation power of the atmosphere, the quicker the soil will dry between water applications and the smaller the time-averaged K_c will be for any particular period.

Magnitude of the wetting event

As the amount of water available in the topsoil for evaporation and hence the time for the soil surface to dry is a function of the magnitude of the wetting event, $K_{c\ ini}$ will be smaller for light wetting events than for large wettings.

Depending on the time interval between wetting events, the magnitude of the wetting event, and the evaporation power of the atmosphere, $K_{c\ ini}$ can vary between 0.1 and 1.15.

Time Interval between Wetting Events

In general, the mean time interval between wetting events is estimated by counting all rainfall and irrigation events occurring during the initial period that are greater than a few millimetres. Wetting events occurring on adjacent days can be counted as one event. The mean wetting interval is estimated by dividing the length of the initial period by the number of events.

Where only monthly rainfall values are available without any information on the number of rainy days, the number of events within the month can be estimated by dividing the monthly rainfall depth by the depth of a typical rain event.

The typical depth, if it exists, can vary widely from climate to climate, region to region and from season to season. Table presents some information on the range of rainfall depths. After deciding what rainfall is typical for the region and time of the year, the number of rainy days and the mean wetting interval can be estimated.

Table. Classification of Rainfall Depths

Rain Event	Depth
Very light (drizzle)	≤ 3 mm
Light (light showers)	5 mm
Medium (showers)	≥ 10 mm
Heavy (rainstorms)	≥ 40 mm

Where rainfall is insufficient, irrigation is needed to keep the crop well watered. Even where irrigation is not yet developed, the mean interval between the future irrigations should be estimated to obtain the required frequency of wetting necessary to keep the crop stress free.

The interval might be as small as a few days for small vegetables, but up to a week or longer for cereals depending on the climatic conditions. Where no estimate of the interval can be made, the user may refer to the values for $K_{c\ ini}$ of Table.

Example. Estimation of interval between wetting events

Estimate, from mean monthly rainfall data, the interval between rains during the rainy season for a station in a temperate climate (Paris, France: 50 mm/month), dry climate (Gafsa, Tunisia: 20 mm/month) and tropical climate (Calcutta, India: 300 mm/month).				
Station	monthly rain (mm/month)	typical rainfall (mm)	number of rainy days	interval between rains
Paris	50	3	17	~ 2 days
Gafsa	20	5	4	weekly
Calcutta	300	20	15	~ 2 days

Determination of Kc ini

The crop coefficient for the initial growth stage can be derived which provide estimates for $K_{c\ ini}$ as a function of the average interval between wetting events, the evaporation power ET_o, and the importance of the wetting event.

Light wetting events (infiltration depths of 10 mm or less): rainfall and high frequency irrigation systems

When wetting during the initial period is only by precipitation, one will usually use to determine $K_{c\ ini}$. The graph can also be used when irrigation is by high frequency systems such as microirrigation and centre pivot and light applications of about 10 mm or less per wetting event are applied.

SOIL EVAPORATION REDUCTION COEFFICIENT (K_R)

Soil evaporation from the exposed soil can be assumed to take place in two stages: an energy limiting stage, and a falling rate stage. When the soil surface is wet, K_r is 1. When the water content in the upper soil becomes limiting, K_r decreases and becomes zero when the total amount of water that can be evaporated from the topsoil is depleted.

Maximum Amount of Water that can be Evaporated

In the simple evaporation procedure, it is assumed that the water content of me evaporating layer of the soil is at field capacity, θ_{FC} shortly following a major wetting event and that the soil can dry to a soil water content level that is halfway between oven dry (no water left) and wilting point, θ_{WP}. The amount of water that can be depleted by evaporation during a complete drying cycle can hence be estimated as:

$$TEW = 1000\ (\theta_{FC} - 0.5\ \theta_{WP})\ Z_e$$

where

TEW total evaporable water = maximum depth of water that can be evaporated from the soil when the topsoil has been initially completely wetted [mm],

θ_{FC} soil water content at field capacity [$m^3\ m^{-3}$],

θ_{WP} soil water content at wilting point [$m^3\ m^{-3}$],

Z_e depth of the surface soil layer that is subject to drying by way of evaporation [0.10-0.15m].

Where unknown, a value for Z_e, the effective depth of the soil evaporation layer, of 0.10-0.15 m is recommended. Typical values for θ_{FC}, θ_{WP} and TEW are given in Table.

Typical Soil Water Characteristics for Different Soil Types

Soil type (USA Soil Texture Classification)	Soil water characteristics			Evaporation parameters	
	θ_{FC}	θ_{WP}	(θ_{FC} - θ_{WP})	Amount of water that can be depleted by evaporation	
				stage 1 REW	stages 1 and 2 TEW* (Z_e = 0.10m)
	m^3/m^3	m^3/m^3	m^3/m^3	mm	mm
Sand	0.07 - 0.17	0.02 - 0.07	0.05 - 0.11	2 - 7	6 - 12
Loamy sand	0.11 - 0.19	0.03 - 0.10	0.06 - 0.12	4 - 8	9 - 14
Sandy loam	0.18 - 0.28	0.06 - 0.16	0.11 - 0.15	6 - 10	15 - 20
Loam	0.20 - 0.30	0.07 - 0.17	0.13 - 0.18	8 - 10	16 - 22
Silt loam	0.22 - 0.36	0.09 - 0.21	0.13 - 0.19	8 - 11	18 - 25
Silt	0.28 - 0.36	0.12 - 0.22	0.16 - 0.20	8 - 11	22 - 26
Silt clay loam	0.30 - 0.37	0.17 - 0.24	0.13 - 0.18	8 - 11	22 - 27
Silty clay	0-30 - 0.42	0.17 - 0.29	0.13 - 0.19	8 - 12	22 - 28
Clay	0.32 - 0.40	0.20 - 0.24	0.12 - 0.20	8 - 12	22 - 29

Fig. Soil Evaporation Reduction Coefficient, K_r

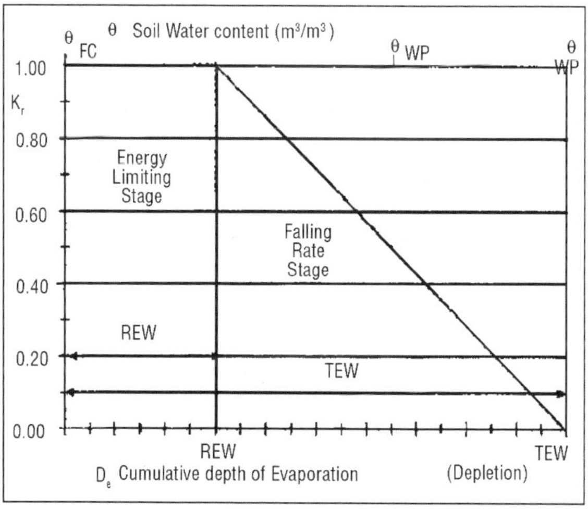

Stage 1: Energy Limiting Stage

At the start of a drying cycle, following heavy rain or irrigation, the soil water content in the topsoil is at field capacity and the amount of water depleted by evaporation, D_e, is zero. During stage 1 of the drying process, the soil surface remains wet and it is assumed that evaporation from soil exposed to the atmosphere will occur at the maximum rate limited only by energy availability at the soil surface. This stage holds until the cumulative depth of evaporation, D_e, is such that the hydraulic properties of the upper soil become limiting and water cannot be transported to the soil surface at a rate that can supply the potential demand. During stage 1 drying, $K_r = 1$. The cumulative depth of evaporation, De, at the end of stage 1 drying is REW (Readily evaporable water, which is the maximum depth of water that can be evaporated from the topsoil layer without restriction during stage 1). The depth normally ranges from 5 to 12 mm and is generally highest for medium and fine textured soils.

Stage 2: Falling Rate Stage

The second stage (where the evaporation rate is reducing) is termed the 'falling rate stage' evaporation and starts when D_e exceeds REW. At this point, the soil surface is visibly dry, and the evaporation from the exposed soil decreases in proportion to the amount of water remaining in the surface soil layer:

$$K_r = \frac{TEW - D_{ei-1}}{TEW - REW} \text{for} D_{ei-1} > REW$$

where

K_r dimensionless evaporation reduction coefficient dependent on the soil water depletion (cumulative depth of evaporation) from the topsoil layer (K_r = 1 when $D_{e,\ i-1} \leq$ REW), $D_{e,\ i-1}$ cumulative depth of evaporation (depletion) from the soil surface layer at the end of day $_{i-1}$ (the previous day) [mm],

TEW maximum cumulative depth of evaporation (depletion) from the soil surface layer when $K_r = 0$ (TEW = total evaporable water) [mm],

REW cumulative depth of evaporation (depletion) at the end of stage 1 (REW = readily evaporable water) [mm].

SOIL PHYSICAL PROPERTIES OF CROP GROWTH

Soil physical properties affect root and shoot growth directly and indirectly, the latter for example through poor drainage causing pores to fill with water and plants to suffer from anaerobiosis. Root growth has been described under various soil physical conditions, but relationships have only rarely been established between features such as crop yield, root growth and soil pore size distribution or conductivity, a more aggregate measure.

The difficulty in establishing simple relationships does not mean that soil has little influence upon root growth; rather it points to the complexity of the interactions and the internal homeostasis which plants maintain. Roots both elongate and proliferate and spread laterally as they grow and age. Concomitantly some roots die and others become suberised and function as conduits, but not as absorbers, of water and nutrients. Roots can elongate downwards as fast as 8 cm/d, as for example, soybean growing in a silt loam in a rhizotron. Deep-rootedness and maximum rooting depth reflect soil properties (for example, roots will not grow through pores that they cannot deform to a larger diameter than the root).

However, relationships are not often reported. Maximum rooting depth varies with species and soil type. For example, wheat roots penetrated to 0.8 m in heavy-textured soils and to 1.2 m in a loamy sand but it is often found that a variety will have a consistent rooting depth across similar soil types in a particular year or in one soil across several years. Angus *et al.* found that rice and six dryland crops (mung bean, cowpea, soybean, groundnut, Maize and sorghum) extracted different amounts of stored soil water (ranging from 100 mm for rice to 250 mm for groundnut) and that extraction was, in part, related to rooting depth.

The spread of roots with age can be related to the growth (increase in weight) of the whole plant, and to accumulated temperature or growing day-degrees (GDD); indeed, there is some evidence that temperature influences the direction of newly-appeared roots as well as the rate of appearance and extent of growth. Clearly, however, there are factors other than plant size, temperature and soil which influence root proliferation. Otherwise the plants sown at three different times of year in the same soil in Figure would align their root growth along a single growth-GDD relationship. These other factors, of which day length is probably particularly important, tend to mask underlying relationships between growth and soil structure.

Tillage can affect root length, though in this case its effects took three years to develop. Measurable differences in soil porosity developed under two tillage treatments: in the first year both root growth and water infiltration (K approximately 5 mm/h) were the same under minimum tillage and conventional tillage. By the third year, when differences were measured between roots, infiltration rates were 84 mm/h in minimum tillage and 0.2 mm/h under conventional tillage. Despite the differences in root growth there were no substantial differences in grain yield, reflecting the overall constraint of climate in the semi-arid environment.

One of the clearer associations between soil porosity and plant growth is described by Tisdall and Cockroft and Tisdall. Soil was ameliorated by deep ploughing, inserting gypsum at depth, and incorporation of straw and green manure in the topsoil. This treatment gave 10-fold increases in the number of earthworms and a 4-fold increase in soil pores. Infiltration rates also increased 10-fold relative to untreated soil. Root growth was not measured, but fruit yields from peach trees increased from 18 to 75 t/ha.

Increases in soil density or strength retard root penetration and thus limit the volume of soil exploited by the crop and the water available. It is difficult to quantify the relationships between these soil parameters and plant growth. In the cases of bulk density and strength, particularly, a gross measure of either for an undisturbed mass of soil can give only a remote indication of what a root encounters. A determination of gross bulk density does not assess whether a root is growing within a pore (in which case it may deform surrounding soil before its radial environment reaches the density or strength of the gross soil) or if it is growing within the soil material, in which case it has already exerted a radial force equivalent to that measured for the gross soil.

This problem of scale-what is measured in a gross estimate cannot assess the micro-environment of the root, to which it (and through hormonal signals, the plant top) responds—does not invalidate some general hypotheses about plant behaviour in response to soil compaction and structural arrangement. Roots stop growing when they are unable to deform their micro-environment. This probably occurs at suctions about 60 kPa when they cannot generate adequate internal turgor. It is thought that root elongation declines curvilinearly with increasing either bulk density or shear strength. In the case of bulk density, little effect is noted until a 'critical value' and root elongation ceases within a further 10 per cent increase in bulk density. As explained earlier, these values vary with soil type. Root elongation declines asymptotically with increasing soil strength though the actual critical values would be expected to vary with soil type, water content and the method of measurement.

Studies of the effect of bulk density or strength on other plant processes, particularly germination and shoot elongation, also suffer from the same technical problem of scale. Shoots are able to explore macropores without being subject to the gross values of the soil they are in. The actual local values which inhibit shoot elongation appear to be quite small, for example, 0.76 kPa. These values from controlled situations contrast sharply with gross field values and with the large number of inconsistent correlations which arise from attempting to correlate crop performance with grossly-measured soil structure. Perhaps, most encouragingly, studies of genetic variation in the sensitivity of crops to soil strength suggest that there is appreciable range in plant sensitivity. The relative ranking of genotypes is, however, the same when under near-critical stress as when growing with virtually no mechanical stress. Genotypes suited to stressful situations may be selected, therefore, by screening at a single soil strength.

The above comments relate to the direct effects of soil physical properties on crop growth. Two further points may be made. First, changing soil properties may not cause, or at least not immediately cause, measurable differences in plant performance. This neither invalidates an association between the soil and the plant, nor the need to remain concerned about changes in soil properties. It is common but not often reported, that changes in a cropping system and soil characteristics first affect aspects other than the crop. It is only when a system has become significantly degraded, and there are associated environmental impacts, that reduced plant performance is noticed. This is illustrated in a study at Ibadan, Nigeria, where no-till Maize in rotation with cowpea gave similar yields to conventionally cultivated crops. The no-till fields had only one-tenth the run-off and erosion of those conventionally cultivated. Likewise, alley-cropping of the annuals between six year-old hedgerows of tree legumes (*Leucaena leucocephala, Gliricidia sepium*) caused increased crop yields, but within the alley-cropping systems the yields of Maize were very similar despite substantial variations in run-off and soil loss.

The second point about soil-crop relations is that the crop also affects the soil through ground-cover, depth of rooting and other properties. The crop attributes that most influence soil physical properties are speed of establishment and development of foliage cover. Rapid establishment and growth minimizes topsoil structural decline and soil erosion by wind and water. Thereafter, deep-rooting directly affects soil structure, particularly if deep-rooted crops, such as safflower, are grown in rotation as a 'biological plough' to create macropores and these are minimally disturbed before the next crop is sown. A general list of desirable crop characteristics. Not all these are universally applicable (nor indeed, accepted by all scientists) because some attributes such as stolons and rhizomes, on the one hand, can provide advantageous ground cover and bind the soil, but they are undesirable if the species presents weeding or other problems.

The list includes both herbaceous and tree species. Ezenwa considers the attributes important for the choice of trees in Sahelian areas as the ability to:

- Enrich a microsite by depositing litter;
- Fix nitrogen and not be allelopathic;
- Avoid competition;
- Withstand stress, either water or wind;
- Provide alternative produce such as fodder, fuel or fruit.

CROP GROWTH STAGES

As the crop develops, the ground cover, crop height and the leaf area change. Due to differences in evapotranspiration during the various growth stages, the K_c for a given crop will vary over the growing period. The growing period can be divided into four distinct growth stages: initial, crop development, mid-season and late season. The general sequence and proportion of these stages for different types of crops.

Initial Stage

The initial stage runs from planting date to approximately 10 per cent ground cover. The length of the initial period is highly dependent on the crop, the crop variety, the planting date and the climate. The end of the initial period is determined as the time when approximately 10 per cent of the ground surface is covered by green vegetation. For perennial crops, the planting date is replaced by the 'greenup' date, *i.e.*, the time when the initiation of new leaves occurs. During the initial period, the leaf area is small, and evapotranspiration is predominately in the form of soil evaporation. Therefore, the K_c during the initial period ($K_{c\ ini}$) is large when the soil is wet from irrigation and rainfall and is low when the soil surface is dry.

The time for the soil surface to dry is determined by the time interval between wetting events, the evaporation power of the atmosphere (ET_o) and the importance of the wetting event. General estimates for $K_{c\ ini}$ as a function of the frequency of wetting and ET_o. The data assume a medium textured soil. The procedure for estimating $K_{c\ ini}$ is presented in Chapter.

Table. Approximate Values for $K_{c\ ini}$ for Medium Wetting Events (10-40 mm) and a Medium Textured Soil

Wetting Interval	Evaporating Power of the Atmosphere (ET_o)			
	Low 1-3 mm/day	Moderate 3-5 mm/day	High 5-7 mm/day	Very High > 7 mm/day
less than Weekly	1. 2- 0.8	1.1 - 0.6	1.0 - 0.4	0.9 - 0.3
Weekly	0.8	0.6	0.4	0.3
Longer than once Per week	0.7 - 0.4	0.4 - 0.2*	0.3 - 0.2*	0.2*- 0.1*

Values derived from Figures.

Note: (*) Note that irrigation intervals may be too large to sustain full transpiration for some young annual crops.

Crop Development Stage

The crop development stage runs from 10 per cent ground cover to effective full cover. Effective full cover for many crops occurs at the initiation of flowering. For row crops where rows commonly interlock leaves such as beans, sugar beets, potatoes and corn, effective cover can be defined as the time when some leaves of plants in adjacent rows begin to intermingle so that soil shading becomes nearly complete, or when plants reach nearly full size if no intermingling occurs. For some crops, especially those taller than 0.5 m, the average fraction of the ground surface covered by vegetation (f_c) at the start of effective full cover is about 0.7-0.8. Fractions of sunlit and shaded soil and leaves do not change significantly with further growth of the crop beyond f_c » 0.7 to 0.8. It is understood that the crop or plant can continue to grow in both height and leaf area after the time of effective full cover. Because it is difficult to visually determine when densely sown vegetation such as winter and spring cereals and some grasses reach effective full cover, the more easily detectable stage of heading (flowering) is generally used for these types of crops.

For dense grasses, effective full cover may occur at about 0.10-0.15 m height. For thin stands of grass (dry rangeland), grass height may approach 0.3-0.5 m before effective full cover is reached. Densely planted forages such as alfalfa and clover reach effective full cover at about 0.3-0.4 m.

Another way to estimate the occurrence of effective full cover is when the leaf area index (LAI) reaches three. LAI is defined as the average total area of leaves (one side) per unit area of ground surface.

As the crop develops and shades more and more of the ground, evaporation becomes more restricted and transpiration gradually becomes the major process. During the crop development stage, the K_c value corresponds to amounts of ground cover and plant development. Typically, if the soil surface is dry, $K_c = 0.5$ corresponds to about 25-40 per cent of the ground surface covered by vegetation due to the effects of shading and due to microscale transport of sensible heat from the soil into the vegetation. A $K_c = 0.7$ often corresponds to about 40-60 per cent ground cover. These values will vary, depending on the crop, frequency of wetting and whether the crop uses more water than the reference crop at full ground cover (*e.g.*, depending on its canopy architecture and crop height relative to clipped grass).

MID-SEASON STAGE

The mid-season stage runs from effective full cover to the start of maturity. The start of maturity is often indicated by the beginning of the ageing, yellowing or senescence of leaves, leaf drop, or the browning of fruit to the degree that the crop evapotranspiration is reduced relative to the reference ET_o. The mid-season stage is the longest stage for perennials and for many annuals, but it may be relatively short for vegetable crops that are harvested fresh for their green vegetation.

At the mid-season stage the K_c reaches its maximum value. The value for K_c ($K_{c\ mid}$) is relatively constant for most growing and cultural conditions. Deviation of the $K_{c\ mid}$ from the reference value '1' is primarily due to differences in crop height and resistance between the grass reference surface and the agricultural crop and weather conditions.

LATE SEASON STAGE

The late season stage runs from the start of maturity to harvest or full senescence. The calculation for K_c and ET_c is presumed to end when the crop is harvested, dries out naturally, reaches full senescence, or experiences leaf drop. For some perennial vegetation in frost free climates, crops may grow year round so that the date of termination may be taken as the same as the date of 'planting'.

The K_c value at the end of the late season stage ($K_{c\ end}$) reflects crop and water management practices. The $K_{c\ end}$ value is high if the crop is frequently irrigated until harvested fresh. If the crop is allowed to senesce and to dry out in the field before harvest, the $K_{c\ end}$ value will be small. Senescence is usually associated with less efficient stomatal conductance of leaf surfaces due to the effects of ageing, thereby causing a reduction in K_c. The variation in K_c for different crops as influenced by weather factors and crop development.

COMPARISON OF CONSTRUCTED CURVES WITH MEASUREMENTS

Because the ET_c data for the dry bean crop at Kimberly, Idaho were measured using a precision lysimeter system during 1974 by Wright, the actual K_c measurements can be compared with the constructed K_c curves, where actual K_c was calculated by dividing lysimeter measurements of ET_c by daily ET_0 estimated using the FAO Penman-Monteith equation.

As illustrated in the graph, the mid-season length for the general, continental climate overestimated the true mid-season length for dry beans in southern Idaho, which averaged only about 30 days rather than 40 days. This illustrates the importance of using the local observation of 30 days for mid-season period length rather than the general value.

The final, best estimate for the K_c curve for the dry bean crop in southern Idaho is plotted (lower curve in graph) using K_c values of 0.15, 1.19, and 0.35 and the actual observed lengths of growth stages equal to 25, 25, 30 and 20 days. Note the impact that the error in estimating mid-season length has on the area under the K_c curve. This supports the need to obtain local observations of growth stage dates and lengths.

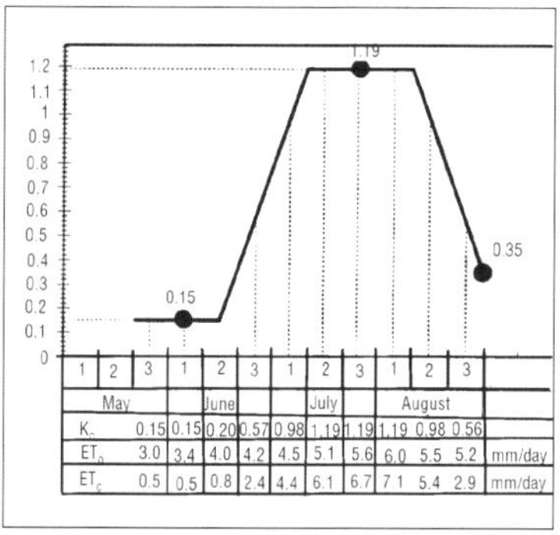

	1	2	3	1	2	3	1	2	3	1	2	3	
	May				June			July		August			
K_c			0.15	0.15	0.20	0.57	0.98	1.19	1.19	1.19	0.98	0.56	
ET_0			3.0	3.4	4.0	4.2	4.5	5.1	5.6	6.0	5.5	5.2	mm/day
ET_c			0.5	0.5	0.8	2.4	4.4	6.1	6.7	7.1	5.4	2.9	mm/day

Fig. K_c curve and Ten-day Values for K_c and ET_c derived from the Graph for the Dry Bean Crop Example

The value calculated for $K_{c\,mid}$ (1.19) appears to have underestimated the measured value for K_c during portions of the mid-season period at Kimberly. Some of this effect was due to effects of increased soil water evaporation following four irrigations during the 1974 mid-season which increased the effective K_c. Where the basal K_{cb} + K_e approach is introduced and demonstrated for this same example.

The 0.15 value calculated for $K_{c\,ini}$ agrees closely with measured K_c during the initial period. Measured K_c during the development period exceeded the final K_c curve during days on or following wetting events. The day to day variation in the lysimeter measured K_c is normal and is caused by day to day variations in weather, in wind direction, by errors in prediction of R_n and ET_o, and by some random errors in the lysimeter measurements and weather measurements. First five days of that decade, K_c = 0.15, while during the second part of the decade K_c varies from 0.15 to 0.36 at the end of day 10. The K_c for that decade is consequently: 5/10 (0.15) + 5/10(0.15+0.36)/2 = 0.20.

Numerical Determination of K_c

The K_c coefficient for any period of the growing season can be derived by considering that during the initial and mid-season stages K_c is constant and equal to the K_c value of the growth stage under consideration. During the crop development and late season stage, K_c varies linearly between the K_c at the end of the previous stage ($K_{c\ prev}$) and the K_c at the beginning of the next stage ($K_{c\ next}$), which is $K_{c\ end}$ in the case of the late season stage:

$$K_{ci} = K_{cprev} + \left[\frac{i - \Sigma(L_{prev})}{L_{stage}}\right](K_{cnext} - K_{cprev})$$

where

i day number within the growing season [1.. length of the growing season],

$K_{c\ i}$ crop coefficient on day i,

L_{stage} length of the stage under consideration [days],

Σ (L_{prev}) sum of the lengths of all previous stages [days].

Equation 66 applies to all four stages.

Example. Numerical determination of K_c

Determine K_c at day 20, 40, 70 and 95 for the dry bean crop (Figure 36).		
Crop growth stage	Length (days)	K_c
initial	25	$K_{c\ ini}$ = 0.15
crop development	25	0.15... 1.19
mid-season	30	$K_{c\ mid}$ = 1.19
late season	20	1.19 .. $K_{c\ end}$ = 0.35

At i = 20:	initial stage, $K_c = K_{c\ ini}$ =	0.15	-
At i = 40	Crop development stage,		
For:	Σ (L_{prev}) = L_{ini} =	25	days
and:	$L_{stage} = L_{dev}$ =	25	days
From Eq. 66:	K_c = 0.15 + [(40 - 25)/25](1.19 - 0.15) =	0.77	-
At i = 70:	mid-season stage, $K_c = K_{c\ mid}$ =	1.19	-
At i = 95	late season stage,		
For:	Σ (L_{prev}) = $L_{ini} + L_{dev} + L_{mid}$ = (25 + 25 + 30) =	80	days
and:	$L_{stage} = L_{late}$ =	20	days
From Eq. 66:	K_c = 1.19 + [(95-80)/20](0.35-1.19) =	0.56	-
The crop coefficients at day 20, 40, 70 and 95 for the dry bean crop are 0.15, 0.77, 1.19 and 0.56 respectively.			

Alfalfa-Based Crop Coefficients

As two reference crop definitions (grass and alfalfa) are in use in various parts of the world, two families of K_c curves for agricultural crops have been developed. These are the alfalfa-based K_c curves by Wright and grass-based curves by Pruitt and those reported in this chapter.

The user must exercise caution to avoid mixing grass-based K_c values with alfalfa reference ET and vice versa. Usually, a K_c based on the alfalfa reference can be 'converted' for use with a grass reference by multiplying by a factor ranging from about 1.0 to 1.3, depending on the climate (1.05 for humid, calm conditions, and 1.2 for semi-arid, moderately windy conditions, and 1.35 for arid, windy conditions):

$K_{c\ (grass)} = K_{ratio}\ K_{c\ (alfalfa)}$

where

$K_{c\ (grass)}$ grass-based K_c (this handbook),

$K_{c\ (alfalfa)}$ alfalfa-based K_c,

K_{ratio} conversion factor (1.0... 1.3).

A reference conversion ratio can be established for any climate by using the $K_{c\ mid}$ = 1.20 listed for alfalfa and then adjusting this $K_{c\ mid}$ for the climate using Equation. For example, at Kimberly, Idaho, the United States, where RH_{min} = 30 per cent and u_2 = 2.2 m/s are average values during the summer months, a reference conversion ratio between alfalfa and grass references using Equation is approximately:

$$K_{ratio} = 1.2 + [0.04(2.2 - 2) - 0.004(30 - 45)]\left(\frac{0.5}{3}\right)^{0.3} = 1.24$$

where

h = 0.5 m is the standard height for the alfalfa reference.

Transferability of Previous K_c Values

The values for $K_{c\ mid}$ and $K_{c\ end}$ listed in Table are for a large part based on the original values presented in FAO Irrigation and Drainage, with some adjustment and revisions to reflect recent findings. Similarly adjustments in $K_{c\ mid}$ to compensate for differences in aerodynamic roughness and leaf area, as introduced in Equation are derived from the K_c values given for different wind and RH_{min} conditions in the concerned K_c Table, with some upward adjustment to better reflect increased ET_{crop} values under high wind and low RH_{min} when applied with the FAO Penman-Monteith equation.

The K_c's from FAO-24 were based primarily on a living grass reference crop. The FAO Penman-Monteith equation presented in this publication similarly represents the same standardized grass reference. For that reason K_c values are in general not very different between these publications except under high wind and low RH_{min}. The modified Penman was found frequently to overestimate ET_o even up to 25 per cent under high wind and low evaporative conditions and required often substantial local calibration. K_c values derived from crop water use studies which used the FAO-24 Penman equation to compute grass reference crop evapotranspiration, can therefore not be used and need to be adjusted using ET_0 values estimated from the FAO Penman-Monteith equation. Similarly crop water requirement estimates based on the FAO-24 Modified Penman equation will need to be reassessed in view of the found differences between the FAO-24 Penman and the FAO Penman-Monteith reference equations.

INFLUENCE OF SOILS ON LAND USE

Soil plays a central role in the effective operation of landuse systems. In agricultural systems it has an influence on the yield of produce from arable, pastoral and forested land, particularly through its control over conditions within the root zone (rhizosphere). It also affects the nature and timing of mechanical operations and animal stocking levels.

In addition, soil has an influence on the behaviour of fertilizers and pesticides, and other related agricultural chemicals; consequently the susceptibility of certain crops to pests and diseases is influenced by soil characteristics. In non-agricultural landuse systems, the role of soil is just as important as in agricultural systems.

For example, in the engineering and construction industries, the success of mechanical operations, and the stability of resulting structures, depends on the mechanical characteristics of soils and their parent materials. Soil characteristics also have an important bearing on waste disposal practices, and on land reclamation and restoration procedures.

Many of the soil characteristics which influence landuse practices are interrelated to a greater or lesser extent. Hence, statistical associations between individual soil properties and landuse parameters, such as crop yield, are often unclear. Nevertheless, it is possible to examine some of the broader influences of the more important soil characteristics on landuse potential.

Soil Physical Factors

Texture is one of the most stable soil properties and has a major influence on landuse capability. Clay soils display relatively good water and nutrient retention but are often characterised by poor drainage and aeration. Furthermore, because such soils are often wet, they respond only slowly to changes in air temperature. Consequently, in clay soils of temperate environments, crop germination and emergence may be delayed in spring, although the onset of autumn frost is also delayed. Clay soils are frequently difficult to cultivate and are thus said to have a 'heavy' texture. Workability and trafficability are particularly restricted during the autumn and winter periods. As clay soils are characterised by the development of large aggregates, however, they tend to be relatively resistant to erosion.

In contrast to clay soils, sandy soils display relatively poor retention of water and nutrients, and tend to be freely or excessively drained, and well aerated. Consequently, they respond rapidly to changes in air temperature. Sandy soils are generally easy to cultivate and they are said to have a 'light' texture.

Aggregation is poor in such soils, however, and they may be susceptible to erosion, particularly if they are dominated by fine sand and contain little organic material. Similarly, silty soils are prone to crusting and capping, which results in reduced infiltration, increased frequency of overland flow and increased risk of erosion. In an agricultural context, medium-textured soils, especially the finer silt and clay loams, are the most desirable. Such soils display optimum conditions in terms of water and nutrient retention, drainage and aeration, response to changes in air temperature, resistance to erosion and ease of cultivation.

In terms of landuse capability, soil structure is one of the most important soil characteristics, and is best developed in soils which contain appreciable quantities of clay and humus. Aggregates provide the basis for nutrient and water retention, while inter-aggregate pore networks facilitate aeration, rapid drainage of excess water, and extension of plant root networks.

Similarly, well-developed soil structure plays a crucial role in the reduction of erosion risk. Plant growth tends to be most effective in soils where crumb and granular structures predominate; optimum growth conditions are often associated with crumb aggregates of 1-10 mm diameter.

Soil structure should not be confused with the *tilth* of the soil. Provided the appropriate soil constituents are present, structure will develop in response to natural processes, such as wetting and drying and freeze-thaw cycles. In contrast, tilth is created by tillage operations, such as harrowing and discing, which break down large aggregates to form a finer seedbed suitable for the germination and emergence of crops.

The production of a good tilth requires great skill and practice. If the soil is too wet, then compaction may lead to impeded drainage and poor development of plant root networks.

Furthermore, the resulting structural damage may increase the susceptibility of soils to erosion. These problems may be overcome to some extent through the adoption of minimum or zero-tillage options, although these are not suitable for all soil types and conditions.

The porosity and pore size distribution of soils are influenced directly by texture and degree of structural development. They are also closely associated with bulk density and have a major influence on water retention and aeration characteristics of soils.

The beneficial effects of ploughing on soil structure and macroporosity in a farm-scale study of poorly drained clay soils in southern England. Here, macroporosity was found to be consistently higher in ploughed soils, even though they were traversed by agricultural machinery, than in direct-drilled soils where the use of such machinery was limited.

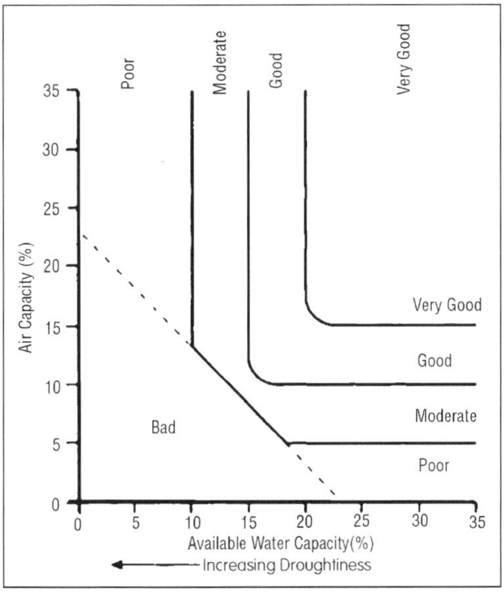

Fig. Classification of the Structural Quality of Topsoils

The content of plant-available water is a vitally important soil characteristic in relation to cropping potential. The amount of plant-available water depends largely on texture, porosity and pore size distribution, degree of structural development, and organic content, and is at its maximum in soils that are near field capacity.

In coarse-textured, sandy soils, amounts of plant-available water are small as water drains rapidly, under the influence of gravity, from relatively large pore spaces. Such soils tend to be drought susceptible (droughty) and irrigation water is often needed to supplement soil water reserves. In fine-textured, clayey soils, amounts of plant-available water are also small.

Here, much of the water is held. in very small pore spaces at tensions greater than permanent wilting point (> 15 bars) and is thus unavailable to plants. Such soils tend to suffer from restricted drainage and are susceptible to waterlogging. In these circumstances, removal of excess water is encouraged through drainage practices. Amounts of plant-available water are greatest in soils with a silt loam texture.

When the water content of a soil is less than field capacity, it is said to be in a state of soil moisture deficit (SMD). The potential SMD is calculated on the basis of a growing season and represents the accumulated sum of potential evapotranspiration minus precipitation.

A cereal crop which yields 10 t ha^{-1} of dry matter, for example, could transpire 2,000-5,000 t ha^{-1} of water during a growing season, which is equivalent to 200-500 mm of rainfall. This value makes no allowance for drainage losses, surface runoff or evaporation directly from the soil surface.

Thus, in drier parts of eastern and southern Britain, for example, where mean annual rainfall is near 500 mm, crop yields may be adversely affected and irrigation is required to supplement soil water reserves on arable land. When the water content of a soil is above field capacity, cultivation using heavy machinery, and excessive animal stocking, may cause compaction, particularly in clay soils. In most instances, the field capacity period sets broad limits to good ground conditions for use of agricultural machinery and for animal stocking.

An understanding of the hydrological characteristics of soils is important not only on agricultural land, but also on land used for urban and industrial activities. For example, drainage and aeration problems are commonly experienced on land which has been restored following opencast mining programmes.

Usually, these problems are overcome through the installation of appropriate drainage systems. Similarly, the behaviour and migration of solutions are of critical importance in relation to contaminated land, and land which is used for waste disposal purposes.

The mechanical characteristics of soils have a major influence on their trafficability, and determine the nature and timing of mechanical operations. If agricultural machinery is used on soils which are wetter than their plastic limit, severe structural damage, compaction and the development of cultivation pans may occur. These problems will in turn result in impeded drainage and increased susceptibility to waterlogging, and under these conditions crop performance may be severely restricted.

Problems may also arise in soils with unfavourable shrink-swell characteristics. Expansion of swelling clays on wetting and shrinkage on drying result in churning and cracking of the soil. These changes cause disruption of soil structure and tilth, and are likely to have an adverse effect on crop yields. Vertisols in particular undergo swelling and shrinkage, in response to seasonal wetting and drying cycles, and are thus very difficult to cultivate. In pastoral systems, the mechanical characteristics of soils have an important bearing on the timing and intensity of animal stocking. Once again, under wet conditions, particularly when the soil is above its plastic limit, animal trampling and associated compaction results in poaching and puddling of the soil.

Similar mechanical problems also occur in forested systems where heavy machinery is used in land preparation prior to planting, and during felling operations. In many areas of the world, forestry is practised on land which is unsuitable for arable cultivation. Soils are often developed on steep slopes, or are highly organic or have high water contents, and therefore tend to be mechanically weak.

Under these conditions, severe disturbance may result in compaction, increased rates of organic matter breakdown and enhanced soil erosion.

On land used for urban and industrial activities, the mechanical characteristics of soils, particularly load-bearing capacity and consistency limits, have a major influence on the stability of buildings and their foundations. Similarly, if the substrate contains large amounts of swelling clays, stability may be adversely affected. This has been illustrated in many parts where, in recent years, drought has resulted in severe shrinkage of clay soils and parent materials, and consequent subsidence of buildings.

Soil temperature has a major influence on plant growth at all stages. At very high temperatures, however, the main limitation to growth is intense evaporation, and the resulting moisture stress, rather than temperature alone. Under these conditions, plants may suffer from physiological drought. Here, plants are able to cope with daytime moisture deficits, induced by intense evapotranspiration, by making best use of night-time moisture excesses.

Soil Biochemical Factors

Soil organic matter plays a fundamental role in agricultural land use, largely through its influence on water content, nutrient status and structural stability. The humus fraction in particular has a very high water-holding capacity and can retain up to four times its own dry weight of water; about 50 percent of this is likely to be plant-available.

In fact, the presence of 5 percent humus will increase the plant-available water content of a sandy loam by more than 50 percent, and that of a clay loam by about 30 percent, above levels in comparable organic-free soils. In addition to its water retention properties, organic matter is often added to soils as a mulch to reduce water losses by evaporation. Organic matter is also an important source of essential plant nutrients, particularly nitrogen.

The humus fraction in particular has a very high CEC and is therefore able to retain nutrients such as base cations, which are available for plant uptake. In a mechanical context, organic matter plays a crucial role in the development of soil structure. For this reason, soils with appreciable quantities of organic material are less susceptible to erosion than comparable soils with low organic contents.

The organic content of soils varies dramatically in response to differences in land use. Contents are often lowest in soils under intensive arable cultivation, as relatively little organic matter is returned to the soil after harvesting. Moreover, with the advent of inorganic fertilizers, the importance of organic material as a source of nutrients has declined.

In southern and eastern areas of Britain, where arable cultivation is widespread, soil organic contents are often below 2 percent by weight, although values vary somewhat depending on the type of crop rotation system adopted, and position within the rotation cycle. In soils under permanent pasture, levels of organic material are considerably greater than in arable soils, and are often in excess of 10-15 percent.

Furthermore, soils under continuous arable cultivation often suffer a rapid decline in organic content over time, unlike those under pasture where organic contents remain relatively high. In forested soils, particularly those under needleleaf trees, organic contents can be very high. In many parts of upland Britain, for example, needleleaf plantations are often established on soils which are peaty in character.

The size and diversity of soil organism populations are just as important as organic contents in terms of land use. Earthworms in particular are known to have a beneficial effect on nutrient levels, and on structural stability, in agricultural soils.

The number and diversity of soil organisms are often greatest in soils under permanent pasture where relatively high organic and moisture contents are maintained. In soils under arable cultivation, however, populations are often restricted as the food supply is relatively low. Similarly, in forested soils, particularly where needleleaf plantations are widespread, soil organism populations are restricted due to high levels of soil acidity and the prominence of unpalatable organic material.

Plant nutrient status is defined in terms of a number of soil properties, including exchangeable base content, nitrogen and phosphorus contents, CEC, percent base saturation and acidity. In fertile, nutrient-rich soils, CEC is dominated by exchangeable bases and the percentage base saturation is high. In acid, nutrient-poor soils, however, CEC is dominated by H^+ and Al^{3+} and base saturation is low.

Here, it may be increased by liming, where adsorbed H^+ and Al^{3+} ions are replaced by Ca^{2+} ions. Similarly, levels of K^+ ions in the soil may be increased by the addition of NPK (nitrogen-phosphorus-potassium) fertilizers. It should be noted, however, that prolonged dominance of the soil exchange complex by one or two specific cations is detrimental to the availability of other cations to plants.

Such competition between different ions for plant uptake is known as 'antagonism'. High levels of Na^+, for example, are common in saline and sodic soils of semi-arid environments. Na^+ is toxic to many plants and tends to inhibit crop performance. Furthermore, high levels of Na^+ encourage aggregate dispersion and consequent breakdown of soil structure.

In addition to the plant macronutrients, amounts of soil micronutrients, or trace elements, such as molybdenum, boron and copper, may need to be controlled. Deficiencies or toxic excesses are often evident in crops and grazing animals, and can be regulated either through soil amendments or by direct treatment of crops and animals.

Many anions are just as important as cations, in plant nutrition. Such anions include nitrate, phosphate and various forms of sulphur. As AEC is restricted in most soils, particularly in temperate environments, anion retention is limited.

Consequently, losses through leaching tend to be rapid and levels which are sufficient for productive arable cultivation can only be maintained through fertilizer additions, or perhaps by mulching. The close relationship between nutrient status and organic content of soils, in a study of changes in the nitrogen content of arable soils in response to different manurial treatments; nitrogen contents were found to be considerably greater in soils which received larger inputs of farmyard manure (FYM).

Acidity plays a crucial role in the availability of plant nutrients. If pH is too high (> 8.0), or too low (<5.0), availability is likely to be restricted for certain nutrients. For most arable crops, growth is optimal at pH 6.0-7.0. Soil acidity is, however, rather unstable and susceptible to change. Excessive addition of nitrogen fertilizers, for example, may result in soil acidification.

Grassland vegetation on pastoral land, particularly that in humid, cool temperate environments, grows most effectively at soil pH values of 5.0-6.0, while needleleaf trees often grow most effectively where soil pH values are even lower (3.5-4.5); few other productive crops can tolerate this level of acidity.

Soil Type

Because soil type is defined by a combination of numerous soil characteristics, its influence on land use will be more general and less direct than that of an individual soil property, such as organic content, texture or acidity. Consequently, the relationship between soil type and land use is observed most clearly in geographical investigations of relatively large areas, particularly at the regional or national scale where a range of soil types and land uses may be identified.

At the more local scale, individual soil properties are likely to show much greater spatial variation than soil type, and therefore to have a much greater influence on land use. The broad geographical relationship between soil type and land use will be examined with reference to soil orders of the Soil Survey Staff classification system; approximate major soil grouping equivalents according to the FAO-Unesco classification.

Entisols

These are shallow, immature soils which are found in a variety of situations, including steeply sloping sites with colluvial parent materials, and floodplain sites on alluvial deposits. They may also exist in cold or exposed environments where soil formation is restricted. Entisols often have a low water-holding capacity and low nutrient status, and can rarely be used for productive arable cultivation.

They may be used for grazing although the quality of pasture is likely to be limited unless a pasture improvement programme is adopted. Entisols are rarely deep enough to sustain productive forestry, although shallow-rooted species may grow well in such soils.

Inceptisols

These are deeper and rather better developed than Entisols, although the morphological expression of soil development may not be particularly strong. They often have a relatively low nutrient status but with careful management they can be highly productive; this might include maintenance of organic matter levels and protection against erosion.

Inceptisols are widespread in their occurrence and are particularly common in floodplain areas of the River Ganges in India, and in many other parts of southern Asia where they are often used for high yielding rice production. They are also common in the Amazonian region of Brazil, and in parts of the USA they are used extensively for sugar-cane production. In the Sahelian zone of Africa, however, where Inceptisols are common, severe drought combined with widespread overgrazing has resulted in severe erosion problems.

Alfisols

These are particularly common in humid temperate environments and are widespread in western Europe, including eastern and southern areas of Britain, and in eastern parts of North America. Alfisols, particularly those with a fine loamy texture, are often highly fertile and contain significant quantities of mull humus. They also have a high water-holding capacity and stable soil structure, and therefore tend to be well-suited to productive arable cultivation, scoring highly in terms of landuse capability. In Britain, they are often used for intensive cropping of both winter and spring cereals and exhibit few limitations to agriculture. In low-lying areas, however, particularly where profiles have a pronounced clay enriched B (argillic) horizon, waterlogging may present problems during the autumn and winter months.

In such circumstances, trafficking is restricted to prevent compaction and the formation of cultivation pans, and installation of artificial drainage is often necessary.

Mollisols

Mollisols are particularly common in relatively dry temperate environments, under prairie or steppe grasslands of continental interiors, such as central North America and Eurasia, and Argentina. These soils are rich in mull humus and have a well-developed crumb structure in their upper horizons. They also contain high levels of calcium and large earthworm populations, and are potentially very fertile.

They are, however, regularly exposed to drought conditions and, because they are often developed in loess parent materials, they tend to be susceptible to wind erosion, particularly if the surface organic-rich layer is degraded or removed. If carefully managed, using appropriate irrigation and crop rotation practices, Mollisols are highly productive and are widely used for arable cultivation in areas where population densities are relatively low.

In such areas, cereals, especially wheat, are widely grown; collectively, these areas constitute the world's 'bread basket'. In Argentina, however, extensive rearing of beef cattle is also practised in areas dominated by Mollisols.

Histosols

These soils may develop in a variety of situations from upland bog and moorland sites, to lowland fen and carr environments. Some Histosols are highly acidic and nutrient-poor, while others are nutrient-rich, and all tend to be waterlogged for long periods of time. Acid, nutrient-poor Histosols are of limited agricultural use, while nutrient-rich peats can be of great value for arable farming, once they have been drained.

However, drainage leads to rapid decomposition (oxidation) of organic matter which has a detrimental effect on structural stability, water-holding capacity and nutrient status. Furthermore, conservationists are particularly concerned about the destruction of wetland habitats, caused by extensive drainage of these soils.

Spodosols

These often have a peaty surface horizon consisting predominantly of mor humus and poorly decomposed plant remains. In addition, they are often coarse-textured, highly leached and acidic in character.

For the most part, Spodosols have a low water-holding capacity, which renders them drought susceptible, and also have a low nutrient status. Consequently, they are rarely used for productive agricultural purposes except in the case of forestry.

Spodosols are widespread in cool temperate, humid regions, such as the boreal forest zones of North America and Eurasia, particularly where soil parent materials are low in base-rich, weatherable minerals. In many parts of eastern USA, between the Great Lakes and the Atlantic seaboard, much of the virgin land colonised during the late eighteenth and early nineteenth centuries was underlain by Spodosols. This was of low fertility, and crop yields were so low that much of it had been abandoned by the mid-nineteenth century.

Ultisols

Ultisols are at the 'ultimate' or end stage of weathering and contain few weatherable minerals. These highly weathered and leached soils are acidic and of low nutrient status, and are dominated by kaolinitic clays. They are mature soils which have developed over a very long time period, in humid warm temperate to tropical environments. Ultisols occur most commonly in southeastern parts of the USA, eastern and southern parts of Brazil, and in parts of southeast Asia.

Although the inherent fertility of these soils is rather low, they can be utilised for productive arable cultivation if carefully managed. In southeast USA, Ultisols used to be subjected to a type of shifting cultivation regime, based largely on tobacco and cotton, the soils being cropped for a few years and then abandoned as fertility declined. However, this practice became unnecessary with the development of effective crop rotation systems which incorporated the use of lime, manures and legumes.

Oxisols

Oxisols are very highly weathered and leached, and have developed over long time periods. In addition, they are often acidic and of low nutrient status. They are characterised by high contents of oxides and hydroxides of iron and aluminium, and occur most commonly in humid tropical environments, particularly in Africa and South America.

In rainforest areas of central Africa and Amazonia, where rainfall is evenly distributed throughout the year, Oxisols are used largely for shifting cultivation. Here, small-scale plots of land are cleared for cultivation and then abandoned after a few years due to declining fertility and weed invasion.

The land then reverts to forest-fallow and nutrient levels are gradually restored. In a regional context, however, shifting cultivation can only sustain a relatively low population density. In savanna and savanna-woodland areas, where rainfall is more seasonal in character, Oxisols are not well suited to shifting cultivation, and pastoral activities are more common. Plots of land used for arable cultivation can then benefit from applications of manure derived from grazing animals. In spite of their low fertility, Oxisols can be highly productive with intensive management.

Vertisols

Vertisols are characterised by the presence of swelling clays and are subject to severe cracking on drying, and to swelling on wetting. Such physical disturbance often leads to tillage and trafficability problems, and to restricted water availability. In spite of these problems, however, these soils often have a high nutrient status and are highly fertile.

Vertisols occur most frequently in warm temperate to tropical environments, and are particularly common in Australia, India and Sudan. They also occur in southern coastal regions of the USA where they are developed in marine clays. Cultivation of these soils is difficult due to their adverse physical characteristics. They tend to be very hard when dry and highly plastic when wet. Consequently, the timing of various cultivation operations is critical. In India, for example, planting of summer crops, which should ideally take place just before the monsoon rains, was virtually impossible in soil which was severely cracked and hardened following the long dry season.

Similarly, once the monsoon rains commenced, planting was also difficult as the soils soon developed plastic characteristics. Thus, planting had to take place just after the monsoon rains had finished. This practice was rather inefficient as the crops were unable to take full advantage of available water in the soil and depended primarily on the presence of stored water. This problem of timing has now been overcome using a different approach to tillage, developed from research at the International Crop Research Institute of the Semi-Arid Tropics (ICRISAT). Soils were tilled immediately after the harvest of the winter crops, before the soil became too dry. Farmers could then establish the summer crop in loose dry soil just before the monsoon rains commenced.

Aridisols

These develop in arid and semi-arid environments, where soil water is severely restricted and productive agriculture is virtually impossible without irrigation. Poor management of Aridisols often results in soil degradation by wind erosion, salinisation and sodification. High levels of sodium in such soils are of particular concern. Sodium is toxic to plants, although the degree of tolerance varies from species to species. High levels of sodium can also lead to aggregate dispersion and consequent breakdown of soil structure.

Effective management of soil salinity and sodicity depends on sensible irrigation and drainage practices, together with evaporation control, selective chemical treatments, and the use of salt tolerant crops. With careful management, Aridisols are able to support a range of arable and pastoral activities. In the irrigated 'western range region' of the USA, for example, grazing of cattle and sheep is widely practised, together with cultivation of arable crops.

Andisols

These develop in parent materials of volcanic origin, notably ash and related deposits. They tend to have a relatively fine texture, with large amounts of fresh, weatherable minerals and low bulk density. They also contain significant quantities of organic matter and have a high ion exchange capacity. Consequently, they are often very fertile and tend to be intensively cultivated, particularly in tropical environments where they are renowned for their flexibility.

In some instances, however, aluminium saturation and low phosphate availability may present problems. Moreover, Andisols are commonly developed on rather steep slopes in mountainous regions, such as parts of southeast Asia, central America, east Africa and North Island New Zealand, and are fairly localised in their occurrence. Consequently, these soils may be particularly susceptible to erosion if poorly managed.

AGRICULTURAL LAND USE

ARABLE CULTIVATION

Methods of arable cultivation have evolved in different ways in response to the vast array of contrasting global environments. In many humid tropical regions, where soils are of low fertility, shifting cultivation, sometimes known as *slash and burn,* is commonly practised, although this often sustains only a small fraction of the population in such regions. Essentially, shifting cultivation involves a three-phase rotation.

This commences with the clearance of existing vegetation, often by burning which helps to improve soil nutrient status, particularly in tropical ecosystems where much of the nutrient stock is stored in the biomass rather than in the soil. Following clearance, the land is cultivated, but for only a few cropping cycles (cultivation phase), as the decline in soil fertility is rapid and weed invasion is a major problem.

The period of cultivation is influenced to some extent by rainfall, with wetter areas being cropped for shorter periods than drier areas. Generally, however, a cropping period of about one to five years is commonly adopted. After cultivation, the land is abandoned and former vegetation is allowed to re-colonise (bush-fallow phase). This phase is essential for the restoration of soil nutrient levels, and its length again depends to a large extent on rainfall. Usually, the recommended period is five to twenty years, although it may be slightly longer in wetter areas and a little shorter in drier areas.

An example of shifting cultivation is the Chitemene system of northern Zambia. Here, felling of trees is limited and crops are grown in ash gardens created by burning lopped branches. Similarly, the Jhum system practised in Bengal, India comprises five-, ten- and twenty-year cycles with a long fallow and a variety of mixed cropping. A related practice is the Khoriya system of Nepal.

In temperate areas, land has often been more closely settled, and competition for land has been greater, than in tropical regions. Furthermore, land in temperate areas cannot be cropped for significant periods of the year, due to the restricted growing season, and consequently more intensive cropping systems have evolved.

One of the oldest recorded systems in Britain, practised as early as the Celtic period, was a three-year rotation of autumn cereal, spring cereal and fallow. This was largely replaced during the eighteenth century by other rotations such as the Norfolk four-course system. Here, grass-clover pasture, root crops and cereals were alternated, and arable practices were closely integrated with livestock rearing. The nitrogen for crop growth could therefore be fixed in the soil by clover, which is a legume, and released from grass residues and manure.

During the nineteenth century, other cropping practices evolved, largely in response to depressed cereal prices and the increasingly accepted belief in the beneficial effects of grass as a soil 'conditioner'. At this time arable-pasture (or arable-ley) rotations, with a relatively long period of ley (up to three years), became increasingly common. It was during this period that mixed farming became widespread, with livestock being reared on farm-produced grass, cereals and root crops.

This type of farming was based on a nutrient supply from farm-derived organic materials such as farmyard manure (FYM), which consists of animal dung and urine mixed with straw, slurry (dung suspended in urine and washing water) and compost. Sewage sludge from local sewage treatment works has also been used on agricultural land. Unfortunately, however, this material often contains high levels of heavy metals, such as lead, zinc and cadmium, derived from industrial effluent.

Green manure crops, sometimes known as *catch crops,* have been an additional source of nutrients. Here, a quick-growing leafy crop is ploughed into the soil, releasing nitrate for the benefit of the following cash crop. This approach reduces the risk of nitrate leaching, but its effects are short-lived and it contributes little to soil organic matter levels.

In recent decades, due largely to increasing population pressure, demand for food and competition between different land uses, there has been a move away from the more traditional agricultural practices outlined, in both tropical and temperate regions. During the Second World War in particular, the demand for food increased and agricultural activity intensified.

Arable and livestock rearing became increasingly separated so that FYM was no longer available where it was needed. At this time, greater emphasis was placed on cereal monoculture at the expense of arable-ley rotations. This change became feasible as the traditional sources of nitrogen were replaced by inorganic nitrogen fertilizers.

However, excessive use of such fertilizers led to increased soil acidification, and to nutrient enrichment of runoff from agricultural land with consequences for surface water quality and aquatic biota.

In addition to the chemical changes, the intensity and scale of mechanisation in arable cultivation has increased dramatically. This has contributed to increased soil compaction which in turn has led to impeded drainage and restricted development of plant root networks. In an attempt to overcome these problems, minimum and zero-tillage options have been developed, where the use of heavy agricultural machinery is minimised.

An example of such a practice is direct drilling where crops are sown directly into the existing vegetation cover. This is now practised in many parts of the USA, particularly in the central corn/maize belt, and in many parts of tropical Africa, South America and Australia. The advantages of minimum and zero-tillage include improved topsoil organic content, aggregate stability, moisture retention, erosion control and workability. In addition, energy requirements are reduced and excessive heating of the soil in hot environments is prevented. In a comparison of zero-, minimum and conventional tillage practices in northern Italy, Expert reported an improvement in a number of soil characteristics, particularly those relating to aggregate stability, with reduced tillage.

However, the disadvantages of minimum and zero-tillage include increased bulk density, reduced total porosity and slow warming of the soil during the spring. Moreover, phosphorus and potassium become concentrated in the surface layers (0-5 cm) of the soil which may restrict their availability to plants if the soil dries out. This is in contrast to tilled soils where the nutrients are more evenly mixed throughout the topsoil (0-20 cm).

In addition, accumulation of volatile fatty acids around fermenting crop residues may inhibit seed germination, and the increase in perennial weeds is difficult to control with herbicides alone.

In many tropical and subtropical areas, particularly in the developing world, low external input agriculture is becoming increasingly common as a component of sustainable development. Here, farmers use crop varieties which are adapted to the main soil constraints, while minimising the use of purchased inputs, and maximising nutrient recycling.

This is very different from the high input systems from which much of the world's food supply is derived, but which involve intensive use of agricultural chemicals to overcome soil constraints. This system was developed as part of the transition process between shifting and continuous cultivation.

The principal features include slash and burn clearance, rotation of acid-tolerant upland rice and cowpea cultivars, maximum residue retention, zero-tillage, and absence of lime and fertilizers. As yields declined, due to nutrient deficiencies and increased weed pressure, a kudzu fallow was grown for one year.

Subsequent options included pasture, agroforestry and fertilizer based continuous cultivation. It was suggested that this system helped to preserve some agro-ecosystem diversity, while contributing towards sustainable production and income for farmers.

All of the developments are clearly reflected in increased crop yields, particularly in the developed world. In western Europe, for example, the amount of grain produced per head of population increased from about 250 kg in 1950 to nearly 500 kg in 1990, an increase of 2.5 percent a^{-1}.

To a large extent, these increases have been achieved through the intensification of agriculture, rather than by increasing the area of land under cultivation. In the developing world, however, production has increased only slightly, often by increasing the area of marginal land under cultivation, and in some areas production has declined.

In Africa, for example, the amount of grain produced per head of population decreased from approximately 150 kg in 1950 to about 120 kg in 1990, a decrease of 1.6 percent a^{-1}. This decrease, coupled with rapid population growth and the extension of cultivation onto marginal land in many areas, is clearly not sustainable in the long term.

Pastoral Activities

Grazing as a landuse practice involves the rearing of animals mainly for transport, meat, milk or wool. In many areas, natural and semi-natural grassland and scrub are used for grazing. These include parts of the prairies and steppes of the temperate continental interiors of North America and Eurasia, the pampas of South America, and the savanna and rangeland areas of Africa and Australia.

Even some of the most fragile ecosystems are used to support grazing livestock, including semi-arid regions of Africa and Asia, and tundra areas of northern Scandinavia.

In other parts of the world, natural and semi-natural vegetation has been cleared or at least substantially modified to allow good quality grassland to become established. The quantity and quality of herbage can be improved either by replacement of existing vegetation, in conjunction with fertilizer and lime treatments, or by improvement of existing grasses.

Although the former approach is often considered to be the most effective, it takes longer to achieve and is more costly than the latter. Once grazing has commenced, re-establishment of pre-existing vegetation is prevented by the consumption of new shoots.

Until the 1960s, almost all hill land improvement in Britain involved ploughing and reseeding, which proved to be appropriate for the brown earth and podzolic soils most commonly associated with grazing prior to this time. More recently, however, with the extension of pasture improvement onto peaty gleyed podzols and humic gleys, the more gradual improvement of existing grasses appeared to be a more appropriate option.

Consequently, in order to achieve the most effective solution, in economic and biological terms, an understanding of soils, vegetation and grazing strategy is essential. One of the most widely used pasture improvement strategies in the British uplands still involves the replacement option. Soil pH is raised to about 5.5 or more by liming (*c.* 7 t ha^{-1} ground dolomitic limestone with high levels of calcium and magnesium).

Soil nutrient levels are also increased by fertilizer treatments and a seed mixture of grasses and white clover. Clover, being a legume, helps to fix nitrogen in the soil, a process which is enhanced by inoculation of clover seed with nitrogen-fixing bacteria such as *Rhizobium* spp.

Sowing is usually carried out in April or May, with limited cultivation and light grazing initially. Grazing can also be controlled effectively using appropriate fencing strategies. Pasture quality must be maintained through the application of dressings of lime and fertilizer, particularly at wetter sites. In addition, livestock should be monitored for problems relating to trace element deficiency or toxicity.

A number of environmental problems have been associated with pastoral activities, including erosion, compaction, salinisation and acidification of soils, and leaching of organic wastes into potable water supplies.

The decline in soil organic content and nutrient levels in the grazed plots was explained in terms of a number of interrelated causal factors. Removal of the protective vegetation cover by grazing livestock and by burning, together with soil compaction caused by animal trampling, resulted in increased soil exposure and enhanced surface runoff.

These increases in turn resulted in accelerated soil erosion, organic matter decomposition and leaching losses. It appeared that the return of organic matter and nutrients to the soil by grazing livestock was insufficient to counteract the deterioration in soil herbage quality. Suggested responses to these problems included rotational grazing, incorporating the planting of a supplementary fodder of legumes in grazing areas. This practice may reduce the frequency of overgrazing, particularly during the dry season.

Hay storage for use during the latter part of the dry season was also encouraged. In addition, it was suggested that the timing and amount of burning should be reviewed. Burning helps to reduce pest levels and to restrict bush invasion, but it also destroys soil organic matter. It was suggested that burning should be carried out early in the dry season, before the ground becomes tinder dry, rather than late in the season when it can be particularly intense and destructive.

Another problem with grazing in dryland environments is that it has traditionally been transhumant, with pastoralists moving to floodplain areas in dry seasons and away from them in wet seasons. Commercial irrigated farming in floodplain areas has prevented this, leading to the over-use of other, more marginal areas for grazing.

Forestry

Many areas under natural and semi-natural forest and woodland have been, and are currently being, exploited for their timber, often using large-scale, mechanised logging operations. Such areas include the tropical forests of central Africa, Amazonia and southeast Asia, and the boreal forests of North America and Eurasia. It should be noted, however, that reliable estimates of deforestation rates are difficult to establish.

In recent years, satellite imagery has been used to assess the extent of deforestation, particularly in tropical forest areas. Ideally, this approach should be used in conjunction with ground verification otherwise estimates may be misleading and possibly erroneous. For example, areas of deforestation have sometimes been determined from aerial estimates of recent fire damage.

However, this can lead to overestimates of deforestation rates, as smoke plumes from recent fires are likely to cover a far larger area than that which has been deforested. Conversely, underestimates may result if forest areas are thinned rather than clear-felled; in this situation low or intermediate canopy trees may be removed, while the upper canopy remains intact.

The reasons for deforestation are largely economic. Softwoods from areas under boreal forest are used in the paper and construction industries, while tropical hardwoods are used in the manufacture of furniture and for construction purposes. In many developing countries, forest and woodland areas are also heavily exploited for woodfuel, which is the main source of energy, and for agricultural land.

Clearly, the exploitative decline in forest and woodland resources, is not sustainable in the long term and a number of environmental problems, including soil erosion, salinisation and biodiversity losses, have developed as a result. In addition, improved road access into deforested areas is likely to encourage population influx and increased shifting cultivation. A variety of approaches to these problems have been suggested.

These focus on the sustainable use of forests, which involves selective logging of several areas each year, followed by forest closure and regrowth. This approach allows for sustainable timber production, conservation of soil and water resources, and maintenance of ecosystem diversity. It also ensures the survival of indigenous populations within forest areas.

An example of sustainable forest management is agroforestry, where trees and arable crops are combined. This has been particularly successful in tropical and subtropical regions where the degree of shading is less than in temperate areas. The trees are usually planted in rows, between which a variety of crops are established.

Although the trees may compete with crops for nutrients, there are a number of benefits associated with the incorporation of trees into cropping systems. As well as providing shelter, trees help to control runoff and erosion. They also help to improve soil nutrient status through the addition of organic material and maintenance of soil structure. Moreover, the trees are an important source of timber, woodfuel, animal fodder and fruit crops.

The most successful agroforestry projects use indigenous tree species which are best adapted to local soil and climatic conditions. In India, for example, species of *Leucaena* are considered to be particularly beneficial as they are able to fix nitrogen in the soil. In contrast, eucalyptus trees are less favoured because they have a high water demand and their leaf litter is toxic.

Studies by the Soil Conservation Institute of the Indian Council of Agricultural Research have shown that eucalyptus trees planted close to crops may reduce crop yields by 10-15 percent. In spite of these problems, eucalyptus trees grow rapidly, providing a useful source of woodfuel, and consequently they are often incorporated into agroforestry systems, but are usually planted in stands well away from crops.

Another approach to sustainable forestry and woodland management is the promotion of small-scale social forestry programmes, where local communities are involved in the establishment of tree nurseries. Farmers are then encouraged to transplant saplings from the nurseries onto their own land. Multipurpose tree species, which provide animal fodder, fruit and woodfuel, and which grow rapidly, are particularly favoured.

Since the late 1970s, social forestry schemes, established at the local *panchayat* (village lands) scale, have been successful in parts of Nepal, where they have been supported by the FAO, UN Development Programme, World Bank and Community Forestry Development Programme (CFDP). Sensible forestry and woodland management has also been facilitated by switching energy sources from woodfuel to cheap and sustainable alternatives such as bio-gas, which is derived from the decomposition of animal and human excrement and related waste products.

In many temperate regions, forestry programmes often utilise rapidly growing softwood species. Here, forestry practices are usually relegated to steeply sloping uplands which are of limited agricultural value. In the last few decades, many areas of the British uplands have been converted from moorland and rough pasture to needleleaf forestry.

The main species planted include Sitka spruce *(Picea sitchensis),* Scots pine *(Pinus sylvestris),* Lodge-pole pine *(Pinus contorta),* Japanese larch *(Larix kaempferi)* and Norway spruce *(Picea abies).* These species are well adapted to the cool climatic conditions, poor soils and steep slopes often found in upland areas, and are ready for harvesting within a few decades.

Although expansion of the forestry industry may be appropriate in economic terms, the environmental consequences may be serious, especially if expansion is poorly managed. In particular, coniferous afforestation has been closely associated with soil and surface water acidification.

As the soils in many planted areas are often wet and acidic, drainage and nutrient amendments are frequently required before planting can commence. Drainage ditches are usually constructed parallel to the direction of slope, and frequently feed into deep check drains which run across the slope. These serve to reduce rates of water flow and soil transport into adjacent streams.

Following these land pre-treatments, saplings are planted on the ridges between drainage ditches and fertilizers may be added to stimulate growth in the important early phase of tree development. About ten to twenty years later, thinning takes place. This practice encourages stronger growth, thus reducing the susceptibility of plantations to damage by windthrow which is a serious problem in exposed areas. The trees are usually harvested after about forty to sixty years, towards the end of the optimum growth phase. After this, the economic rate of return diminishes as growth rate slows. The harvesting process is highly mechanised and may be carried out in one of two main ways.

First, only the trunk of the tree is removed, whilst the thin top and branches (brash) are left to protect the soil (lop and top felling). Second, all parts of the tree are removed (whole-tree harvesting) and the soil is left with minimum protection. In a study of whole-tree clear-cutting on soils it was observed increases in bulk density caused by mechanical compaction; for high-value timber, however, this can be alleviated by helicopter logging.

Decreases in pH, CEC and percent base saturation were also observed, particularly in the upper soil horizons. In a similar study, compared the effects of conventional clear-felling (lop and top) and whole-tree harvesting of Sitka spruce on future productivity of forestry at a site in North Wales. Soil potassium reserves were used as an indicator of productivity, and although both methods of felling resulted in the return of some potassium to the soil, amounts were greater following conventional clear-felling.

MAJOR CROP PLANTS

The genus Allium (Family: Alliaceae) includes the common garlic and the onion which form the very basis of Asian and Mediterranean food preparations. Allium ampeloprasum or sweet leek (Levant garlic) is found all over Europe, Asia Minor and North Africa. Its presence in the Channel Islands and western France is thought to represent relics of former cultivation rather than recent introduction. According to Vavilov, the primary center of origin of onion is in Central Asia and that of leek in the Near East and the Mediterranean.

Allium ascalonicum or shallot is closely related to A. deserti-syriaci from the deserts of Syria and Iraq. It has also been described as a variant of A. cepa, the common onion. Allium kurrat or salad leek originated in Sinai and is grown in the Nile delta, Palestine and Arabia for its leaves. Allium porrum or leek has its greatest diversity in Asia Minor and is derived from A. ampeloprasum, its wild progenitor. And finally, A. sativum or garlic has its primary center in Central Asia but a secondary center in Southwest Asia. Wild Allium species often invade cereal fields all over Southwest Asia. They can be distinguished quite simply by their deep purple or light green inflorescence.

Pistacia vera (Anacardiaceae) or pistachio is native to the Near East and Southwest Asia where it has been cultivated for over 4000 years. Kernels of naturally occurring forms are known from sites long before that. Secondary centers of diversity have developed in Iran and Syria. Relic populations of former extensive cultivation are found in Turkmenistan and Kyrgyzstan. Another species, P. terebinthus, is used as an alternative pollinator and as rootstock on which young male and female pistachio trees can be grafted.

Carthamus tinctorius (Asteraceae) or safflower has its origin in the Near East according to Ashri and Knowles since the cultivated form is similar to two closely related species: C. flavescens found in Turkey, Syria and Lebanon (usually as a weed in wheat fields), and C. palaestinus found in the deserts of Western Iraq and Palestine. In these areas introgression between the wild and cultivated forms may still occur, leading to an exchange of genes. One species in the group, C. leucocaulos, is distinctly different from other species. Stems are smooth, head size is small, coloured flowers are light purple or light pink instead of purple, and self-fertility is high. It is endemic to the Aegean Islands, but has been introduced to southern Europe, Australia and the USA.

Carthamus divaricatus, which is endemic to Libya, is the only species which is morphologically distinct. It has horizontal branches; strongly divaricate outer involucral bracts; yellow, purple and white flowers; yellow pollen, and dark purple striped anthers. Carthamus oxyacanthus, found in continental areas from western Iraq through to northwestern India, and extending northward to southern former USSR, in all areas is a very serious weed. Two other species also found are C. gypsicolus, similar to C. oxyacanthus and restricted to the former USSR, and C. curdicus, with characteristics of both C. gypsicolus and C. flavescens and restricted to northern Iraq. All the above have 12 pairs of chromosomes. There are two species with 32 pairs of chromosomes, C. turkestanicus and C. baeticus. Carthamus turkestanicus ranges from Turkey in the west to northern Pakistan and Kashmir, India in the east, with presumably introduced populations in Ethiopia. Carthamus baeticus may be found around the entire Mediterranean Sea and in its islands. All wild species of safflower are spiny weeds, some of them very serious because they occupy fields sown to other crops. Others are more prevalent in roadsides and waste places. They are characterized by yellow flowers, although white flowers with yellow pollen grains are not unknown. At an early stage in its evolution, C. tinctorius spread to Egypt, Ethiopia, southern Europe, South Asia and the Far East where distinctive types have evolved.

Several wild Beta spp. (Chenopodiaceae), which are relatives of Beta vulgaris or beetroot, are widespread in this region. Among them are B. corolliflora found in Turkey, Azerbaijan, Georgia and in parts of Iran. It is known for its resistance to frost and is often used in crosses with cultivated species. Beta intermedia is said to have evolved as a result of a cross betweenB. lomatogona and B. trigyna, two other wild forms. Beta lomatogona is itself a weed found all over Asia Minor and B. trigynais found around the Black Sea with a few samples found as far as the Caspian Sea, Ukraine and Hungary. Beta macrorrhizais found at higher altitudes in the mountains of Iran, Turkish Armenia near Lake Van and the Caucasus. It is a cold-resistant species, sweet tasting with a white pulp. The sugar beet, B. vulgaris, has its origins in the western Mediterranean.

Spinacia oleracea (Chenopodiaceae) or spinach is a native of Southwest Asia but has become well established almost all over the world, having been introduced in Europe, the Americas and the Far East.

Cichorium intybus (Compositae) or chicory is a perennial salad plant native to the parts of southern Europe bordering on the Mediterranean. It was cultivated in Greek and Roman times and has become naturalized in parts of North America.Cichorium endivia or endive is indigenous to the eastern Mediterranean. Cynara scolymus (Compositae) or artichoke was probably developed from its wild progenitor C. cardunculus, a native of the Mediterranean whose stalks and leaves are edible. There are several words for the plant in the Berber language, which indicates the possibility of its origin in the Maghreb. It moved in to Europe in the Middle Ages.

Lactuca sativa (Compositae) or lettuce evolved as a natural selection from its closest wild progenitor, L. serriola, which is a prickly biennial of the temperate regions. Lactuca sativa originated in the eastern Mediterranean about 7000 BP, most likely in Egypt when the climate of that part of northeastern Africa was not as dry as it is today. Lettuce was a common oilseed crop in Upper Egypt as far back as 6500 BP as depicted by rosettes of a tall, large vegetable with subulate leaves painted on the walls of tombs at Thebes. The presence of latex in primitive lettuce also marked the plant as a fertility symbol. Unfortunately, the available illustrated record found in ancient monuments is not complemented by dry or carbonized remains of leaves or seed of this crop in Egypt.

Early human selection for non-shattering seed heads, late flowering, non-prickly leaves, decrease in latex content, and hearting character (the tendency of leaves to congregate in layers in a heart-shaped head) is said to have led to its domestication as a leafy salad vegetable. Lettuce reached China around the 7th century AD where a special cultivar of the asparagus type was developed and reported by Fuchs in 1543 from Sinkiang. The local name of this variety is 'usum' or 'on-sen'. According to Vavilov this form of lettuce is practically unknown in Southwest Asia.

Cornus mas (Cornaceae) or Cornelian cherry is found all over Caucasia and Asia Minor as an underbrush of deciduous forests. The fruits are edible and are used to produce an alcoholic beverage, vin de Cornouille. The fruits are of great antiquity and have been reported from archaeological sites over 10,000 BP in 1991 from the PPNB site of Nevali Çori, near Urfa, in southeastern Turkey. Corylus avellana (Corylaceae) or European hazelnut has its primary center in the Caucasus where a number of otherCorylus species are also found. However, it was domesticated relatively recently in 17th century Italy. Corylus colurna or Turkish hazelnut or cob nut is cultivated in Turkey for export and also used as rootstock for C. avellana since it possesses resistance to several diseases. Hybrids between the two are called 'trazels'. Corylus maxima or filbert nut tree is native to West Asia, southeast Europe and the Caucasus.

Brassica oleracea (Cruciferae) or cabbage has a secondary center in Asia Minor where convar. oleracea and capitata are found. There are wild as well as cultivated forms. Some of the wild inedible forms are also used as ornamentals. Brassica oleracea var. botrytis or cauliflower is a native of the eastern Mediterranean

Cucumis melo (Cucurbitaceae) or musk melon (cantaloupe) has its origins in Africa but has an important secondary center in the Near East where some of the sweetest fruits are found. Similarly, Cucumis sativus or cucumber, a native of India, also has a secondary center in this region and cucumbers are part of all Near Eastern cuisines from Turkey through to Iran and Afghanistan. In Iran tender cucumbers are eaten with yogurt (mást-khiyár) on a daily basis.

Ricinus communis (Euphorbiaceae) or castor was known to the ancient Egyptians as far back as 6000 BP. Its oil was used as an illuminant. Although it originated in Africa, it grows wild in the Yemen and all of the Near and Middle East. It was taken at an early date to India. Today castor oil has great commercial value in the paint and lubricant manufacture industries. The oil also has been used as a purgative (home remedy) since ancient times.

Castanea sativa (Fagaceae) or chestnut is found from southern Italy to Asia Minor and western Georgia in the former USSR. It was introduced to southern Europe by the Greeks and the Romans, and then transported to northern Europe and Britain. Nuts are roasted, the shells removed and the insides eaten or used in confectionery.

A number of Aegilops spp. (Gramineae) or goatgrass are native to West Asia and the Near East. Aegilops was known to the ancient Greeks and the name is related to its supposed healing properties of an eye disease from which goats suffer. Ae. columnaris is spread from Turkey through Iraq, Iran and Caucasia. It is a common weed found in and on the borders of fields. Ae. crassa is also found in Turkey, Syria, Palestine, Iraq, Iran and Afghanistan. Ae. cylindrica is found in the Balkan peninsula, Crete, Turkey, Caucasia, Armenia, Azerbaijan, Iran, Iraq and Afghanistan. It is a weed in fallow fields and along slopes of hillsides. Ae. kotschyi is found from North Africa across Palestine, Iraq, Iran, Afghanistan and Caucasia

Ae. laurentii has a wider distribution, *i.e.* from southern Europe, former USSR, Turkey, Palestine, Iraq and Iran. Ae. mutica is restricted somewhat to Armenia and Anatolia in Turkey and Ae. ovata is just the opposite, being found all over the Mediterranean, Palestine, Syria, Lebanon, Turkey, Iraq, Iran and Afghanistan. A weed of cultivated wheat fields, it has a male-sterilization action on the nucleus of Triticum aestivum and T. turgidum. Ae. speltoides has its primary center in southern Turkey and northern Syria and Iraq. It is less common in places bordering the above. It is often found growing with wild wheat (T. boeoticum) in southern Turkey and northern Syria. Ae. triaristata is found in the Mediterranean, West Asia, Iraq, Iran and southern parts of the former USSR. Ae. triuncialis is all over the Mediterranean area, Turkey, Palestine, Syria, Lebanon, Iraq, Iran, Turkmenistan and Afghanistan. It has been suggested that it is a hybrid between female Ae. caudata × male Ae. umbellulata that originated in West Asia. Ae. umbellulata, on the other hand, occurs on moist steppe, dry slopes of hills, and is a weed in cultivated fields in the Greek islands, Turkey, northern Syria, Iraq, northwestern Iran and Transcaucasia. It is resistant to leaf rust and hence used often in wheat breeding. Some Aegilops spp. cross readily with others, *e.g.* Ae. sharonensis, whereas others are cross-pollinators which readily produce hybrids. Aegilops spp. that occur on the borders of cultivated Triticum fields often exchange genes although the hybrids are mostly sterile.

Hordeum vulgare (Gramineae) or barley is one of the oldest crops to be domesticated in this region, from its wild progenitorH. spontaneum. The wild progenitor of barley is found in the eastern Mediterranean, West Asia, and as far as Turkmenia and Afghanistan. Domestication of barley is said to have taken place around 9000 BP in the Fertile Crescent. However, grains of barley have been found in Egyptian tombs dating from 15,000 BP.

But it would be difficult to distinguish between wild H. spontaneum and its domesticated form from looking at carbonized remains. Other finds include those discovered in Cayönü in Turkey dating from 9000 BP. The domesticated barley spread through the Mediterranean to Rome and beyond as well as eastwards through Iran and Afghanistan into India and from India to China through Tibet. There are several forms but they all have a brittle rachis and very long awns. For some time H. agriocrithon, a brittle rachis six-row form first reported from Tibet, was thought to be the progenitor of six-row barley.

But later this theory was discounted as such brittle rachis forms were also found where H. spontaneum came into contact with six-row barley. These forms mature about two weeks before the cultivated six-row barley and occur as weeds in barley fields all over Tibet. They are easily distinguished from cultivated forms by their purple to black spikes. However, H. agriocrithon, unlike H. spontaneum, has not yet been found growing by itself in the wild state away from cultivated fields.

Avena sativa (Gramineae) or oats. Two hexaploid oats have been recognized: A. sativa and A. byzanthina. Avena sterilis,the wild oat, is recognized as the progenitor with its center of diversity in the Near East and the Mediterranean. Wild oats are perpetual invaders of cereal fields in Southwest Asia and elsewhere. Other wild species found in this region are: A. damascena (found in an area 60 km north of Damascus in Syria), A. clauda, A. longiglumis, A. moroccana (found in Morocco and sometimes mistaken for A. sterilis) and A. strigosa (black oat, found all over the Mediterranean).

Secale cereale (Gramineae) or rye has its primary center of origin in northeastern Turkey and northwestern Iran. A secondary center has been suggested in Afghanistan from where the crop may have migrated to central Asia and Europe. A number of genetic variants exist, one of which (S. segetale) evolved into the annual wild rye weed. Hybridization between cultivated wheat and rye has resulted in the triticales which have resistance to several diseases. Cultivated rye is known from the Neolithic age in Austria, but it seems to have become widespread in Europe only after the Bronze Age.

Secale montanum (Gramineae) or mountain rye is a perennial species found in the central Atlas Mountains of Morocco, Sierra Nevada Mountains of Spain, Italy, Sicily, Yugoslavia, Greece, Lebanon, Turkey, Iran and Iraq. It is highly polymorphic and crosses readily with other Secale spp., producing several weedy forms and varieties. Other species include S. silvestreand S. vavilovii, which occurs on the slopes of Mt. Ararat and along the banks of the Araks River.

Sorghum halepense (Gramineae) or Johnson grass, a relative of the cultivated sorghum (S. bicolor), is found from the Mediterranean up to Pakistan. It is a rhizomatous perennial first identified near Aleppo (Halep) in Syria. It occurs as a weed on the borders or in cereal fields. It is a source of disease resistance for cultivated sorghum with which it crosses readily.

Triticum aestivum (Gramineae) or common wheat or bread wheat has its primary center in Transcaucasia and adjacent areas and is by far the most important crop that this region has given the world. Natural cross-fertilization still takes place between wild and cultivated Triticum spp. and many authors believe that wheat is still evolving. It is said to have arisen as a result of a cross between T. dicoccum (emmer wheat) and Ae. squarrosa. This cross must have taken place after emmer evolved from its wild progenitor, T. dicoccoides, in the area south of the Caspian Sea. Several other hexaploid forms eventually evolved and they are as follows: T. vavilovii, a species peculiar to Armenia, T. spelta (spelt wheat) is still cultivated in parts of Spain, T. macha (macha wheat), T. compactum (club wheat, was grown in Syria until the last century and volunteers can still be observed in cultivated fields of T. aestivum) and T. sphaerococcum (dwarf shot wheat with globoid grains native to India, where it is still grown in small pockets). A secondary center of diversity for hexaploid wheat developed in the Hindu Kush Mountains where the crop encountered a physical barrier to further migration, *i.e.* the theory of peripheral diversity.

Triticum monococcum (Gramineae) or einkorn was domesticated in the Fertile Crescent from one of its wild progenitors, T. boeoticum or T. urartu. Its earliest recorded archaeological find is from the site at Ali Kosh in Iraq dating from 8500 BP.Triticum boeoticum subsp. thaoudar, the wild progenitor with two grains and two awns, has been reported growing in wild stands from Diyarbakir in Turkey and remains have been found from the site at D'jade on the Euphrates where it was grown and harvested in the green state. Triticum monococcum, although considered obsolete, is still cultivated as a feed for poultry and swine in the mountainous villages in Italy, Spain and Turkey and elsewhere.

Triticum timopheevi (Gramineae) consists of two subspecies: araraticum and timopheevi. The former grows wild in the northeastern arc of the Fertile Crescent. The latter species, however, is part of what is termed the Zanduri wheat, cultivated in Georgia in the former USSR. This type of wheat is formed by diploid einkorn, tetraploid timopheevi and hexaploidzhukovskyii wheats. Neither subspecies crosses readily with other wheats.

Triticum dicoccoides (Gramineae) or wild emmer is found all over the Fertile Crescent. Recently wild stands have been reported from Suweida province south of Damascus in Syria. Wild emmer has been used in crosses with cultivated tetraploid wheat to improve disease resistance of the plant and protein content of the grain. Several cultivated tetraploid wheat forms may have been derived from wild emmer: T. dicoccum (emmer wheat), T. turgidum (poulard wheat), T. carthlicum (Persian wheat), T. durum (macaroni wheat), T. polonicum (Polish wheat) and includes T. ispahanicum, T. turanicum (Khurasan wheat) and T. turgidum subsp. turgidum (English wheat).

Triticum dicoccum (Gramineae) or emmer wheat is, according to some authors, the oldest cultivated wheat, being domesticated in the Fertile Crescent around 10,000 BP. Until recently, it was cultivated in large quantities in

Ethiopia (the secondary center of diversity for tetraploid wheats), Iran, eastern Turkey, Transcaucasia, Volga Basin, Yugoslavia, Czechoslovakia, Italy, Spain and India. It was seen from frescoes and wall paintings at Thebes and Luxor that T. dicoccumwas the favoured cereal used for bread-making in Egypt. But today its cultivation is on a decline and it can be found only in small pockets of traditional farming communities in the above areas. In India it is called 'khapli' wheat and is now grown only in limited areas in the state of Rajasthan.

Triticum durum (Gramineae) or macaroni wheat is cultivated all over West Asia and North Africa (WANA) and the Mediterranean. In fact, 80% of all the durum wheat grown in the world is cultivated in this region. It is rarely grown outside the WANA region in the Old World. Its price in the world market is higher than bread wheat because of its great demand in the making of pasta dishes in Italy and elsewhere. Ancient texts speak of the possibility of storing durum wheat over long periods because of the low water content of its grains.

Triticum zhukovskyi (Gramineae) or Zanduri wheat is cultivated in western Georgia in the former USSR. It is a hexaploid with a unique genome formula. The designation 'Zanduri', however, is reserved for a mixture of T. monococcum, T. timopheevisubsp. and T. zhukovskyi.

Cicer arietinum (Leguminosae) or chickpea, also referred to as garbanzo, occurs in two forms: the Mediterranean type(plants with green stem, white flower, owl-head shaped and salmon-white seeds) and the desi type (plants with anthocyanin in the stem and leaves and small seeds with an angular shape and darker colour). Cicer reticulatum, a wild annual species found in Turkey, could be the wild progenitor since it produces fertile progenies when crossed with cultivated chickpea. Lens esculentum (Leguminosae) or lentil.

The genus Lens comprises two species:

1. L. culinaris (syn. Lens esculentum),which includes three subspecies - culinaris (cultivated lentil), and the wild orientalis and odemensis, and
2. L. nigricans, with subspp. nigricans and ervoides.

The cultivated lentil has been domesticated from its wild progenitor subsp. orientalis. Wild lentils occur from western Europe (Spain and Portugal) to central Asia. According to Harlan there is a microcenter of diversity for lentils in Turkey.

Lathyrus sativus (Leguminosae) or grass pea was domesticated in West Asia with a center of diversity in the Mediterranean region.

Lathyrus tingitanus (Leguminosae) or Tangier pea has a microcenter of diversity in Morocco where it is cultivated as a winter annual.

Lupinus albus (Leguminosae) or white lupin is found wild in western Turkey but may have been domesticated in the Balkans. Other lupins from the Mediterranean area are: L. angustifolius, L. cosentini, L. luteus, L. pilosus and L. termis. Egyptian lupin has been found in Palestine and Egypt since ancient times. The seeds contain an alkaloid which has to be neutralized before consumption.

Medicago sativa (Leguminosae) or alfalfa evolved around the Caspian Sea where it occurs as a wild plant. It is believed that it was the first plant to be cultivated as a forage crop to provide feed for horses and cattle. The Greeks came across alfalfa being used to feed the horses of the mighty Persian cavalry of Darius III around 350 BC. The Greeks quickly adopted the crop for their own use and it spread later into southern Italy and Europe and subsequently to other parts of the world. Several wildMedicago species occur in West Asia: Medicago cancellata, which is used as a gene source for adaptation of M. sativa to poor soils, Medicago dzhawakhetica, found in Georgia and parts of Asia Minor, and M. glutinosa, found in the Caucasus and also used as a gene source for crop improvement.

Pisum sativum (Leguminosae) or pea was domesticated from the wild P. humile, the latter being common in northern Iraq, Jordan, Syria, northwestern Iran, Palestine, Turkey and Cyprus. The crop reached the Greeks through the Black Sea and went eastwards to India and China via the Himalayan trade routes passing through Tibet. Pisum elatius is another wild form which is found all over the Near East and in southern Italy and Sardinia. It can also be found in scattered places in North Africa.

Trifolium ambiguum (Leguminosae) or Caucasian clover is found in Caucasia, Crimea and Turkey. It is a very valuable fodder plant. Material taken from Crimea to Australia gave rise to cultivars which can withstand long periods of flooding.

Trifolium israeliticum (Leguminosae) is a forage species reported from northern Palestine.

Trigonella fbenum-graecum (Leguminosae) or fenugreek probably also evolved in West Asia, but is cultivated in southern Europe, North Africa and India as a fodder crop. In India the seeds are also eaten and used in traditional medicine.

Vicia sativa (Leguminosae) or common vetch has a great variation in form, chromosome number and karyotype. Zohary and Plitmann described seven subspecies as follows: sativa, macrocarpa, nigra, cordata, incisia, amphicarpa and pilosa.All these subspecies are found in the Mediterranean and the Near East, extending up to southern and central Europe. Other useful Vicia species include V. ervilia or bitter vetch, with its primary center in the Mediterranean. It was known to be cultivated for food in Turkey and Greece. Vicia narbonensis or the narbonne vetch has its primary center of origin in eastern Georgia and a secondary center of origin in the Mediterranean. Narbonne vetch occurs as a weed in barley and wheat fields in Transcaucasia as does V. villosa or hairy vetch which is found in central Europe, the Mediterranean region and West Asia.

Vicia faba (Leguminosae) or faba bean probably originated in Southwest Asia or the Mediterranean. Its wild progenitor has not been identified with certainty although several theories have been put forward but discounted for one reason or another. A center of diversity has been established in the Mediterranean region. Several finds of the small-seeded var. minuta have been found in the archaeological remains. The oldest documented record of an archaeological find of large-seeded faba bean comes from Iraq around 1000 AD which perhaps accounts for the late development of the large-seeded types we see today.

Linum usitatissimum (Linaceae) or flax (linseed) was probably domesticated almost around the same time (ca. 8200 BP) as wheat and barley in the Near East from where it spread to other parts of the Old World along trade routes. Its wild progenitor,L. bienne, does not occur in Central Asia which Vavilov thought to be a center of origin of flax. Linum bienne has two main races: a continental winter annual found in the semi-arid foothills of Iraqi Kurdistan and Iran, and an Atlantic-Mediterranean form found along the coastlines. The latter is a perennial also sometimes referred to as L. angustifolium. It has a very high oil content. Hence, during the domestication process, according to some authors, two forms emerged, one for fibre (flax) and another for its oil (linseed).

Gossypium herbaceum (Malvaceae) or short-staple cotton, a native of southern Africa, was probably introduced to Ethiopia and thence to Arabia and Balochistan. It is widely cultivated in Syria and Egypt. In Iran it evolved into a special race which spread to western India where it was the first cotton to be cultivated. However, owing to its short staple it was not much in demand compared with the Egyptian cotton G. barbadense which is also later in maturity. Another cotton species, G. incanum or hillcoat cotton, became established in the Yemen, no doubt having arrived there from Ethiopia. It is extremely drought resistant as is G. areysianum found in southern Arabia. Another theory suggests that the northwestern part of the Indian subcontinent was probably the cradle of cotton cultivation, from which an 'ennobled' plant diffused to other parts of Asia Minor, the Near East, Africa and Europe. The earliest archaeological finds of cotton are from Mohenjo-Daro in the Indus Valley 2300 BC.

Cannabis sativa (Moraceae) or hemp originated perhaps near the Caspian Sea. The first use of this plant was for fibre. Hemp is reckoned to be one of the oldest cultivated non-food crops. Evidence of the use of hemp in China goes back to Neolithic times. The crop reached West Asia around 2000 BC and was taken to Europe around 1500 BC. It became well established in the Mediterranean region.

Ficus carica (Moraceae) or common fig, one of the most ancient fruits of the Near East, arrived there from South East Asia, probably along the trade routes. However, it developed a secondary center in the Mediterranean and Asia Minor. By 6000 BP figs were cultivated in Egypt. It also grows wild in these regions and all over tropical Asia.

Nelumbo nucifer (Nelumbonaceae) or the Indian lotus is not native to India but to wetlands of northern Iran. It spread to China, Japan and India where it is often cultivated for its rhizomes and fruits. It was also grown for some time in the eastern Mediterranean region but is not found there today.

Olea europaea (Oleaceae) or olive is one of the longest-living fruit-bearing trees in the world. It is one of the relics of an age when the Mediterranean had a somewhat tropical climate, *i.e.* during the mid-Tertiary period. Excavations in Palestine and elsewhere in the Mediterranean indicate that it is closely associated with the history of civilization, being the source of edible fruits and oil for over 10,000 years. There is little doubt that the olive tree was first domesticated in the eastern Mediterranean where wild Olea europeae subsp. oleaster forms thrive. The Phoenicians, from such ancient cities as Ugarit on the Syrian coast, introduced the olive around 5000 BP to the west and a secondary center of diversity developed in the areas bordering the Aegean Sea. From there the tree spread to southern Italy and as far as Spain where groves known to be 1000 years old still exist.

Phoenix dactylifera (Palmae) or date palm is another tree plant whose cultivation goes back to ancient times in the Near East. This tree and its symbol are synonymous with Arabia. Its present natural distribution includes North Africa, the Middle East and the western coast of India. At one time in the past the genus Phoenix was widespread in Europe but retreated southward as the climate changed. Its cultivation goes back to Neolithic times. The earliest archaeological record dates back to 6500 BP from Egypt. Traces of date palm stones from 5000 BP have been found at achaeological sites in western India. Its wild ancestor is not known although it crosses readily with the two other species of palms - P. dactylifera and P. reclinata. Phoenix atlantica, or false date palm, could also be an ancestor or an intermediate form. Phoenix atlantica var. maroccanaalso produces edible fruits.

Papaver somniferum (Papaveraceae) or opium poppy is a native of West Asia. It is the source of opium which is obtained from dried latex which drips from cuts made in the unripe capsular fruits. Papaver setigerum, the wild progenitor, is found from Persia to the Balkans. Arab traders are credited with spreading the narcotic use of opium. The plant had reached China by the 18th century and was the principal cause of the Opium Wars.

Fagopyrum esculentum (Polygonaceae) or buckwheat was introduced into China, Tibet and northern India from Central Asia. It is cultivated as a grain crop and is highly nutritious. It is a popular food in Japan.

Rheum rhaponticum (Polygonaceae) or rhubarb of Europe and North America originates from Armenia and Azerbaijan around the Caspian Sea where several other species of Rheum are grown. In China and Tibet R. palmatum and R. officinalehave been used in traditional medicine as laxatives since 2700 BC.

Punica granatum (Punicaceae) or pomegranate is the traditional fruit of the central Iranian plateau. It is also one of the most ancient fruit trees to be domesticated and is known to have been grown in the Hanging Gardens of Babylon. It has some cultural significance as well and is found in the central courtyard of every home in the central Iranian plateau. The field genebank maintained by the Agricultural Research Center at Yazd in central Iran has over 700 different types, some of which go back to antiquity. Its only related species is P. protopunica which is endemic to the island of Socotra (Yemen) in the Indian Ocean. Cydonia oblonga (Rosaceae) or quince is native to southern Daghestan and Azerbaijan. It is cultivated in the Near East as a garden tree. Its fruits are used to make jam.

Malus pumila (Rosaceae) or apple and Pyrus communis (Rosaceae) or pear have their primary center of origin in Asia Minor, the Caucasus and Central Asia. The secondary centers of origin are more towards the east, perhaps in western China. However, both apples and pears have existed in Europe from prehistoric times.

Prunus armeniaca (Rosaceae) or apricot was thought by Vavilov to have its secondary center of diversity in the Near East. It is believed to be native to western China.

Prunus amygdalus, syn. Amygdalus communis (Rosaceae) or almond, has its primary gene center in Central Asia as well as the Near East region and is a fruit species of great antiquity. Syria has several cultivars. Amygdalus fenzliana or fenzel almond is found in Anatolia and is grown as an ornamental. Amygdalus persica or peach has its center of origin in China but also a secondary center in Afghanistan and northeastern Iran. The landraces in the native habitats have better taste but also have some bitter forms and are not uniform in shape and size and hence not preferred by commerce.

Prunus avium (Rosaceae) or sweet cherry comes from Asia Minor and wild trees can be found in West Asia and North Africa. Domestication is said to have occurred in different places, leading to several ecotypes. It is interesting to observe that the sweet forms were domesticated whereas the bitter ones were left to grow unprotected in the wild. Another Prunus spp. native to the Near East is P. cerasifera or cherry plum which grows wild in Caucasia, Iran, Asia Minor, the Altai and Central Asia. It could be one of the parents of P. domestica or the garden plum. Prunus spinosa or blackthorn or sloe is found all over the Near East, North Africa and the southern Mediterranean.

Pyrus spp. (Rosaceae) or pears. The Near East is the main geographic center of origin for this genus. Pyrus caucasica, P. syriaca and P. takhtadzhiana are some of the other species of Pyrus found in this region. Hybridization among the wild species is common and the progenies are appreciated as ornamentals, rootstocks for cultivated species, and increasingly in recent years as donors of genes of disease resistance in fruit-breeding programs, particularly in California.

Coffea arabica (Rubiaceae) or Arabic coffee has its primary center in southwest Ethiopia, where it was chewed before being used for making a beverage. From Ethiopia it migrated to the Yemen. The discovery that coffee could be brewed into a beverage was made in Arabia around the 15th century.

A single coffee plant shipped by Dutch plant explorers to Amsterdam from the East Indies in 1706 was nurtured in the botanical garden of that city. Amsterdam was one of the major world centers of maritime trade in those days. Some of the seedlings from this single plant were taken via France to the Caribbean island of Martinique and from there to South America. Hence the genetic base of the Latin American coffee is indeed very narrow and devastating attacks of diseases are common. The African form C. canephora often provides genes for disease resistance. Besides Latin America, coffee is also cultivated in Kenya, India, Java and other parts of the East Indies. Coffee is said to be the second most valuable commodity in world trade after petroleum products.

Citrus medica (Rutaceae) or citron is probably the only citrus native to West Asia, although the northeastern part of India (Assam and the Indo-Burma border area) has been mentioned as a center of origin of citrus species. Citrus medica var.ethrog is used by people of the Jewish faith at the Feast of Tabernacles. Citron was grown in Sardinia and the area around Naples after the 3rd century AD.

Cuminum cyminum (Umbelliferae) or cumin is a native of the Levant whose fruits are used as spice in cooking, for flavouring soups, and are an essential ingredient of curries. Its essential oil is also used in perfumery. It is also grown extensively in the drier parts of western India.

Daucus carota (Umbelliferae) or carrot. Wild forms occur in West and Central Asia. However, a secondary center of diversity has developed in Turkey and unique landraces can even be found in Syria. There is evidence to believe that the carrot was domesticated in Afghanistan from where it spread all over the world, introgressing with the indigenous wild forms to give rise to local cultivars. Two main groups have been recognized: the Asian group with purple and yellow roots, and the Western group with mainly orange roots. In Turkey the two groups seem to converge and give rise to hybrids. In the Near East countries uniform non-hairy hybrid varieties from the West have become popular, but the natives still prefer the hairy purple type with non-uniform roots.

Foeniculum vulgare (Umbelliferae) or fennel is native to the Mediterranean where it has been cultivated since early times. All parts of the plant are aromatic. The essential oil is used as a flavouring agent.

Pastinaca sativa (Umbelliferae) or parsnip is a native biennial of West Asia and has been consumed as a root vegetable since Greek times. Feral forms are known with tough, dry roots. In the tropics, where it has not gained much popularity, parsnip grows well only at higher altitudes.

Pimpinella anisum (Umbelliferae) or anise was domesticated in the Orient and cultivated in the Near East for its aromatic fruits which are used to flavour strong liqueurs, such as Syrian 'arak' and the Greek 'ouzo'. It was known to the ancient Egyptians, Hebrews, Greeks and Romans and was valuable for medicinal purposes during the Middle Ages.

Celtis australis (syn. C. excelsa) (Ulmaceae) or hackberry is native to the eastern Mediterranean and one of the tree species whose fruits were gathered by early farmers from ancient times.

Vitis vinifera (Vitidaceae) or grapevine is perhaps as old as time itself and was one of the first fruits to be domesticated from the wild in Central Asia, the Near East and the Mediterranean region. The wild progenitor, subsp. sylvestris, is found in areas bordering the Mediterranean Sea except in Libya and Egypt. Its primary center is probably in Armenia and northern Iran. Natural hybrids between the wild and cultivated forms are often found in Tajikstan and as a result newer cultivars come into existence all the time.

6

Plant and Crop Production

CROP PRODUCTION

The present human population of the world is 5.7 billion, increasing at an annual rate of 1.6 %.

The rapid population growth in general and the fact that most of the population growth takes place in developing countries rises the question how these additional people can be fed. Basically, there are two different possibilities to increase food production: Enlarging crop yields on already cultured land, or expanding the area of cultured land on previously uncultured land.

Reviewing the development of crop production since 1965 reveals a massive increase. The differences in production between 1965 and 1988-90 (average of three year period) of the eight most important food and cash crops. The production of rice, wheat and maize, which are the most important crops for human nutrition, doubled during the last 25 years. The production increase was due to higher yields per area. Higher yields per area were possible because of the introduction of high-yielding varieties, improved cultivation techniques, fertilizer use and chemical crop protection measures-a development which has been termed "green revolution".

The intensification of crop production improved the level of food consumption in most developing countries during the last 30 years despite a world population growth of 2.4 billion in the same period. However, the intensification also led to a number of disadvantageous side-effects like erosion, deterioration of soil quality, salinization, and others. Every year, about 1 % of the cultured area has to be replaced because of these reasons. Especially in developed countries, the use of agrochemicals is a growing public concern. The amount of crop protection products used has risen steadily during the past 30 years. Presently, more than 25 billion US$ per year are spent for herbicides, fungicides, insecticides, soil disinfectants and growth regulators. In Western Europe, the most crop protection products are used per area of arable land, followed by the USA. The expenditure per area in these and the other principal farming regions of the world.

Despite the use of pesticides, not only the absolute amount of losses, but also relative losses have increased by about 3 to 10 % in the past 30 years (only in coffee the relative losses were reduced by about 5 %). This is thought to be due to the higher susceptibility of the high-yielding varieties, the development of resistance to chemical pesticides in pest and pathogen populations, the use of fertilizers which not only promote growth but also increase susceptibility, and the extension of production into regions where the pressure from pests, pathogens and weeds is higher. The losses have increased predominantly in developing countries, where appropriate crop protection measures often cannot be taken because of their high price and/or lack of knowledge. Plant genetic engineering may contribute substantially to the development of new pest and disease resistant varieties by increasing the speed of conventional breeding and by transferring useful genes from various plants and other organisms into crops. The principal methods used to achieve these goals are described.

The present state of the art of genetic engineering for pest and disease resistance and applications in development or already realized will be presented. The safety issues raised and the regulatory framework for field releases and market introduction complements the topic. As not only plants, but also other organisms like bacteria or viruses are genetically engineered in order to prevent crop losses, these approaches are al so briefly described.

CROP IMPROVEMENT

Protoplasts are naked cells that lack cell walls. They are spherical with a plasmolysed cell content and are contained within a plasmalemma. In principle, each individual protoplast can reform a cell wall, and later initiate either a callus through sustained divisions, or an embryo, defined as a somatic embryo. In banana they are obtained from in vivo tissues or in vitro cultures.

This phenomenon is termed totipotency and denotes the recovery of a whole organism from a single cell, and is also applicable to protoplasts. In theory all cells are totipotent, but in practice it depends on the past cellular environment. Usually the morphogenetic competence is retained at the unicellular stage that corresponds to a protoplast. However since protoplasts are not exposed to the stabilizing and inductive influence from neighbouring cells, they may have lost their plant regeneration capacity. In fact, factors such as the genotype or species of the manipulated plant, and the ontogenetic state of the explant source, exert a powerful effect on the regeneration potential of protoplasts.

Consequently the development of appropriate in vitro conditions for protoplast regeneration is complicated. Notwithstanding this, successful efforts have been made in isolation, cultivation and regeneration of protoplasts, since Nickell and Torrey pointed out their merit for crop improvement.

Various tissues and an increasing number of plant species and genotypes have been successfully used in protoplast culture, but so far sufficient quantities of protoplasts for practical applications are not routinely met. While the value for agriculture of protoplasts still needs to be demonstrated, they are an invaluable tool for studies on permeability of ions and solutes, photosynthesis, phytohormones, phytochrome, and maintenance of totipotency. Moreover, protoplasts are useful for the uptake of foreign genetic material and to produce somatic hybrids through protoplast fusion. In addition, protoplasts are an excellent system for studies on cell genetics and even for plant virology.

These studies rely on isolated, clean and healthy protoplasts, which requires the appropriate choice of osmoticum, hydrolysing enzymatic solution, and pretreatment of the donor explant either in vivo or in vitro. This involves the right choice of explant age, cold treatment, phytohormone pretreatment, light intensity or photoperiod, and subculture rhythm, which affects both the internal metabolic status of the cells and cell-wall composition. Prior to cultivation isolated protoplasts need to be freed from enzymatic remains and debris, which are considered to be toxic. Culture in or on solid medium is considered to be more advantageous than in liquid medium because the development of a single protoplast into a colony can be followed up much better. For development, different factors such as medium composition are crucial, especially the nature and quantity of growth hormones.

Other important factors are the physical environment, plating density, and embedding conditions or the use of feeder layers, which all contribute in one way or another to cell-wall regeneration and cell division. Once cell divisions start, the level of auxin(s), as well as the colony density, usually need to be reduced to avoid overcrowding. This is done by subculturing, which gives the additional benefit that an adequate nutrient supply is maintained. Finally, for rooting, plants are usually transferred to cytokinin-free medium, possibly containing some auxin, and exposed to high light intensities, usually in illuminated plant growth chambers. During recent decades, plant biology workers have recognised the potential of protoplasts in many experimental systems, since an efficient enzymatic method for protoplast isolation was first established by Cocking.

SIGNIFICANCE AND USE OF PROTOPLASTS IN BANANA

Banana is now easily amenable to in vitro culture, and plants are regenerated from various explants through organogenesis, embryogenesis, anther culture and even from cultured protoplasts. This has created opportunities for other biotechnological applications. Cell suspensions, for example, opened the way for gene transfer and improved the production, the quality, and therefore the manipulation of cell suspension-derived protoplasts. Through genetic engineering it has become possible to confer new traits on banana plants, using either particle bombardment or Agrobacterium-mediated transfer.

Protoplasts facilitate the direct transformation of plant cells by DNA microinjection and electroporation. However, many characters of agricultural interest are multigenic or ill-defined, and current transformation methodologies allow the integration of only a few foreign genes. Protoplast fusion, however, allows the transfer of several useful characters, even if detailed genetic or molecular knowledge of genes encoding for these desired characters is lacking. Protoplast fusion is therefore a complementary tool to increase nuclear and cytoplasmic variability and to confer desirable agronomic traits.

Important banana cultivars are susceptible to many pests and diseases, particularly Mycosphaerella fijiensis and Fusarium oxysporum f.sp.cubense, nematodes and insects. Very interesting traits of resistance have been identified in this genus for most serious diseases. However cross-breeding is very difficult in the genus since most edible cultivars are sterile even after profuse pollination by wild pollen-fertile genotypes. Protoplast fusion is therefore an option, as it can overcome such sexual barriers which occur frequently in banana, and which cannot be bypassed even through embryo rescue. Therefore somatic hybridisation between wild and cultivated banana is expected to produce hybrids combining agronomic traits with genetic resistance to pests and pathogens.

Protoplast fusion has the added benefit that it creates the possibility of generating asymmetric fusions, whereby selected cytoplasmic organelles, chromosomes or chromosome fragments from an irradiated protoplast donor could be combined with the genome of an acceptor protoplast. This strategy is of especial interest in banana, where the subspecies balbisiana is considered to be a source of multiple resistance and therefore could be partially utilised to improve banana cultivars, of which the most important component is the subspeciesacuminata. Protoplast fusion also makes the creation of synthetic triploid banana genotypes possible by, for example, combining haploid protoplasts derived from anther culture with protoplasts from a diploid improved cultivar.

Chimerism and somaclonal variation are factors that seriously limit rapid clonal propagation of banana. By their nature, protoplasts avoid chimerism because they originate from a single cell, and protoplasts may be use to dissociate chimeric plants. On the other hand the cellular heterogeneity of protoplast populations can be useful for isolating somaclonal variants with improved characters such as increased yield or pathogenic resistance, so that somaclonal variation may also be exploited to provide new sources of genetic variability.

Banana protoplast isolation and culturing is nowadays routine, although protoplast research on banana started 15 years later than on model plants. Now it is regrettably underused because of the strong emphasis on molecular and genomic studies in banana. Nevertheless, protoplasts have much to offer for non-conventional banana breeding, since they overcome sexual incompatibilities at interspecific and even at the intergeneric level, and allow the incorporation of multigenic traits such as yield, resistance to stress, pests and diseases. Finally protoplast techniques may improve banana when other classical methods have failed.

WORK ON ISOLATED CELLS AND PROTOPLASTS

Early Experiments

The first attempts at banana protoplast culture were made in 1984 by Bakry. Various explants (leaf and sheath from in vitro banana plants, callus from floral explants, and bract and tepals from in vivo banana plants) of diploid and triploid banana plants belonging to the AA, BB, AAA and AAB groups were tested.

Enzyme solutions were composed of a modified CPW (Cell and protoplast washing solution) salt solution containing 0.7 M mannitol, and a mixture of various enzymes at 0.1 to 5% (w/v) such as:

- *Pectinases*: Macerozyme;
- *Hemicellulases*: Hemicellulase
- *Cellulases*: cellulase R-10 'Onozuka', Cellulysin and Driselase.

Table. Cell and Protoplast Washing (CPW) Solution

Component	Concentration (mg/l)
KH_2PO_4	27.2
KNO_3	100
$CaCl_2.2H_2O$	150
$Mg\ SO_4.7H_2O$	250
KI	0.16
$Cu\ SO_4.5H_2O$	0.025

The enzyme solution was adjusted to pH 5.6 with KOH and filter-sterilised (0.22 mm). Small explants of about 1-5 mm^3 were incubated for 9 h with shaking (80 r.p.m.) at 27°C in the dark. The mixture was sieved through an 83 mm metallic sieve, followed by dilution with KMC salt solution. After centrifugation at 100 g for 5 min, the pellet was diluted in a KMC solution. Whatever the plant material, enzymatic solution and plant preconditioning (light/dark), only isolated plasmolysed cells were obtained.

Positive results were obtained when nodular calli were used as starting material. Floral explants cultivated on MS medium complemented with 500 mg/l casein hydrolysate, 2 mg/l IAA, and 2 mg/l BAP lead to calli covered with nodular bodies.

These calli were embryogenic. With the enzyme mixture 2.5% cellulase R10 'Onozuka', 0.2% hemicellulase, 0.3% pectolyase Y23 and 0.6% macerozyme, Bakry produced the first living banana protoplasts. After calcofluor staining, protoplasts were found to be cellulose-free, perfectly round in shape, variable in size and having a dense cytoplasmic content. Although the yield was quite low the protocol was reproducible.

This pioneering work was confirmed by Cronauer and Krikorian in 1986 (using proliferating shoot tips of the cultivar 'Lacatan' [AAA group]) and by Da Silva Conceicao in 1989. Later on regular progress was made by several scientists. Initially banana calli from protoplasts were developed in 1992, followed by plant regeneration directly from protoplasts, protoplast transformation and somatic hybridisation.

Explants used as a Source of Protoplasts

In banana, in vivo explants are not a good source of protoplasts. Although Matsumoto *et al.* obtained protoplasts from in vivo bracts, in most cases initiation started from in vitro cultures. Leaf explants, slices of shoot tissue, roots or callus were the first choice as starting material for protoplast isolation because of their convenience. In fact, protoplasts can be obtained in banana from almost any tissue, including young leaves, sheaths, bracts, roots, and callus, but yields depend on the explant source.

Yields are quite low with most tissues and range from 0.1 to 2.8 × 106 protoplasts per gram, but the protoplasts are unable to divide. However, cell suspension-derived protoplasts gave high yields which ranged from 4 to 5 × 105 per gram with 'Long Tavoy'; from 1.1 to 3 × 105per gram with malaccensis; 2 to 20 × 106 per gram with 'Maça' to 6.6 × 107 per millilitre of packed cell volume (PCV) with 'Bluggoe'.

Essential was the embryogenicity of the suspension. The first sustained cell divisions in protoplast culture leading to calluses were observed by Megia *et al.* in 1992 with M. acuminata cell suspension-derived protoplasts. Megia *et al.* and Panis *et al.* improved the protocol. This was confirmed by Matsumoto *et al.* and Assani *et al.* who produced plants from such 'embryogenic protoplasts'. Consequently embryogenic cell suspensions are the starting material of choice.

Banana Genotypes Tested

Bananas are either triploid, diploid or tetraploid in order of importance for food production, but protoplasts have been obtained so far from triploid, diploid and haploid genotypes. The cultivars and wild types tested; the letter A denotes the contribution of the subspecies acuminata and B the contribution of the subspecies balbisiana.

Triploid Genotypes

- Musa cavendishii L.cv North banana (AAA);
- 'Lacatan' (AAA)
- 'Nanicao' (AAA)
- 'Bluggoe' (ABB)

- 'Maça' (AAB); 'Grande Naine' (AAA); 'Gros Michel' (AAA); 'Currare Enano' (AAB); 'Dominico' (AAB).

Diploid Genotypes

- 'Pisang Lilin' (AA)
- Musa acuminata ssp. burmanica type 'Long Tavoy' (AA)
- Musa acuminata ssp. malaccensis (AA)
- SF 265 (AA); IRFA 903 (AA); Col 49 (AA); 'Colatino Ouro' (AA); 'Pisang Klutuk Wulung' (BB); 'Pisang Klutuk' (BB).

Protoplast Extraction and Purification

Enzymes

Various enzyme mixtures are used depending on the author and banana explant (CS = Cell Suspension; P = Plant explant; C = callus).

They are commonly composed of:

- Cellulases (Cellulase Onozuka R10, Cellulase Onozuka RS, Cellulysin, Macerase, Driselase)
- Pectinases (Pectinase; Macerozyme R10; Pectolyase Y 23)
- Hemicellulase (Rhozyme HP -150; Glusulase)

Protoplast Isolation Media

It is standard practice to incubate the plant material in the enzyme solution for a digestion period of 1-24 h either with shaking (40 r.p.m.) or no shaking, generally at 27°C in the dark. The best source of protoplasts is from embryogenic cell suspension cultivated in the light (30 or 65 mE m^{-2} s^{-1}) or darkness and regularly subcultured. Initiation should start 3-5 days, 7-10 days or 2-15 days after the last subculture, followed by sieving through a 200-400 mm mesh just before enzyme incubation.

Protoplast Purification

Protoplasts are purified from debris by sieving (100 mm and subsequently or directly through 56/32/25 mm sieves), which is preferred to flotation. With flotation purification, a 21% sucrose solution and centrifugation at 120 g is used.

Protoplast Washing

After sieving, protoplasts are generally repeatedly washed to remove enzymes. Protoplast pellets obtained by centrifugation (50/66/90 g) are washed with protoplast isolation medium without enzymes; cell suspension culture medium without growth regulator plus 10% mannitol; or 3% $CaCl_2$ plus 0.5% KCl solution; or 3% $CaCl_2$ 0.5% KCl solution and 10% mannitol; or 0.6 M D-mannitol, 0.1 mM $CaCl_2$, 0.5% PVP-40 and 3.5 mM MES.

Protoplast Quality Control

Complete cell wall degradation is confirmed both by the spherical shape of the released protoplasts and by the absence of fluorescence after staining with calcofluor white. Evans blue or FDA staining are commonly used for viability assessment.

Protoplast Culture Media

For banana protoplast culture, derived MS medium (supplemented with 2,4-D, or 2,4-D plus zeatin) or N-medium composed with N6 salt, vitamins, organic acid and sugar alcohol are commonly used. Due to difficulties in obtaining favourable protoplast development, many culture systems have been tested. Liquid and solid media (derived from MS) were initially tried for protoplast cultivation, but it was found that liquid cultures are not suitable for Musa protoplasts. In contrast, solid media supported sustained divisions.

To overcome the low growth response, other media were tested, such as semi-solid media, protoplasts embedded in solid or semi-solid media, nurse cultures, and nurse cultures with a feeder layer. Cocultivation in banana protoplast cultures was considered as it was reported that coculturing protoplasts of recalcitrant species, especially monocotyledons, with a reliable feeder culture, induced cell division. The banana protoplast literature describes various combinations of media and culture systems but, whatever the protocol, a high plating density (10^5-10^6/ml) was crucial, as well as the use of feeder layer, or nurse culture combined with a feeder layer. Protoplasts are cultured at 27 ± 1°C in the dark.

Culture in Liquid Media

For culturing protoplasts in liquid media, protoplasts with a density of 105-106 per millilitre are suspended in liquid media, with or without shaking. The following options are open:

- N6 salt and KM vitamins, organic acids and sugar alcohols according to Kao and Michayluk, 6.3% (w/v) glucose and 4% (w/v) sucrose as osmoticum, 260 mg/l KH_2PO_4 and 100 mg/l 2-(N-morpholino) ethanesulphonic acid MES;
- N_6 salts, KM vitamins, organic acids and sugar alcohols, vitamins of Morel, 0.4 M glucose, 117 mM sucrose, 0.5 mM MES, 1.9 mM KH_2PO_4;
- N_6 salts, KM vitamins, organic acids and sugar alcohols, vitamins of Morel, 0.4 M glucose, 117 mM sucrose, 1.9 mM KH_2PO_4, 2.28 μM zeatin, 0.90 μM 2,4-D and 5.4 μM NAA;
- MS modified medium, 5 mM 2,4-D and 10% mannitol;
- 1/2 MS, 10% mannitol;
- MS medium, 5 mM 2,4-D, 1 mM zeatin, 0.55 M mannitol;
- MS minor nutrients, 1/2 major nutrients, MS vitamins, 10 mg/l ascorbic acid, 20 g/l sucrose, 5 μM 2,4-D; 0.55 M D-mannitol;
- MS preconditioned medium, 5 mM 2,4-D, 10% filter-sterilised medium from a one-week-old cell culture.

Culture on Solid Medium

For culture on solid medium protoplasts are directly plated at high density on a gelatinous medium or on a membrane which covers the gelatinous medium.

Such media are composed of:

- MS minor nutrients, 1/2 major nutrients, MS vitamins, 10 mg/l ascorbic acid, 20 g/l sucrose, 5 μM 2,4-D, 0.275 M D-mannitol, 0.2% gelling gum; or
- MS 1/2 salt solution, 5% mannitol, 5 μM 2,4-D, 0.8% agarose or 0.2% Gelrite.

Culture in Semi-solid Medium

For culture on semi-solid medium protoplasts are cultivated on:

- 0.6% agarose, N_6 salt, vitamins of Morel, vitamins of Kao and Michayluk;
- 0.6% agarose, 0.35 M glucose, 40 g/l saccharose, 260 mg/l KH_2PO_4, 3 mM MES, and Pichloram, BAP, NAA, L-asparagine and L-proline at various concentrations;
- Protoplasts at twice final desired concentration in culture medium, are mixed with the same medium containing 0.3/0.8% agarose, 5-10% mannitol, 0-10 μM 2,4-D.

Culture Embedded in Solid Medium

For embedment in solid medium protoplasts at twice the desired concentration (2×10^6) are placed in a twofold concentrated culture medium: MS minor nutrients (× 2), MS major nutrients, MS vitamins (× 2), 20 mg/l ascorbic acid, 40 g/l sucrose, 10 μM 2,4-D, but with 0.55 M D-mannitol.

This is followed by mixing with an equal volume of solution containing a gelling gum at 0.6%, 0.55 M D-mannitol, previously melted and maintained at 55°C.

Culture in Alginate

Protoplasts are cultured in a solid medium, or in calcium alginate beads according to Schilde-Rentschler *et al*. For this purpose, the alginate solution (2.8% alginic acid, 0.4 M mannitol) is first autoclaved. Protoplasts are added to the alginate solution to obtain a concentration of 1.0×10^6 protoplasts per millilitre.

Later, drops of the protoplast-alginate mixture are polymerised in 0.4 M mannitol/50 mM CaCl2 solution, and transferred into protoplast liquid culture medium in small Petri dishes. Pieces of solid medium containing protoplasts are transferred into liquid medium:

- MS medium, 10 mg/l ascorbic acid, 20 g/l sucrose, 5 μM 2,4-D, 0.275 M D-mannitol, with or without 0.3% activated charcoal;

- 'A medium' consisting of N6 salts, KM vitamins, organic acids and sugar alcohols, vitamins of Morel, 117 mM sucrose, 1.9 mM KH_2PO_4;
- 'B medium' consisting of the 'A medium' containing 2.28 μM zeatin, 0.90 μM 2,4-D and 5.4 μM NAA.

Nurse Cultures

Living cells as nurse in liquid medium: Protoplasts can be maintained according to one of the following procedures:

- Nurse cell suspensions are maintained in N6 salt with MS vitamins, 10 μM L-proline, 300 mg/l casein hydrolysate, 5 μM 2,4-D, and 3% (w/v) sucrose at pH 5.8;
- 1 ml PCV of nurse cells is freshly subcultured and mixed with 50 ml of MS minor nutrients, 1/2 major nutrients, MS vitamins, 10 mg/l ascorbic acid, 20 g/l sucrose, 5 μM 2,4-D and 0.275 M D-mannitol. Embedded protoplasts prepared as above are then immersed in the nurse solution. Cultures are placed on a rotary shaker at 120 r.p.m.

A MillicellTM-HA μm pore size) may be used as a physical barrier, to separate protoplasts from nurse cells during culture.

IMPROVING CROP PERFORMANCE

The dramatic gains in agricultural productivity seen in the second half of the last century are often linked uniquely to increased mechanisation and the widespread introduction of fertilizers and agrochemicals.This 'Green Revolution' would not have been possible without the huge genetic improvements made to the crops. Plant breeding has contributed around half of the threefold increase in UK wheat yields recorded from 1947 to 1992 and yields have continued to improve since then. Plant breeding can directly improve the performance of crops in different ways.

PRODUCTIVE YIELD

Developing crop varieties, which convert more of their biomass into productive yield, is the single biggest contributor to improved crop output. The introduction of shorter-strawed cereals is a striking example of how this has been achieved, by transforming more of the crop's productive energy into valuable grain.

Physical Characteristics

Changing a crop's physical structure can also contribute to increased yields. For example, the development of semi-leafless varieties of field peas helped to boost intrinsic yield. It also stimulated the growth of stabilising tendrils, which significantly improved the crop's standing ability so reducing crop losses at harvest.

Disease Resistance

Genetic resistance to disease enables crops to realise their yield potential– it can also mean reduced use of agrochemicals. Plant breeding has significantly improved the genetic resistance of crops against the threat of viral and fungal infection. Key examples include resistance to blight in potatoes, rhizomania in sugar beet, and barley yellow mosaic virus in cereals. The challenge of breeding resistant varieties is constant because new strains of disease develop naturally.

Timing of Maturity

Plant breeding technology has brought major improvements in the uniformity with which crops ripen ready for harvest. This not only reduces potential crop losses at harvest (as in the case of pod shatter in oilseed rape), but has also improved growers' ability to mechanise harvesting operations. In the field vegetable sector, for example, the labour intensive (and unpleasant!) task of picking crops such as cauliflower, broccoli and Brussels sprouts has been transformed by the development of improved varieties.

Other Agronomic Factors

By improving crops' ability to cope with a range of other agronomic pressures, advances in plant breeding continue to underpin progress in agricultural productivity.

They include:

- Genetic resistance to pests, such as nematode resistance in potatoes
- Shorter crop life-cycle, to expand a crop's geographical growing areas
- Stress tolerance, such as frostresistance in field vegetables, to extend the seasonal availability of home-grown fresh produce.

IMPROVING CROP QUALITY

Alongside progress in keeping Britain's farmers competitive through improved productivity, developments in plant breeding have also brought significant gains in crop quality. Gone are the days when Europe encouraged farmers to concentrate solely on maximising output, assured that their crops would find support under the Common Agricultural Policy.

Today, as production-based support and barriers to international trade continue to be reduced, Britain's growers must compete with the best in the world, on quality and cost of production. Consumer expectations of food are also more exacting. This is reflected in tighter quality specifications along the food chain. Plant breeders have responded with a continuous stream of new varieties, tailored to the needs of specific end-markets.

Cereals

Britain's historical dependence on North American imports of wheat for bread making has been reversed over the past 40 years thanks to improvements in baking technology and the development of high protein, hard milling varieties of wheat.

Until the mid-1960s, imported wheat typically accounted for up to 65% of our bread making requirements. Today as much as 90% of our bread is produced from homegrown varieties. In barley, improvements in malting quality have underpinned a fourfold increase in the volume of beer brewed per tonne, from around 2,000 liters in 1950 to nearly 8,000 liters today.

The quality of malting barley varieties grown in Britain is internationally recognized, with export markets from the near Continent to the Far East.

Oilseed Rape

Oilseed rape has become a major homegrown source of vegetable oil and animal feed within the past 30 years. The rapid expansion of this important break crop from 50,000 hectares in the mid-1970s to around 500,000 hectares today can be linked to two major breakthroughs in plant breeding. Varieties for food use were first developed in 1970. These contain reduced levels of erucic acid in the oil, making it more suitable for human consumption. This was followed in the late 1980s by 'doublelow' varieties, which also offered reduced levels of glucosinolates in the residual meal to improve quality for animal feed.

Potatoes

In the potato sector, plant breeders have responded to increasing consumer demands for quality and choice. Fresh produce counters now boast a bigger range than ever, from traditional main-crop varieties through to baby new, baking and salad potatoes. Breeding programmes have also targeted processing and catering markets, with specific varieties used to produce crisps, instant mash and chips.

Sugar Beet

Higher root yield and increased sugar content have combined to double sugar production per hectare in the past 50 years. The introduction of monogerm seed, allowing seeds to be planted individually rather than in clusters, ensures genetic improvements are fully realised in the field.

Pulses

As interest in home-grown sources of protein has increased, breeders have developed varieties of peas and beans for specialist markets–freezing and canning for human consumption, flaking for pet food, and tannin-free varieties for animal feed.

Creating a New Variety

The creation of each new variety is a complex, costly and skilled operation. It is also painstakingly slow–today's breeding programmes are already looking ten years ahead to the needs of farmers, consumers and the environment at the end of the next decade and beyond. Techniques vary between crop species, but the scientific principles of plant breeding remain unchanged from Mendel's first discovery that selected parent plants can be cross-pollinated to combine desired characteristics. Genes–units of hereditary material that are transferred from one generation to the next, determine plant characteristics. Since each plant contains many thousands of genes, and the breeder is seeking to combine a range of traits in one plant, such as high yield, quality and resistance to disease, developing a successful variety is an extremely lengthy process–up to 12 years in the case of cereals, even longer for potatoes. Plant breeding has been compared to playing a fruit machine–not with three reels, but several hundred. The skill of the plant breeder lies in improving his chances of hitting the jackpot by combining all the desired characteristics in the same variety.

TILLAGE AND PLANT PRODUCTION

Loosening the soil with the aim of improving conditions for the crops plays a crucial part in arable farming. One purpose of turning the soil is to eliminate plants competing with the crop. The extent to which the soil structure is changed depends on the form of tillage, which may involve;

- loosening the soil with hooks or tines,
- turning the soil with the plough, or
- crumbling and breaking up the soil with a powered rotary cultivator or harrow.

If greater drive power becomes available, this may lead to selection of implements (*e.g.* rotary cultivator rather than plough) which modify the soil structure to a greater extent. In addition, less suitable (marginal) sites may be "put to the plough". Both of these steps increase the risk of soil degradation, which may involve reduction of pore volume, water absorption capacity and water storage capacity, the danger of puddling and crusting and a loss of organic matter. Crusting impedes water penetration and plant growth.

The optimum degree of soil moisture for tillage lies within narrow limits. If the soil is tilled when it is too wet it will be compacted, while tilling a soil which is too dry will result-depending on its clay content-in formation of clods or pulverisation. Compaction may assume sizeable dimensions if heavy tractors and implements are used.

Among other things, compaction of the soil has effects on plant growth, soil organisms and the availability and breakdown of plant nutrients and pesticides. On slopes, soil layers above the compacted horizon may slip.

Loosening of the soil and introduction of organic matter have positive impacts on the soil fauna. By contrast, compaction and puddling, frequent disturbance of the soil by tillage and application of pesticides and fertilisers all adversely affect the development of soil organisms. Measures for preventing soil erosion and compaction include the following in particular:

- The soil should if possible have permanent vegetation cover in the form of living crop plants (permanent cropping, intercropping, alley cropping) or dead plants (mulch).
- Crops should be sown directly into the remnants of the preceding crop (without the soil being turned).
- Crop residues should be left on the surface and should not be ploughed under.
- Coarse soil structures are to be created or preserved through appropriate crop rotation measures and use of suitable implements.
- Contour ridges or terraces are to be created on slopes. This may be a somewhat complicated operation; there may be no vehicle access to the land in question.
- Windbreaks are to be planted at right angles to the direction of the prevailing wind.
- Organic matter is to be preserved and increased if possible.
- The pressure to which the soil is subjected when vehicles are driven over it is to be reduced as far as possible by using small/light-weight tractors and machines or larger tyres.
- Where possible, the soil should be driven over and tilled only when the moisture content is optimum.
- Shallow and deep tillage implements are to be used alternately over the course of time.

Emphasis must be placed on promoting a willingness to abandon excessive exploitation of the soil in favour of sustainable, site-appropriate forms of cultivation.

Sowing/planting, Crop Tending and Fertilising

Sowing and planting, which are performed after the soil has been tilled, are intended to create optimum conditions for the growth of the seed or young plant. After tillage, the soil may be totally or partly without cover until the crops have fully developed. It is thus exposed to the risks of erosion and to puddling as a result of heavy rainfall or severe evaporation.

With large cropping areas, mechanical aids are virtually essential for distributing chemical pesticides. Such equipment calls for highly skilled operators. Unsuitable, defective or incorrectly operated equipment can result in overdosage of fertilisers and pesticides, which will have negative effects on soil, plants and water as well as on the equipment users. Application of highly concentrated liquid pesticides using the ULV (ultralow-volume) technique may lead to severe air pollution, with drift causing contamination over wide areas.

Pesticide users can be exposed to serious health hazards by touching or inhaling the chemical substances. It is often difficult to empty the containers completely and water used to rinse them out can contaminate surface water and drinking water. Pesticides and the equipment used to apply them are often stored improperly and are frequently kept in the same place as food for want of any other lockable rooms. Under unfavourable conditions, mechanical weed control measures (hoeing) can destroy the soil structure and encourage erosion; despite this fact, they should still be preferred to chemical methods.

Agricultural engineering can make a major contribution to ensuring that pesticides and fertilisers are used and applied correctly. Apart from selection of the right equipment, which is determined in part by the formulation of the agents used (*e.g.* powder or liquid), correct operation is equally important. If the time of application is appropriately chosen and economic-threshold concepts are used, quantities can be cut, drifting reduced and risks thereby minimised. Protective clothing, including face masks, is to be provided.

Under the climatic conditions prevailing in the tropics and subtropics, however, wearing of such clothing imposes considerable physical stress.

The stresses and risks for the equipment operators can be reduced to a large extent if the work is appropriately organised (*e.g.* operators proceed in the direction in which the wind is blowing).

Harvesting, Threshing, Processing, Preservation, Storage

The technical aids used in harvesting and threshing are intended to enable the work to be performed more easily and rapidly, besides minimising losses and risks. Harvesting and processing of "dry goods" (*e.g.* grain, burned-off sugar cane) can give rise to dust emissions which affect only a small area but are highly intensive. Such emissions affect people working and living in the vicinity, as well as animals. They can be reduced by taking technical measures at their source or their effects mitigated through the wearing of face masks and protective clothing.

Threshing and processing may yield by-products (glumes, hulls etc.). However, the quantities occurring at farm level do not constitute a serious environmental hazard, since the farm can generally utilise such products itself.

When crops are harvested, various substances are removed from the natural cycling systems on the site concerned. Efforts should be made to ensure that at least the by-products are returned to the soil either directly, after use for other purposes (*e.g.* as livestock fodder) or after composting.

Technical measures for preserving and storing produce at farm level seldom have any environmental impacts, unless chemicals are used. Crop drying calls for an energy supply, however, and in certain cases this may lead to over-use of local forest resources. Efforts to alleviate this problem should focus on ways of reducing energy consumption (heat sources).

Supplying and Distributing Water

To supplement the environmental brief Irrigation, attention must be drawn here to a number of important interrelationships and areas of overlap between these two sectors. The water application and distribution system (gravity-flow method with open channels, pressure method with pipes or hoses) has a considerable influence on mechanisation:

- Channels determine the field size and bridges are needed to cross them.
- Small embankments and ditches in the field are damaged when driven over.
- Pipes have to be removed before a field can be tilled or urgent plant protection measures carried out.

Aspects of Animal Production

Pasture farming has always been the traditional form of livestock husbandry. The use of technical aids is confined to protective and security measures (pens etc.); the same applies to the keeping of small animals (*e.g.* poultry, rabbits, bees). It is not until livestock farming is intensified, with livestock kept in confinement, that technical aids become increasingly important. In industrialised countries where intensive animal husbandry is practised (particularly in Europe), technical equipment has come to play just as vital a role in livestock farming as it does in crop growing.

In poorly ventilated livestock housing, heat, dust and gases-particularly ammonia-can create stresses for both man and animal.

Sizeable quantities of ammonia can escape into the air when animal excrement is stored and applied to fields as manure. In industrialised countries, ammonia is one of the principal factors causing the gradual death of the forests in regions where intensive livestock husbandry is practised. One effective countermeasure is to make sure that solid or liquid manure is immediately worked into the soil.

Improper storage and spreading of animal manure can lead to over-fertilising (eutrophication) of both surface water and groundwater.

Protective measures must often focus first of all on bringing about changes in awareness, for only when this has been achieved can technical measures contribute to reducing negative environmental impacts. Animal excrement must be regarded and treated as a valuable fertilizer and not as waste. It is also important that this fertilizer be spread over the cropping area precisely in line with the plants' nutrient requirements. Only if such measures can be realised will it be possible to develop intensive livestock farming methods that are environmentally sound in the long term.

Evaluation of Environmental Impacts

Among the negative consequences of cultivation-magnified by use of mechanical aids-it is erosion that has by far the greatest significance worldwide. While there are numerous ways of reducing erosion (*e.g.* crop-growing measures such as mulching, technical measures such as terracing and planting of windbreaks), standards for evaluating the effect of erosion are largely confined to criteria for recording and assessing the removal of soil. Specific cultivation bans or requirements are occasionally imposed in the catchment areas of reservoirs particularly at risk from sediment.

Manufacturers of tractors and agricultural machinery in industrialised countries are called upon to fulfil widely varying national environmental requirements. These include;

- standards and guidelines on design and durability;
- provision of safety devices, protective circuits etc., particularly for motorised vehicles and machines;
- provision of special equipment if the vehicles use public roads (danger to other road users);
- emission standards (exhaust emissions, noise).

Institutions performing official functions, such as agricultural-machinery testing stations, carry out type approval testing whose results are binding on manufacturers. Compliance with requirements is relatively easy to monitor.

It is far more difficult to ensure that regulations are observed by users. Safety devices may be removed, protective clothing and masks not worn or emission standards, speed limits and the like disregarded.

VARIETY REGISTRATION OF EXISTING CROP PLANTS

Although the variety registration of existing crop plants has not resulted in adverse effects in humans, the conventional assessments carried out by plant breeders was not considered to be sufficient to ensure food safety of GM crop plants. As a result, Europe issued a regulation on novel foods and novel food ingredients that sets the legal framework for the market introduction of genetically modified organisms (GMOs). This came into force. The Novel Foods Regulation establishes a system of mandatory pre-market approvals of novel foodstuffs, including GM plants.

The accompanying EU guideline gave a clear indication of the types of data that are needed to form the basis of an assessment. In particular, four central safety issues have been set forth related to food and feed safety:

- The nutritional and toxicological consequences of inserted gene products;
- The potential of pleiotropic (unintended) effects in the host organism due to the insertion event;
- The allergenicity of expressed proteins and novel foodstuffs;
- The potential of gene transfer to human and animal gut flora.

This means that in transgenic insect-resistant Bt tomatoes encoding the C-terminal truncated Bt2 gene derived from a Bacillus thuringiensis (Bt) strain, IAb5, the novel protein has to be evaluated on its own merit, while the safety and wholesomeness of the 'remaining' transgenic tomato is assessed separately. Whereas, for example, in antisense RNA exogalactanase tomatoes, which have improved rheologic characteristics due to downregulation of the endogenous exogalactanase activity, the safety assessment is primarily focused on the 'remaining' novel fruit. In general, there is basic agreement on the safety issues to be addressed. One particular area of concern is the possibility of unexpected or unintended metabolic perturbations due to genetic modification that may alter, for instance, levels of nutrients and health-influencing components. The use of rDNA techniques does not necessarily result in fundamental changes in crop plants compared to the food produced by conventional breeding methods. However, it should be emphasized that a uniform international agreement for the evaluation of GMOs is urgently needed.

For example, an agreement on how to establish in practical terms the similarities or differences between the 'remaining' GM crop plant and the traditionally bred plant seems to be much more difficult to achieve. This chapter is devoted to the description of advanced strategies for the evaluation of potential unintended effects in GM crop plants. Pleiotropic (unintended) alterations in agronomic traits or composition may arise from insertion mutagenesis or as a result of metabolic effects of the novel gene product(s).

The challenge is to gain the ability to discriminate between metabolic alterations due to somaclonal variations, unintended effects or natural diversity. Thereto, emphasis is given to a platform of technologies designed to identify changes in the molecular machinery of crop plants. As is the case for detection methods for GM crops and derived products, there are needs and criteria for sampling strategies, statistical models and manufacturing standard reference matehers recently.

UNINTENDED EFFECTS IN FOOD SAFETY AND WHOLESOMENESS

Strategies for assessing the food safety and wholesomeness of novel foodstuffs are currently in the exploration phase. Above all there is a need for uniform and harmonized quantitative methods for the testing of potential unintended effects in whole foods, such as GM crop plants are. The safety of, for example, synthetic chemicals and food additives is established by assessing separate elements of the compound in question.

Although very successful in case of single chemicals, the evaluation of a whole food product may not be a simple addition of individual assessments of the constituents. Traditional testing protocols involving labouratory animal 90-day feeding trials with (whole) food products are far from ideal. This type of animal feeding study with complex food matrices is complicated by the likelihood of nutritional imbalances leading to dietary problems, confounding factors, an insensitivity for specific endpoints, and the impossibility of using large safety margins, if any. Therefore, in a global context several regulatory, research and industry-based bodies have proposed alternative concepts for the safety evaluation of genetically engineered foods and food ingredients. In particular the Organisation for Economic Co-operation and Development (OECD), the Food and Agriculture Organization (FAO) and the World Health Organization (WHO) have developed concepts and principles for the safety evaluation of GMOs and derived products.

Their reports concluded that risk assessments should be directed at demonstrating that a GM crop plant or derived food product is as safe as a traditional or previously authorized product. Accordingly the EU Competent Authority incorporated these general principles into their Novel Foods Regulation and accompanying guidelines. In this context, novel foods and food ingredients that are considered to be 'no longer equivalent' to traditionally bred plants are subject to labelling as defined in the Council Regulation.

SUBSTANTIAL EQUIVALENCE

The term 'substantial equivalence' appears in the EU Novel Foods Regulation as a major principle in the risk assessment. It emphasizes that, within the context of the Regulation the evaluation of the safety and nutritional value of a GM plant should be focused on a comparative analysis with conventionally bred products. OECD's Group of National Experts on Safety in Biotechnology enunciated this comparative approach towards the safety assessment of transgenic food plants as the concept of substantial equivalence. In their report, it is assumed that existing food organisms possess a long history of safe use (*i.e.* GRAS). Consequently, the most practical approach to the determination of safety is to consider whether the GM plants are substantially equivalent to analogous food product(s), if such exist. Once substantial equivalence of a novel foodstuff has been established, it provides assurance of safety that is equal to or better than that of its comparators.

Further clarification of the concept of substantial equivalence originates from a WHO sponsored workshop. It has been concluded that the establishment of substantial equivalence is not a safety assessment per se, but a dynamic, analytical exercise. Thus, it may mean that analysis of, for instance, gene expression patterns, global changes in protein expression and/or differences in metabolic capabilities (the metabolome) of the novel product in comparison with conventionally bred products should be performed in order to examine equivalence.

This strategy has worked satisfactorily for the assessment of the first-generation crop plants for which sufficient background knowledge is available. However, relatively less experience has been gained with the safety and wholesomeness evaluation of novel food plants for which no substantial equivalence can be established. Even in the case of single gene modification, potential alterations in metabolic pathways are difficult to identify and, consequently, the implications of the genetic modification process on the metabolism of plants are poorly understood. Moreover, data on the mechanisms by which plants regulate, for instance, their response to environmental stress factors and other conditions are remarkably scanty.

Critical Nutrients and key Toxicants

Application of the concept of substantial equivalence appears in practice to lead to various interpretations. Different requirements for risk assessment with regards to the potential of unintended effects exist as a result of genetic modification. Actual approaches entail a consideration of the characteristics of host and donor organism, phenotypic properties and toxicological evaluation and include comparative determinations of any changes in critical (anti-)nutrients and key toxicants for the food source in question.

In general, the analysis of an expanded spectrum of components is unnecessary, but should be considered if there is an indication from other traits that there may be an unintended secondary effect of genetic modification. However, compositional analyses based on single parameters using a list of crop-specific critical (anti-)nutrients and key toxicants as a minimum requirement of screening for possible variations in the content has its limitations.

Limited data are typically available for the less important food components, as in less well-known crops most critical (anti-)nutrients and natural toxicants will be unknown. Moreover, the toxicants to be assessed may be partly influenced by the cellular function of the product encoded by the inserted gene. Furthermore, the natural variation of critical comparators may mask the evidence of secondary effects due to genetic modification.

Thus, there is no criterion that determines whether a difference between a GM plant and its natural counterpart is outside the literature values for the ranges of parameters that express natural diversity. Relevant information on indicators of unexpected effects may not have been observed during the development of an edible transgenic crop species. In the future other compounds could become of importance, such as anti-oxidative, oestrogenic and anti-carcinogenic compounds.

In determining critical nutrients and/or key toxicants, differences in consumption patterns and practices of processing and consumption in various geographical settings and cultures must also be recognized. It is our opinion that conclusions about relative safety and wholesomeness based on a list of selected single components may not be equally valid in a global context and in all times. Therefore, there is an urgent need to set up generic approaches using platform technologies to evaluate undesirable metabolic perturbations in GM food crops.

A Post-genome Challenge

In our view the identification of unintended effects encompasses the ability to interpret and use innovations in molecular genetics research such as genomics, proteomics and metabolomics on a crop-by-crop basis. This will establish a whole data package directed towards a holistic view on possible unintended side effects due to genetic modification. In essence, the focus is shifting towards molecular characterization to understand functional activity.

Vital to this approach will be informative profiles of molecules at different integration levels, *e.g.* mRNAs, proteins including post-translational modifications and molecules at the level of primary and secondary plant products and metabolites that might be useful benchmarks for the detection of unintended effects. The relative importance of these various 'profiling' techniques in establishing data sets for assessing unintended effects is not indicated by their order and will vary from species to species.

Alterations in, for example, expression levels, post-transitional modification, interactions or chemical composition do not per se imply that the product is less safe or even unsafe. Such a platform of technologies may determine whether unintended effects exist between the GM plant and its control(s) or whether there is substantial equivalence apart from certain well-defined expected differences.

Measures of the relative activities of various molecular constituents associated with metabolic capabilities in non- and modified plant cells under different conditions will provide new insights into how metabolism and genetic modification are orchestrated. But it should be emphasized that the fields of plant genomics and proteomics are still in their infancy.

THE FORTRESS FOOD ECONOMY AND PRODUCTION

The Nazi administration has managed the food economy of the Continental fortress by devices that are a part of the ordinary machinery of a planned and regimented economy. Large stock piles of all essential commodities accumulated before the war have been carried over in order to meet temporary emergencies and shortages.

Production in Germany as well as in all the occupied territory is directed according to national production goals, is stimulated by generally profitable prices, its progress is constantly surveyed down to the last farm. The goals are fixed annually, and determine whatever changes in acreage of specific crops are to be made and how large the herds of livestock and the output of animal products shall be.

The delivery of all products from the farm to the proper agencies in the market is subject to public control as to time, volume, and destination of delivery. On the demand side of the market a similar system of public planning and controls operates. After ample reserves are set aside, the available annual output of all foodstuffs is allocated to major groups of claimants according to necessities of warfare and priority rank. While the armed forces claim their allotment of food on the basis of established army diets, the requirements of all civilians are tailored to fit available residual supplies. Civilian consumption is either strictly rationed, or, where this is not feasible, consumption is restricted by curtailment of available supplies. The same principle of public control of demand applies to the utilization of food or feed products and to fertilizer as well.

In other words, not only the people, but all the animals and even plants are put on rations. Any sort of utilization of agricultural produce is subject to central controls. In principle this system is airtight. It almost eliminates the element of surprise and uncertainty in consumption, demand, and utilization, while it still must cope with the fluctuation of yields and crops due to weather and with many other biological factors beyond human influence.

In comparison with the strategy used in the American war food administration, the German administration has made maximum use of the price incentive by fixing such differentials that production goals are enforced.

In August, 1944, at the end of five years of world conflagration, all the available evidence convinces the author that the thesis he held and made public from the outset of the war is still valid, and probably will remain so: the Axis cannot be defeated by food shortage, but must be defeated on the field of battle.

This considered opinion, formed on the basis of years of economic analysis and observation, in no way implies that populations everywhere in Europe or even a majority of the people could maintain anything like their habitual peacetime diet, or that agricultural production could be protected against the effects of total war and could continue to produce as in prewar years. What has happened in more than four years of war behind the blockade cannot be condensed into any simple formula. It requires the careful consideration of the many multiple facets of the food situation. While a great amount of information is at hand on many details, nowhere has an attempt been made to compose a more general picture of the European food economy during the war, sufficiently concise and accurate yet free from detail that would swell it to encyclopedic form. Following is an attempt to fill this gap at least tentatively. It is confined by the requirements of our subject, the reconstruction of world agriculture.

As soon as Germany had accomplished the first part of her bid for Continental empire by military conquest, she began to consolidate the accumulated resources, to reorganize their utilization, and to mold them into an integrated system of planning and regimentation.

Indeed, the nazis did not wait for the completion of conquest, but began to set the wheels of reorganization in motion a few days or weeks after the armies had swept through. It was reliably reported that three days after the Germans had begun to occupy the Sudeten area, German soldiers in uniform were seen plowing the fields. In Poland, the farm-management personnel of the army's *WirtschaftsStabe* followed immediately behind the combat forces and began to direct farming operations. There is much evidence that the same pressure to re-establish farming to full activity was exerted in all occupied areas. The methods used differed from country to country. Moreover, each country was assigned a particular place and task within the food economy of the Continental fortress.

The Continental Food Economy

There is no scheme of the Nazi blueprint at hand, but the major outlines of the plan for the organization and the performance of European agriculture can be drawn. A welter of articles in European newspapers and magazines especially from neutral countries, reports by high Nazi officials, and orders and ordinances provide the evidence. Since they had to face the possibility that the conquest might be disputed at any time, either by invasion and air bombardment from without, or by mass revolt and sabotage from within, the nazis conceived their Continental food economy as a system of concentric rings, corresponding to successive rings of military lines of defence.

Agriculturally these rings proceed from the outermost zones, which may or may not yield some supplies beyond those needed solely for their populations, to zones of increasing importance for the import requirements of the Reich, with the Reich itself forming the centre. Naturally these areas are not evaluated exclusively as sources of agricultural products. Their economic rank is decided by the aggregate of all materials and services to be derived from them, as well as by their military importance. Considerations of food requirements for the conquered people are subordinated to those of securing the essential war supplies for the nazi machine.

Thus the inner ring of the fortress consists of the Reich of 1937, Denmark, Holland, Belgium, the French *départements* east of the Maginot Line, including Alsace-Lorraine, Austria, Czechoslovakia, Poland except for Polish western Ukraine and Polish western White Russia, Lithuania, Latvia, and Estonia.

This central pivot embraces a great range of agricultural conditions. It has a decided slope from west to east in productivity of soil and the skill of men on farms, but it has also a concurrent slope from densely settled and highly industrialized areas in the northwestern corner and the west in general, towards more rural and sparsely populated regions in the east. Therefore the eastern belt from Estonia down to Moravia counts heavily as a food surplus producing area, especially for animal products, although the Corridor and Posnan can also export sizable quantities of potatoes and rye. Austria's and Czechoslovakia's industrial resources, located — at least for several years — well out of heavy bomber range, are of such importance to Germany's war potential that they are considered vital parts of the Reich.

If they need food, they will get it, just as any German province would; and if their normally good capacity to produce food can be expanded, they will receive the means with which to expand it.

Belgium and the adjacent corner of France are in the same class. The Brie-Longwy coal basin and the heavy industries in it are an extension of the Ruhr and the Saar districts. For that reason their food deficits will be covered by the German food administration. At the same time, every effort is being made to step up and adjust to war needs the local production of food, particularly since the climate and soil are good, farmers very efficient, and local produce can find its market without involving a great deal of transportation.

Denmark and Holland are the two most productive food provinces Germany has. They rank equally high or higher with the lower Rhineland because of their unequaled efficiency and their professional versatility in agricultural production. It is very deceptive to measure the agricultural plant of these two nations simply on the basis of cultivated acreage, livestock numbers, or similar statistical data; and it is similarly misleading to focus mainly on the peacetime dependence of these countries on overseas imports of grains and oilseeds. The German General Staff's *Wehrwirtschaft* section knows better. It is aware that the average Dutch or Danish cow yields in marketable dairy products at least twice as much as the average cow in Poland, and that therefore the 6,000,000 Dutch and Danish cattle yield for the market far more milk of an incomparably better grade than Poland's 10,000,000 cattle.

The war food economy in this central ring, marked "I" on the map, is so managed that it is much better controlled than all the other rings. In greater Germany, i. e., the old Reich plus AlsaceLorraine, Austria, Sudetenland, and the annexed northern part of Poland, the so-called *"Warthegau,"* Danzig, and Memelland, agriculture is incorporated in the *Reichsnahrstand* and its production and marketing are ruled by commodity cartels. The system of food planning, made foolproof before the war, was simply extended to the annexed areas, and experienced administrators transferred to them. It soon began to perform as well as it had earlier although in many areas it did not succeed in keeping production up to prewar levels.

Denmark, Holland, and Belgium were left more or less to self government in agricultural affairs with advice or instructions issued only from the top. In Denmark and Holland the most competent administrations, the farmers' co-operatives, were not only in a position to perform any agricultural planning job much more skillfully than any German occupation authority, but they were keenly interested in doing so, because co-operation promised to be the far lesser evil. Belgium, on the other hand, was a net importer and was consequently less experienced in agricultural planning, so a new war food administration had to be organized.

In 1940 the Ministry of Agriculture, and the Ministry of Food Supplies founded in September, 1939, were merged. In October, 1940, this new ministry formed a national food cartel on the model of their *Reichsnahrstand.* Like its German forerunner, this cartel, the *Corporation nationale de l'agriculture et de Palialimentation,* is all-embracing, and comprises producers, distributors, and processors. The ministry and the corporation throughout the war have been responsible for production planning, delivery controls, rationing, and planning of exports and imports. Needless to say, the final decision about how much food can be moved into Belgium, how much can be squeezed out of it, rests with the German command.

In the area east of the Maginot Line running from the Channel to the Swiss border, a different system of managing the food economy was established. It is essentially the one applied in the east. The German administration, in view of the absence of any workable part of a national agricultural or food administration for the thousands of abandoned farms, whose occupants either fled or were evacuated, and faced with the Frenchman's suspicion and animosity towards governmental interference in any way whatsoever, put all the vacant land under the management of its eastern farm management corporation, the Ostland G. m. b. H., a corporation under the jurisdiction of the *Reichsnahrstand* in autumn 1940.

This organization put 2.6 million hectares of land under cultivation, using French prisoners of war for labour and large-scale mechanized farming methods. Merging of fields, plowing down old property lines and consolidating the land for tractors and steam plows was to the German managers simply the difference between allowing land to lie idle or using it for producing food. What it did to the French villages and the people who had lived on the good inherited earth was of no concern to the nazis. The German food administration was satisfied if even between the forts and on top of the Maginot Line food was grown once more.

In the eastern part of the central ring, agriculture had to be similarly managed, but it did not present insurmountable obstacles there, either. The "Corridor" and the province of Posen, which had been German up to 1919, were annexed and integrated into the German agricultural fabric. The rest of Poland was organized as the territory under the "Government General" seated in Cracow.

All farms over 5 hectares, 3,900 of them, were taken over by a farm-management office which established a centralized direction and control. All farms under 50 hectares, the Polish peasant farms, were under the authority of chambers of agriculture, with a German commissioner in charge, but otherwise manned by Polish personnel. The marketing of all products was organized on the German cartel system. A central agricultural office became responsible for supplying all the farmers' needs and the delivery and sale of all farm produce. The existing system of agricultural co-operative associations was much expanded. In the Baltic states a similar system was applied, leaving the production in the hands of the family farmers, but controlling their supplies and sales, as well as planning production from central offices under German management.

One major gap in the central ring remained. Many large-scale farms were left without owners, occupants, or managers. They comprised either large estates, whose owners had been killed, interned, or had fled, or groups of farmer-peasant holdings consolidated by the Soviet government as collective farms. Most of these large-scale farm units were turned over to the Ostland Corporation, mentioned in connection with the French belt east of the Maginot Line. This corporation operated no less than 5,000,000 hectares of tillable land in 1941.

TEMPERATURE CONTROL IN CROP PRODUCTION

MICROCLIMATE

Greenhouse (British "glasshouse," often a misnomer in this plastic age) production is the obvious example of microclimate temperature control. But greenhouse production has its own considerable expertise and literature. Thus, the examples of greenhouse research cited here are included only to illustrate specific situations in which individual control of air and soil temperatures is important. Even outdoors, although climate (including temperature) is usually regarded as beyond the control of man, localized temperature control is sometimes effective on a microclimate scale.

Vancouver, Canada, is a few miles north of the 49th parallel, about 60 miles farther north than Minot, North Dakota, with its legendary harsh winter temperatures. But constant foehn winds off the Pacific Ocean make Vancouver winters mild and wet, although sunshine is scant. When I had a garden there in the late 1940s, a neighbor used to say that I "cheated God" to bring in my lettuce and tomatoes earlier than anyone else.

The bed in which the vegetables grew was banked towards the south at approximately 50 degrees, and the area between the plants was covered with flat stones, gathered from the nearby beach, to maximize soil heating from the weak late winter–early spring sun.

This management was, of course, an extreme example of microclimate modification for crop production. Nevertheless, it was no more than ingenious growers have done to survive inhospitable climates throughout the ages, as with pre-Columbian Andean potato growers. In recent years, the native peoples of the Andean Altiplano have learned to revive the methods of their ancestors, growing potatoes on high narrow beds at the foot of mountain slopes. On freezing nights, the cold air settles between the raised beds without damaging the aerial parts of the plants, whose subterranean portions are protected by the latent heat of the water accumulated in the troughs between the beds, an ancient example of sophisticated microclimate control.

Poinsettia is typical of an ornamental grown for a specific date; unless the plants are marketable at Christmas, their value drops dramatically. Growth of the plants can be sharply reduced by too cool air temperatures. Maintaining temperatures in a greenhouse in very cold weather is very expensive.

However, it has been found that raising soil temperature to 23°C (which is much cheaper to do) could counteract the adverse effects of air temperature as low as 11.5°C.

Sometimes the reverse modification is needed. Flowering of *Alstroemeria* (lily-of-the-Incas) was stimulated by cooling the root zone with 10°C circulating water. There was also an interaction with light, supplementary lighting being essential in winter but harmful in spring and summer.

A beneficial lowering of root zone temperature explained an anomalous result with azaleas pot-grown outside on either clamshell mulch or black polyethylene. Placing the pots close together increased growth of azaleas in black pots but not in white pots. The beneficial effect was traced to a decrease in root zone temperatures by shading when the plants in black pots were placed close together.

Another unexpected root zone temperature effect was traced to the chilling effect of cold greenhouse irrigation water in winter. The effect was noted with roses and chrysanthemums and was sufficient to affect turgidity, stomate opening, and flowering. Such unforeseen deleterious temperature effects are particularly easy to overlook when they involve the temperatures of soil rather than air.

Root zone heating usually involves use of expensive fuel. This potential cost was halved in an ingenious system of pumping comparatively warm water from a well 100 m deep and circulating it through buried pipes.

Temperature, of course, affects more than plant growth. It is sometimes necessary to tread a fine line between temperatures optimum for growth and those that initiate or increase fungal attack. This can be a problem for Florida foliage growers in warm weather, as shown in a study of aerial blight (*Rhizoctonia solani*) infection of Boston fern (*Nephrolepis exalta*). Some plant quality had to be sacrificed if potting medium and air temperatures were to be regulated to restrict development of the pathogen.

Given sufficient irrigation water, many deserts will blossom as the rose. But sometimes the desert sun is too hot, with consequent potential for crop damage. An obvious remedy is to spray the crop with an overhead irrigation system.

The cooling effect of such sprinkling is sharply dependent on initial air temperature. A California study reported the following (the results have been converted from Fahrenheit to Celsius):

Macroclimate temperature	Lowered by
32°C	2–3°C
38°C	5°C
39°C	7.5°C

In addition to other benefits, the water spray at 39°C was reported as being successful in reducing excessive "June drop" of small fruitlets. But such spraying of water in extremely hot weather can cause localized injury due to the "lens effect" of standing drops of water on the leaves. Lens effect injury can be avoided, and better temperature reduction obtained, by using nozzles that emit a fine mist instead of streams of water.

Microclimate is being modified on a very large scale. Whole hectares are commonly covered with plastic sheeting, which may be black, white, or transparent. Plastic covering may be spread over raised beds, with the plants inserted through holes in the plastic; it may lie over individual rows secured along the sides, with or without some form of framing; or it may be used as "floating row covers," supported by the crop itself and rising as it grows. Sometimes such plastic covering serves essentially for weed and soil moisture control. Often, however, some degree of temperature elevation is sought, and air and soil temperatures are commonly included in research reports. The elevation of temperature under plastic film will depend on both the climate and the type of plastic. In sunny climates, temperature rise may be sufficient to provide effective disinfestation of pathogenic fungi.

Sunshading

Another form of large-scale microclimate control is by shading. A practice that started as "slat houses" for orchids and "cloth houses" for high-quality tobacco has developed into very considerable industries, usually growing ornamentals. A high proportion are foliage plants, grown under coarse-woven plastic material developed to give certain "percentages" of shade. Obviously, any modification of insolation (irradiance) also modifies temperature. It is remarkable that although research reports commonly pay considerable attention to the expression of the exact degree of shade, temperature differences are often not mentioned. It can be very helpful to include temperature as a variable, as demonstrated in a study of disease intensity under different levels of shade. Research workers in this field are urged to routinely measure and report the temperature variations that inevitably accompany any modification of irradiance. Shade conditions can be expected not only to lower daytime temperatures but also to raise night temperatures, particularly under cold night–clear sky conditions, in which ground-to-sky radiation can cause a very rapid, possibly harmful, drop in temperature near the ground. Similarly, the use of spectral filters can be expected to modify not only light quality but also temperature.

Freeze Protection

The first, most obvious, and least expensive protection against freeze injury is to select a planting site where injurious freezing is unlikely to occur. Because this is often not possible, freeze protection measures may be necessary. Burning fossil fuels should be regarded as a last resort—the fuels themselves are very expensive, and their use is often environmentally questionable. Only too often, freeze protection methods are ineffective because of ignorance of the following basic thermodynamic and meteorological principles.

1. Cold air will roll down a slope until arrested by some physical barrier, which then forms a "frost pocket."
2. Hot air rises vertically. It cannot be made to move up a slope.
3. Radiated heat travels in all directions uniformly but only in line-of-sight (straight) lines. Thus, to be warmed by irradiation from a heat source (such as an orchard heater), a plant must be able to "see" the heat source. Because radiated heat, like all forms of radiation, is subject to the inverse square law (*i.e.*, intensity decreases proportionately to the square of the distance traveled), radiation warming decreases sharply with distance from the heat source.
4. The total heat content of a mass of air depends not only on its temperature (sensible heat) but also on its latent heat, the two together approximating its total energy content or enthalpy. Thus, total heat content can be very much greater for moist air than for dry air at the same temperature. Putting this in a different way: air masses at the same atmospheric pressure and conditions of 15°F (9.4°C) and 100% RH, 20°F (6.7°C) and 40% RH, and 25°F (3.9°C) or 0% RH all have the same heat content of approximately 5.5 Btu per pound of dry air (ca. 3 kg cal kg1).
5. The latent heat of evaporation is approximately 7.5 times as great as the latent heat of freezing. Thus, when spraying irrigation water for freeze protection (a common practice for Florida strawberries and various other crops), it is essential to freeze at least 7.5 times as much water as is evaporated. In calm or near-calm weather, this is no problem. Continuing to spray after the onset of a brisk breeze, however, can be disastrous. An ingenious application of this principle is to use such evaporative cooling to delay the blooming of fruit trees until the danger of a blossom freeze is over. The blooming of apple trees was delayed by as long as 17 days by use of thermostatically controlled sprinkling whenever prebloom temperatures exceeded 7°C (44.6°F).
6. Smoke from burning oil or other fuel does *not* form a protective shield. It used to be believed (particularly in California) that "smudge pots" could create a low cloud that reflected heat back to the crop below. It is now known that the smoke particles are not in a size range suited to reflect infrared emissions. It is, however, possible to generate very fine water fog with droplets of the appropriate size. An added benefit is that any fog droplets that freeze give off latent heat to the surrounding atmosphere.
7. Freezes are classified as "convection freezes" or "advection freezes." Convection freezes occur with calm air and cloudless skies, conditions in which the earth is radiating heat to the sky, with consequent rapid cooling of the air near the ground. For orchard crops, it is beneficial to have bare ground to radiate ground heat to

the trees. Weeds or cover crops trap such radiated heat at the expense of the trees. Convective conditions commonly result in atmospheric inversions, in which the lower air is colder than that at 10–30 m above the ground. In such conditions, "wind machines" mounted on tall towers or pylons can be beneficial. Helicopters have sometimes been used to achieve the same effect, particularly to prevent dangerously cold air from accumulating in the "frost pocket" hollows.

In an advective freeze, a wind strong enough to disrupt normal convection patterns freezes crops on the exposed higher ground, with much less freeze injury in the valleys and lowlands. Wind machines are worse than useless in an advective freeze, but rows of heaters placed at right angles to the wind direction can benefit crops for a considerable distance downwind. A deadly interaction among temperature, humidity, and wind speed can occur in an advective freeze. Tender leaves and shoots can be killed, not by freezing, but by desiccation, if wind speed is high enough when the temperature approaches the freezing point of plant tissues under conditions of very low humidity (which frequently occur).

For further information on methods and principles of freeze protection, readers are referred to an extensive chapter on freeze protection.

Incidental Effects of Temperature

Old Ecclesiastes said, "Of the making of many books there is no end," and a number of them probably could be written on the incidental effects of temperature. However, only a very few examples can be cited here to indicate how often temperature is an unforeseen or unplanned-for variable.

Temperature can move in mysterious ways, its wonders to perform, through its subtle influence on the activity of growth regulators. As noted earlier, fruit setting in tomato plants is inhibited by too high temperatures. A role for growth regulators in this high-temperature inhibition is indicated by a report that relative levels of gibberellin and auxinlike growth regulators were sharply affected at high temperatures.

On a purely physical basis, temperature can be expected to affect gas diffusion rates, hence rates of photosynthesis and leaf respiration. However, not only can the physical effects of temperature be complicated by the metabolic effects of temperature on rates of photosynthesis and respiration, but such gas exchange is reported to be affected by an interaction between temperature and humidity.

Exact control of temperature is routine, but equivalent accuracy in control of humidity can be difficult, and exact simultaneous control of temperature and humidity can be very challenging indeed. Vegetable transplants usually benefit from hardening by controlled temperature and/or moisture stress before being planted out in the field. This does not appear to be the case for sweet potato transplants, which are vine cuttings rather than seedlings.

Transplants held at 13–18°C were reported to have greatly increased vitality and ultimately higher yields compared with transplants held at an ambient temperature of 26.7°C. (That "26.7°C ambient" temperature is curiously exact and is possibly a translation from "ca. 80°F ambient.")

A particularly intriguing example of an unexpected temperature effect is reported in a study of male sterility in the common bean (*Phaseolus vulgaris* L.). When, in the course of an atypically cool summer, unexpected fertility was noted in supposedly male sterile plants, research was transferred to growth chambers.

PROTECTION OF THE CROP PLANT

Protection of the crop plant against disease follows two main principles, prevention and cure. The whole study of protection will be dealt with here, first genera.

FUNGAL PARASITES

Probably no method of protecting a crop against fungal disease is so attractive as the production and growth of an immune variety of host. If a variety completely immune to all diseases could be evolved in each crop, the practice of agriculture would be considerably simplified. This is hardly possible, but varieties showing heriresistance to a specific disease are not only possible but are now well known. "Conferred immunity" acquired by a plant and lasting for the life of that generation only is also known. When considering an immunity of a crop plant to a disease three main points must be kept in mind. In the first place, immunity to one disease does not confer immunity to another. Resistance to each of the diseases to which the plant is heir must each be bred for or developed separately.

Hence, in writing of an immune variety, what is referred to is a variety which will not readily contract some specified complaint. Secondly, complete and total immunity to a single disease in an otherwise susceptible crop probably does not exist.A rather more accurate view is that some individuals or varieties are more resistant or less susceptible to a particular disease than are others.

If, however, the pressure of infection on these resistant forms is increased past a critical point, or if environmental conditions are highly favourable to the parasite or adverse to the host, then the fungus may gain the upper hand. Thirdly, the fact that in disease there are two components, the host and the parasite, must never be lost sight of.

Just as resistant forms of the host may develop due to genetical alterations in it, so too changes of a hericharacter may take place in the parasite which render it more virulent and capable of breaking down a previously existing immunity. Many disease organisms have developed a number of strains or biological races, each showing different degrees of virulence but all morphologically alike.

So a particular crop plant variety may be immune to one strain of a fungus but be quite susceptible to another strain. Very often the strains of different virulence, or different ability to attack, inhabit different districts or countries. Thus a crop plant variety immune to a definite disease in one geographical area may become infected by that disease when transferred to another area.

So, too, the inadvertent transportation of a strain of fungus from its "home"territory to a new area may cause disease of the crops long resident there and immune to the local form of the pathogen. A case of this was seen when the wheat variety "Bobs"was taken from Australia to India and South Africa. Bred in Australia, "Bobs"had proved to be immune To the black rust there. In India and South Africa the immunity broke down completely.

Either the black rust fungus of India and South Africa is of a different biological race from that of Australia, or growth in the new environment had so altered the wheat that the resistance mechanism failed.

CLIMATE CHANGE ON CROP PRODUCTION

At the basis of any understanding of climate impacts on agriculture lies the biophysical sciences. The rates of most biophysical processes are highly dependent on climate variables such as radiation, temperature, and moisture, that vary regionally. For example, rates of plant photosynthesis depend on the amount of photosynthetically active radiation and levels of atmospheric carbon dioxide (C02). Temperature is an important determinant of the rate at which a plant progresses through various phenological stages towards maturity. The accumulation of biomass is constrained by the availability of moisture and nutrients to a growing plant.

Numerous studies have examined the impacts of past climatic variations on agriculture using case studies, statistical analyses and simulation models. Such studies have clearly demonstrated the sensitivity of both temperate and tropical agricultural systems and nations to climatic variations and changes. In the temperate regions, the impacts of climate variability, particularly drought, on yields of grains in North America and the Soviet Union have been of particular concern because of their effects on world food security.

In the tropics, drought impacts on agriculture and resulting food shortages have been widely studied, especially when associated with the failure of the monsoon in Asia or the rains in Sudano-Sahelian Africa. In the temperate regions, climatic variations are associated with economic disruptions; in the tropics, droughts bring famine and widespread social unrest.

THE BIOPHYSICAL IMPACT OF CLIMATIC CHANCES ASSOCIATED WITH GLOBAL WARMING

It is frequently assumed that global change will bring higher temperatures, altered precipitation, and higher levels of atmospheric CO2.

What might these changes mean for the biophysical response of agricultural crops?

Interactions with thermal regimes. Higher temperatures in general hasten plant maturity in annual species, thus shortening the growth stages during which pods, seeds, grains or bolls can absorb photosynthetic products. This is one reason yield are lower in the tropics. Because crop yield depends on both the rate of carbohydrate accumulation and the duration of the filling periods, the economic yields of both temperate and tropical crops grown in a warmer and CO2-enriched environment may not rise substantially above present levels, despite increases in net photosynthesis.

Because temperature and tropical regions differ in both current temperature and the temperature rise predicted for climate change, the relative magnitudes of combined CO2 and temperature effects will likely be different in the different regions. In the mid-latitudes, higher temperatures may shift biological process rates towards optima, and beneficial effects are likely to ensue. Increases in temperature will also lengthen the frost-free season in temperate regions, allowing for longer duration crop varieties to be grown and offering the possibility of growing successive crops (moisture conditions permitting). In tropical locations where increased temperatures may move beyond optima, negative consequences may dominate.

Both the mean and extreme temperatures that crops experience during the growing season will change in both temperate and tropical areas. Extreme temperatures are important because many crops have critical thresholds both above and below which crops are damaged. Prolonged hot spells can be especially damaging. Critical stages for high temperature injury include seedling emergence in most crops, silking and tasseling in corn, grain filling in wheat, and flowering in soybeans. In general, higher temperatures should decrease cold damage and increase heat damage. Agro-climatic zones are expected to shift poleward as lengthening and warming growing seasons allow new or enhanced crop production (soil resources permitting).

Changes in hydrological regimes. The hydrological regimes in which crops grow will surely change with global warming. While all GCMs predict increases in mean global precipitation (because a warmer atmosphere can hold more water vapor), decreases are forecast in some regions and increases are not uniformly distributed. The crop water regime may further be affected by changes in seasonal precipitation, within-season pattern of precipitation, and interannual variation of precipitation. Increased convective rainfall is predicted to occur, particularly in the tropics, caused by stronger convection cells and more moisture in the air.

Too much precipitation can cause disease infestation in crops, while too little can be detrimental to crop yields, especially if dry periods occur during critical development stages. For example, moisture stress during the flowering, pollination, and grain-filling stages is especially harmful to maize, soybean, wheat and sorghum.

The amount and availability of water stored in the soil, a crucial input to crop growth, will be affected by changes in both the precipitation and seasonal and annual evapotranspiration regimes. Some GCMs predict mid-continental drying the Northern Hemisphere and other GCM predictions have been interpreted to suggest that the rise in potential evapotranspiration will exceed that of rainfall resulting in drier regimes throughout the tropics and low to mid-latitudes. Because the soil moisture processes are represented so crudely in the current GCMs, however, it is difficult to associate much certainty with these projections.

Global climate change is likely to exacerbate the demand for irrigation water. Higher temperatures, increased evaporation, and yield decreases contribute to this projection. However, supply of needed irrigation water under climate change in uncertain. Where water supplies are diminishing, such as the Ogallala Aquifer in the United States, extra demand might require that some land be withdrawn from irrigation.

***Physiological effects of* CO_2.** The study of agricultural impacts of trace gas induced climate change is complicated by the fact that increasing atmospheric CO_2 has other effects on crop plants besides its alteration of their climate regime. These are often called "fertilizing" effects, because of their perceived beneficial physiological nature. Specifically, most plants growing in enhanced CO_2 exhibit increased rates of net photosynthesis. The higher photosynthesis rates are then manifested in higher leaf area, dry matter production, and yield for many crops. In several cases, high CO_2 has contributed to upward shifts in temperature optima for photosynthesis and to enhanced growth with higher temperatures; other studies, however, have not shown such benefits.

CO_2 enrichment also tends to close plant stomates, and by doing so, reduces transpiration per unit leaf area while still enhancing photosynthesis. The stomatal conductances of 18 agricultural species have been observed to decrease markedly (by 36%, on average) in an atmosphere enriched by doubled CO_2. However, crop transpiration per ground area may not be reduced commensurately, because decreases in individual leaf conductance tend to be offset by increases in crop leaf area. In any case, higher CO_2 often improves water-use efficiency, defined as the ratio between crop biomass accumulation or yield and the amount of water used in evapotranspiration. Increases in photosynthesis and resistance with higher CO_2 have been shown to occur at less than optimal levels of other environmental variables, such as light, water, and some of the mineral nutrients.

Temperate crops may benefit more from increasing CO_2 than tropical crops. In crop species with the C_3 pathway characteristic of non-tropical plants (*e.g.*, wheat, soybean, cotton) CO_2 enrichment has been shown to decrease photorespiration, the rapid oxidation of recently formed sugars in the light, a process which lowers the efficiency of overall photosynthesis. C_4 crops, which are particularly characteristic of tropical and warm arid regions (*e.g.*, maize, sorghum, and millet), are more efficient photosynthetically under current CO_2 levels than C_3 plants (because they fix CO_2 into malate in their mesophyll cells before delivering it to the RuBP enzyme in the bundle-sheath cells). Because of this CO_2-concentrating and photorespiration-avoiding mechanism, experimental data show that C4 plants are less responsive to CO_2 enrichment.

The physiological effects of high levels of atmospheric CO_2 described above have been observed under controlled experimental conditions. In the open field, however, their magnitude and significance are still largely untested, and their importance relative to the predicted large-scale climatic effects uncertain. Greenhouse and field-chamber environments tend to be much smaller, less variable, and more protected from wind than field conditions. Furthermore, physiological feedback mechanisms such as starch accumulation or lack of sink (that is, growing, storing, or metabolizing tissue) for the products of photosynthesis may limit the extent to which the "fertilizing" CO_2 effects may be realized. Finally, if trace gas emissions continue to grow unchecked, their climate warming effect is projected to continue even up to 2000 ppm, but the beneficial boost to photosynthesis appears to level off at about 400 ppm for C_4 crops and about 800 ppm for C_3 crops.

Soils. Climate change will also have an impact on the soil, a vital element in agricultural ecosystems. Higher air temperatures will cause higher soil temperatures, which should generally increase solution chemical reaction rates and diffusion-controlled reactions. Solubilities of solid and gaseous components may either increase or decrease, but the consequences of these changes may take many years to become significant. Furthermore, higher temperatures will accelerate the decay of soil organic matter, resulting in release of CO_2 to the atmosphere and decrease in carbon/nitrogen ratios, although these two effects should be offset somewhat by the greater root biomass and crop residues resulting from plant responses to higher CO_2.

In temperate countries where crops are already heavily fertilized, there will probably be no major changes in fertilization practices, but alterations in timing and method (*e.g.*, careful adjustment of side-dress applications of nitrogen during vegetative crop growth) are expected with changes in temperature and precipitation regimes. In tropical countries, where fertilization level is not always adequate, the need for fertilization will probably increase. Sea level rise, another predicted effect of global warming, will caused increased flooding, salt-water intrusion, and rising water tables in agricultural soils located near coastlines. This is particularly crucial in tropical countries such as Bangladesh, with large agricultural regions and high rural population located near current sea level.

Pests. Pests are organisms that affect agricultural plants and animals in ways considered unfavourable. They include weeds, and certain insects, arthropods, nematodes, bacteria, fungi, and viruses. Because climate variables (especially temperature, wind and humidity) control the geographic distribution of pests, climate change is likely to alter their ranges. Insects may extend their ranges where warmer winter temperatures allow their over-wintering survival and increase the possible number of generations per season. Pests and diseases from low latitude regions, where they are much more prevalent may be introduced at higher latitudes. As a consequence of pest increase, there may be a substantial rise in the use of agricultural chemicals in both temperate and tropical regions to control them.

THEORETICAL PRODUCTION ECOLOGY

Theoretical production ecology tries to quantitatively study the growth of crops. The plant is treated as a kind of biological factory, which processes light, carbon dioxide, water and nutrients into harvestable parts. Main parameters kept into consideration are temperature, sunlight, standing crop biomass, plant production distribution, nutrient and water supply.

MODELLING

Modelling is essential in theoretical production ecology. Unit of modelling usually is the crop, the assembly of plants per standard surface unit. Analysis results for an individual plant are generalised to the standard surface, *e.g.* the Leaf Area Index is the generalised surface of all crop leaves per surface unit.

Processes

The usual system of describing plant production divides the plant production process into at least five separate processes, which are influenced by several external parameters.

Two cycles of biochemical reactions constitute the basis of plant production, the light reaction and the dark reaction.

- In the light reaction, sunlight photons are absorbed by chloroplasts which split water into an electron, proton and oxygen radical which is recombined with another radical and released as molecular oxygen. The recombination of the electron with the proton yields the energy carriers NADH and ATP. The rate of this reaction often depends on sunlight intensity, leaf area index, leaf angle and amount of chloroplasts per leaf surface unit. The maximum theoretical gross production rate under optimum growth conditions is approximately 250 kg per hectare per day.
- The dark reaction or Calvin cycle ties atmospheric carbon dioxide and uses NADH and ATP to convert it into sucrose. The available NADH and ATP, as well as temperature and carbon dioxide levels determine the rate of this reaction. Together those two reactions are termed photosynthesis. The rate of photosynthesis is determined by the interaction of a number of factors including temperature, light intensity and carbon dioxide.

- The produced carbohydrates are transported to other plant parts, such as storage organs and converted into secondary products, such as amino acids, lipids, cellulose and other chemicals needed by the plant or used for respiration. Lipids, sugars, cellulose and starch can be produced without extra elements. The conversion of carbohydrates into amino acids and nucleic acids requires nitrogen, phosphorus and sulfur. Chlorophyll production requires magnesium, while several enzymes and coenzymes require trace elements. This means, nutrient supply influences this part of the production chain. Water supply is essential for transport, hence limits this too.
- The production centers, *i.e.* the leaves, are sources, the storage organs, growth tips or other destinations for the photosynthetic production are sinks. The lack of sinks can be a limiting factor for production too, as happens *e.g.* in apple orchards where insects or night frost have destroyed the blossoms and the produced assimilates cannot be converted into apples. Biennial and perennial plants employ the stored starch and fats in their storage organs to produce new leafs and shoots the next year.
- The amount of crop biomass and the relative distribution of biomass over leafs, stems, roots and storage organs determines the respiration rate. The amount of biomass in leafs determines the leaf area index, which is important in calculating the gross photosynthetic production.
- extensions to this basic model can include insect and pest damage, intercropping, climatical changes, etc.

Parameters

Important parameters in theoretical production models thus are:

Climate

- Temperature - The temperature determines the speed of respiration and the dark reaction. A high temperature combined with a low intensity of sunlight means a high loss by respiration. A low temperature combined with a high intensity of sunlight means that NADH and ATP heap up but cannot be converted into glucose because the dark reaction cannot process them swiftly enough.
- Light - Light, also called photosynthetic Active Radiation (PAR) is the energy source for green plant growth. PAR powers the light reaction, which provides ATP and NADPH for the conversion of carbon dioxide and water into carbohydrates and molecular oxygen. When temperature, moisture, carbon dioxide and nutrient levels are optimal, light intensity determines maximum production level.

- Carbon dioxide levels - Atmospheric carbon dioxide is the sole carbon source for plants. About half of all proteins in green leaves have the sole purpose of capturing carbon dioxide.

Although CO_2 levels are constant under natural circumstances, CO_2 fertilization is common in greenhouses and is known to increase yields by on average 24%.

C_4 plants like maize and sorghum can achieve a higher yield at high solar radiation intensities, because they prevent the leaking of captured carbon dioxide due of the spatial separation of carbon dioxide capture and carbon dioxide use in the dark reaction. This means that their photorespiration is almost zero. This advantage is sometimes offset by a higher rate of maintenance respiration. In most models for natural crops, carbon dioxide levels are assumed to be constant.

Crop

- Standing crop biomass - Unlimited growth is an exponential process, which means that the amount of biomass determines the production. Because an increased biomass implies higher respiration per surface unit and a limited increase in intercepted light, crop growth is a sigmoid function of crop biomass.
- Plant production distribution - Usually only a fraction of the total plant biomass consists of useful products, *e.g.* the seeds in pulses and cereals, the tubers in potato and cassava, the leafs in sisal and spinach etc. The yield of usable plant portions will increase when the plant allocates more nutrients to this parts, *e.g.* the high-yielding varieties of wheat and rice allocate 40% of their biomass into wheat and rice grains, while the traditional varieties achieve only 20%, thus doubling the effective yield.

Different plant organs have a different respiration rate, *e.g.* a young leaf has a much higher respiration rate than roots, storage tissues or stems do. There is a distinction between "growth respiration" and "maintenance respiration".

Sinks, such as developing fruits, need to be present. They are usually represented by a discrete switch, which is turned on after a certain condition, *e.g.* critical daylength has been met.

Care

- Water supply - Because plants use passive transport to transfer water and nutrients from their roots to the leafs, water supply is essential to growth, even so that water efficiency rates are known for different crops, *e.g.* 5000 for sugar cane, meaning that each kilogram of produced sugar requires up to 5000 liters of water.
- Nutrient supply - Nutrient supply has a twofold effect on plant growth. A limitation in nutrient supply will limit biomass

production as per Liebig's Law of the Minimum. With some crops, several nutrients influence the distribution of plant products in the plants. A nitrogen gift is known to stimulate leaf growth and therefore can work adversely on the yield of crops which are accumulating photosynthesis products in storage organs, such as ripening cereals or fruit-bearing fruit trees.

PHASES IN CROP GROWTH

Theoretical production ecology assumes that the growth of common agricultural crops, such as cereals and tubers, usually consists of four (or five) phases:

- Initial spread - In this phase, the crop does not cover the field yet. The growth of the crop is linearly dependent on leaf area index, which in its turn is linearly dependent on crop biomass. As a result, crop growth in this phase is exponential.
- Total coverage of field - in this phase, growth is assumed to be linearly dependent on incident light and respiration rate, as nearly 100% of all incident light is intercepted. Typically, LAI is above two to three in this phase. This phase of vegetative growth ends when the plant gets a certain environmental or internal signal and starts generative growth (as in cereals and pulses) or the storage phase (as in tubers).
- Allocation to storage organs - in this phase, up to 100% of all production is directed to the storage organs. Generally, the leafs are still intact and as a result, gross primary production stays the same. Prolonging this phase, *e.g.* by careful fertilization, water and pest management results directly in a higher harvest.
- Ripening - in this phase, leafs and other production structures slowly die off. Their carbohydrates and proteins are transported to the storage organs. As a result, the LAI and, hence, the primary production decreases.

EXISTING PLANT PRODUCTION MODELS

Plant production models exist in varying levels of scope (cell, physiological, individual plant, crop, geographical region, global) and of generality: the model can be crop-specific or be more generally applicable. The emphasis will be on crop-level based models as the crop is the main area of interest from an agronomical point of view.

As of 2005, several crop production models are in use. The crop growth model SUCROS has been developed during more than 20 years and is based on earlier models. Its latest revision known dates from 1997. The IRRI and Wageningen University more recently developed the rice growth model ORYZA2000. This model is used for modeling rice growth. Both crop growth models are open source. Other more crop-specific plant growth models exist as well.

SUCROS

SUCROS is programmed in the Fortran computer programming language. The model can and has been applied to a variety of weather regimes and crops. Because the source code of Sucros is open source, the model is open to modifications of users with FORTRAN programming experience. The official maintained version of SUCROS comes into two flavours: SUCROS I, which has non-inhibited unlimited crop growth (which means that only solar radiation and temperature determine growth) and SUCROS II, in which crop growth is limited only by water shortage.

ORYZA 2000

The ORYZA2000 rice growth model has been developed at the IRRI in cooperation with Wageningen University. This model, too, is programmed in FORTRAN. The scope of this model is limited to rice, which is the main food crop for Asia.

Other Models

The United States Department of Agriculture has sponsored a number of applicable crop growth models for various major US crops, such as cotton, soy bean, wheat and rice. Other widely-used models are the precursor of SUCROS (SWATR), CERES, several incarnations of PLANTGRO, SUBSTOR, the FAO-sponsored CROPWAT, AGWATER and the erosion-specific model EPIC. A less mechanistic growth and competition model, called the Conductance Model, has been developed, mainly at Warwick-HRI, Wellesbourne, UK. This model simulates light interception and growth of individual plants based on the lateral expansion of their crown zone areas. Competition between plants is simulated by a set algorithms related to competition for space and resultant light intercept as the canopy closes. Some versions of the model assume overtopping of some species by others. Although the model cannot take account of water or mineral nutrients, it can simulate individual plant growth, variability in growth within plant communities and inter-species competition. This model was written in Matlab.

SEED STORAGE IN CERTAIN FRUIT CROPS

MANGO

Mango seeds are protected by hard stony endocarp. Seed is ex-albuminous having two fleshy cotyledons with testa and tegmen represented by two papery layers. Polyembryony is observed in certain cultivars. Seed exhibits recalcitrant type of storage behaviour wherein seed loses the viability on drying after extraction and undried seeds cannot be preserved long at low temperatures.

Reduction of seed moisture to below 25% in Alphonso and 32% in Totapuri lowered the germination percentage and seedling vigour (Doijode

1990). Patil *et al.* (1986) obtained 40% of germination in Alphonso stones after 90 days of storage at 25°C in polyethylene bags with charcoal powder. It declined from 75 to 12% after 60 days when stored in open while those stored in polyethylene bags at 8°C did not germinate at all.

The moist storage appears to be successful for preservation of recalcitrant seeds. Here seeds are imbibed with either germination inhibitor or fungicides for preventing germination and pathogens during moist storage. Seed treatment with 1% 8-hydroxy quinoline sulphate and storage at 15°C could preserve the seed viability for one year (Doijode 1995).

Citrus

Citrus seeds are either mono- or polyembryonic in nature. They are largely Utilized for raising rootstocks as well as in breeding programme. Lime (Citrus aurantifolia) seeds are small, oval and polyembryonic with white cotyledons. Lemon (Citrus limon) seeds are ovoid and polyembryonic with 10-15% nucellar embryos and white cotyledons.

Sweet orange (Citrus sinensis) seeds are obovoid, polyembryonic with white embryos. Mandarin (Citrus reticulata) seeds are small polyembryonic with green embryos. Pummelo (Citrus grandis) seeds are large ridged yellowish and monoembryonic. Grape fruit (Citrus parodisi) seeds are white, polyembryonic with white cotyledons.

Seeds of many Citrus species are viable for a short period under ambient conditions. The seeds belong to orthodox group and/or show intermediate storage behaviour and require special procedure for preservation. Seed storage in moisture proof containers at low temperature checks the deterioration by reducing metabolism. Seed storage with calcium chloride in polyethylene bags at 8°C resulted in 100% viability in Citrus karna and rusk citrange and 84% in Citrus jambhiri and Citrus limonia after 150 days (Krishna and Shankar 1974).

Acid lime seeds were successfully stored for 10 years at 5 and -20°C. Grape fruit seeds retained viability for 3 months when stored with charcoal at room temperature and for 4 months at 5-8°C (Chacko and Singh 1968). In Poncirus trifoliata and Citrus grandis, germination was the highest after 80 days in seeds stored in polyethylene bags at 4.4°C and 56-58% RH (Mallareddy and Sharma 1983).

Seeds could be stored in fruit itself for short term preservation. Fruits of some lemon cultivars were successfully preserved for 40 days at ambient temperature and for 60 days at 5°C (Doijode 1984). Bajpai *et al.* (1963) also reported that seed storage in whole fruit of kagzi lime and sweet orange is better than in polyethylene bags.

Rambutan

Seeds are difficult to store under normal conditions (Ellis 1984) and at room temperature they lose their viability within a matter of days. They can be stored for a few weeks in moist saw dust or charcoal to which some aril juice has been added at 21-28°C.

Jackfruit

Seeds are large oblong about 3 × 2 cm in size having thick gelatinous yellow covering and belong to recalcitrant group. Seeds remained viable for 15 days and gave 40% germination after 30 days of storage.

Litchi

Litchi seeds are dark brown, covered with white fleshy, juicy and translucent aril. Seeds are recalcitrant and should not be allowed to dry. When sown immediately after extraction, their germination was 100% which reduced just after one day storage in water and it was 75% on the seventh day and 30% on 15th day after storage. Seeds stored in moisture proof containers sprouted within 15 days and those stored for 10 days in polyethylene bags showed 50.7% germination. However, seeds can be preserved in fruit for short periods. Seed viability was preserved for 24 days by storing fruits in sealed polyethylene bags after treatment with benomyl (0.05%) and 6% wax emulsion.

Mangosteen

Seeds are recalcitrant and lose their viability rather quickly, especially when the thin membrane around the seed is damaged or when seed is placed in ambient or cool atmosphere. Storage of seeds in charcoal or moss at room temperature retains the viability for a few months.

Conclusions

Seeds of several tropical fruit crops show recalcitrant storage behaviour and they do not withstand drying or are able to keep alive at low temperatures. Thus, they are difficult to store for longer period. In this context, more research is needed for extending storage life. However, it is desirable to preserve the germplasm in field as fruit repository. To supplement the conservation process, wherever, as far as possible seeds could be conserved by use of available techniques like moist storage, etc., for certain period.

7

Process of Plant Breeding

Plant Breeding encompasses a broad array of activities. It can be viewed as a pipeline beginning with initial steps such as pre-breeding, etc. and continuing to development of improved germplasm to be used either for further breeding at other venues, for screening in national or other programmes, or for direct use. "Plant Bredding" has been used to describe many types of laboratory activities. Some involve primarily tissue culture techniques such as virus elimination, micropropagation, regeneration from explants, protoplast fusion, etc. Others involve molecular biology. Two major areas of plant molecular breeding are analysis of the nucleic acids or proteins in plants/ pathogens and production of transgenic plants. Management and retrieval of the data obtained from nucleic acid/protein analysis (bioinformatics) is an increasingly important additional area.

TRANSGENIC APPROACHES

Gene transfer may permit development of improved lines when no sources of the desired traits are available in sexually compatible plants. Possible examples include weevil resistance in sweet potato and bacterial wilt resistance in potato. In the former case, sweet potato plants expressing a soybean proteinase inhibitor are already available for testing. Transgenes for bacterial wilt resistance have not yet been identified although various constructs are being tested. For most traits of interest, no suitable transgenes are currently available.

Gene Transfer

Gene transfer also has the potential to be more efficient than traditional breeding for introduction of an important trait, such as virus resistance, into an otherwise well adapted line Obtained by conventional techniques. This may be true even when transfer of resistance via sexual crosses is feasible, particularly for polyploid clonal crops like potato and sweet potato. In this sense, gene transfer can be considered a substitution of a new technology for Conventional breeding.

Thus transformation with suitable transgenes may be a valuable addition to breeding Methodology. It is definitely appropriate to maintain and improve the capacity to transform the potato/sweet potato lines suitable for the mandate areas. This includes work on improving transformation efficiency in the target lines. There are already many types of promoter has been constructed (e.g. tissue-specific) for use in transformation and is working on selectable marker systems that do not involve antibiotic resistance.

Transgenic sweet potato expressing a soybean proteinase inhibitor gene has been produced and will soon be tested for weevil resistance. Ten potato lines with *Bacillus thuringiensis*-mediated resistance to potato tuber moth are also in hand, and potatoes with other potentially useful transgenes are in various stages of pre-field testing. In fact, additional resources will be needed for the technology transfer step, even for the transgenic materials already in hand. For example, it will be important to do food safety tests of the transgenic sweet potato tubers. Other biosafety issues and consumer concerns will also need to be addressed. The current regulatory and public acceptance climate has become much more difficult since the initial days of transgenic research. The intensity of the debate may be greater in the developed world, but developing countries are unlikely to escape it.

Thus initial deployment of transgenic plants will require additional resources, although over time these costs should decrease. The added resources required for deployment of transgenic plants should not come from the conventional breeding programme, which already appears to be stretched very thin. The breeding programme cannot be expected to support itself primarily on outside grants and should have the first call on core funds to maintain at least minimum capacity. Field breeding capacity is needed not only to develop improved materials by conventional techniques, but also to identify transgenic materials that are at least as good as the starting line and also to produce the mapping populations for QTL (Quantitative Trait Loci) studies.

Molecular Markers

Identification of DNA regions associated with phenotypes of interest is an increasing area of research in many crops, including potato and sweet potato. Both phenotypes controlled by single "major" genes and those controlled by multiple genes with quantitative effects (quantitative trait loci or QTLs) have been targeted. The research programme is working both on finding "anonymous" QTLs for resistance and on seeing whether various "candidate genes" known to be related to disease resistance in other systems are correlated resistance.

Some progress has been made on identifying QTLs in a diploid population, although the correlation is seen only with the field response to infection, not with greenhouse or detached leaf assays. Some candidate genes also appear promising. Molecular marker technology has several attractive applications, provided the analytical methods used are convenient, reliable, and inexpensive.

Such applications include the following:

- Characterization of parental lines to allow better informed breeding strategies. For example, it may be possible to determine whether different parental materials carry the same or different resistance alleles and to select those that would contribute the greatest diversity.
- Marker-assisted selection (MAS) has raised many possibilities and hopes, as noted in several of the responses to the questionnaire. A major attraction is the ability to screen seedling populations for traits like pathogen resistance thereby avoiding more laborious later-stage assays. However MAS still faces some serious problems.
  It has not yet been implemented in breeding programmes in any crop (with a few possible exceptions such as those involving major genes in soybean). It will be more difficult to implement in potato and sweet potato than in diploid crops.An economic analysis of the potential efficiencies to be gained by MAS would therefore be appropriate.
- QTLs that are tightly linked to traits of interest might serve as transgenes for direct improvement of otherwise well adapted lines, rather than simply as ways to identify progeny carrying the markers. If known candidate genes are involved, the gene transfer approach would be relatively simple. If cloning of QTLs is required,

Genomics

Genomic/proteomic approaches, which permit simultaneous examination of all the genes/proteins expressed by a plant, are expected to increase understanding of plant metabolism and to result in new strategies for crop improvement. Extensive genomic information for example Solanaceous crops is currently being accumulated in various programmes. Synteny of *Arabidopsis* (soon to be fully sequenced) and tomato is also being examined.

Bioinformatics

- Efficient bioinformatics approaches for accessing nucleic acid and protein sequence data may reduce biotechnology lab costs in the long-run.

Mutation Breeding

In spite of mutational dissection of biological systems, geneticists use mutation in other ways. The aim of breeding is toprepare a better crop plant to make a hybrid and then to select the desired recombinants from the progeny generations. That approach makes use of the variation naturally found between available stocks or isolates from nature. Another way to generate variability for selection is to treat with a mutagen. In this way, the variability is increased through human intervention.

A variety of procedures may be used. Pollen may be mutagenized and then used in pollination. Dominant mutations will be expressed in the next generation, and further generations of selfing reveal recessives. Alternatively, seeds may be mutagenized. A cell in the enclosed embryo of a seed may become mutant, and then it may become part of germinal tissue or somatic tissue. If the mutation is in somatic tissue, any dominant mutations will show up in the plant derived from that seed, but this generation will be the end of the road for such mutations. Germinal mutations will show up in later generations, where they can be selected as appropriate.

Chromosome Mechanics in Plant Breeding

What we know about chromosome mutation can be used in the type of genetic engineering that produces and maintains new crop types in our hungry world. The classic example, performed by E. R. Sears in the 1950s, concerns the transfer of a gene for leaf-rust resistance from a wild grass, *Aegilops umbellulata,* to bread wheat, which is highly susceptible to this disease, to offset a major problem in the wheat industry

The first problem that Sears encountered was that these two species are not interfertile, so the feat of gene transfer seemed impossible. Sears sidestepped this problem with a bridging cross, in which he crossed *A. umbellulata* to a wild relative of bread wheat called *emmer, Triticum dicoccoides.* (Follow the process in Figure below) *A.*

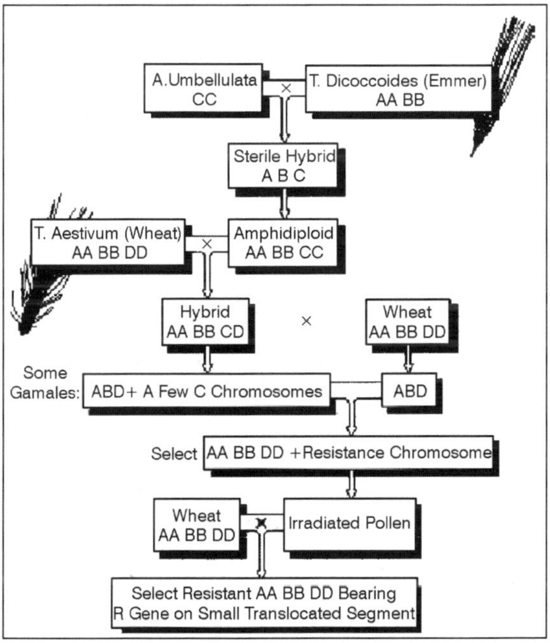

Fig. Summary of Sears's programme for transferring rust resistance from *Aegilops* to wheat. A, B, C, and D represent chromosome sets of diverse origin. *R* represents the genetic determinant (single gene?) of resistance.

Umbellulata is a diploid, $2n = 2x = 14$. We will call its chromosome sets CC. *T. dicoccoides* is a tetraploid, $2n = 4x = 28$, with sets AA BB. From this cross, the resulting sterile hybrid A B C was doubled into a fertile amphidiploid AA BB CC with 42 chromosomes.

This amphidiploid was fertile in crosses with bread wheat *(T. aestivum)*, which is represented as $2n = 6x = 42$, AA BB DD. The offspring, AA BB CD, were almost completely sterile owing to pairing irregularities between the C and D sets. However, backcrosses to wheat did produce a few rare seeds, some of which grew into plants that were resistant to leaf rust, like the wild parent in the first cross.

Some of these plants were almost the desired type, having 43 chromosomes (42 of which were wheat, plus one *Aegilops* chromosome bearing the resistance gene). Thus, in the AA BB CD hybrid, some aberrant form of chromosome assortment had produced a gamete with 22 chromosomes: A B D plus one from the C group.

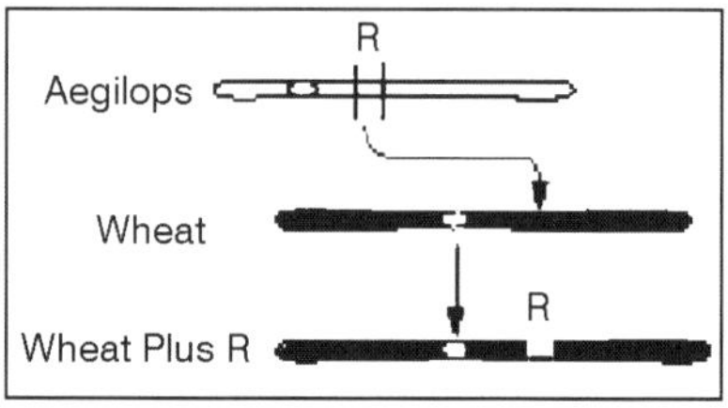

Fig. Translocation of *Aegilops R* Segment to Wheat by Using Radiation as a Means of Breaking the Chromosomes.

Unfortunately, the extra chromosome carried just too many undesirable *Aegilops* genes along with the good one, and the plants were weedy and low producers. So the *Aegilops* gene linkage had to be broken. Sears did so by using irradiated pollen from these plants to pollinate wheat. He was looking for translocations of part of the *Aegilops* chromosome tacked onto the wheat chromosomes. These translocations were quite common, but only one turned out to be ideal—a very small, insertional, unidirectional translocation (Figure above). When bred to homozygosity, the resistant plants were indistinguishable from wheat.

The study of chromosome mutation is of great importance—not only in pure biology, where there are ramifications in many areas (especially in evolution), but also in applied biology. The application of such studies is especially important in agriculture and medicine. Rather curiously, very little is known about the mechanisms of chromosome mutation, especially at the molecular level.

AIM OF PLANT BREEDING

The constant aim of plant breeding is to improve the quality, diversity and performance of agricultural and horticultural crops. The overriding objective is to develop plants better adapted to human needs. The evolution of plant breeding is a classic example of how improved biological understanding has been adapted to provide more effective methods of meeting the demands of a changing world. Plant breeding has existed in its most primitive form since the first farmers saved the seeds of their best plants from one season to the next more than 10,000 years ago. Over the centuries, this selection process has gradually become more scientific, bringing major improvements in the yield, quality and diversity of crops grown worldwide.

In the 19th century Gregor Mendel established the basic principles of plant genetics. Gregor Mendel in the 19th century laid the foundations for traditional plant breeding, in which selected parents are cross-pollinated to combine certain desired characteristics, such as high yield and resistance to pests and disease. He discovered that units of material, which are transferred from one generation to the next, determine inherited traits.

The plant breeder's aim is to reassemble these units of inheritance, known as genes, to produce crops with improved characteristics. In practice, this is a complex and time-consuming process. Each plant contains many thousands of genes, and the plant breeder is seeking to combine a range of desirable traits in one plant to produce a successful variety. Conventional breeding involves crossing selected parent plants, chosen because they have desirable characteristics such as high yield or disease resistance. The breeder's skill lies in selecting the best plants from the many and varied offspring. These are grown on and tested in subsequent years. Typically this involves examining thousands of individual plants for different characteristics ranging from agronomic performance to end-use quality.

Developing a new variety can take up to 15 years for wheat, 18 years for potatoes, even longer for some crops. The scope of conventional plant breeding has increased with improvements in technology. In the laboratory, chemical and mechanical techniques are used to speed up the selection process and remove natural barriers to cross-fertilization, for example between different crop species.

All forms of plant breeding involve the improvement of crops by selection. Since man began farming he has selected seeds from the best plants for the next generation. Over the centuries, this selection process has become more scientific, bringing major improvements in the yield, quality and diversity of agricultural and horticultural crops. Genes-units of hereditary material that are transferred from one generation to the next, determine these characteristics. Since the plant contains many thousands of genes, and the breeder is seeking to combine a range of traits in one plant, developing a successful variety can be an extremely lengthy process-up to 12 years in the case of cereals.

Using biological knowledge derived from cross-pollination, plant breeders have developed physical and chemically assisted ways of enhancing the speed, accuracy and scope of the selection process. Breeding systems involving grafting and hybridization, for example, have extended breeders' capabilities at the whole crop level, while more recent cell-based techniques enable breeders to operate at the level of individual cells and their chromosomes

A major objective in modern plant breeding is the making of crop varieties with the highest possible yield potential. Yield potential is defined as "the yield of a crop when growth is not limited by water or nutrients, pests, diseases, or weeds". For farmers, whose crops are indeed limited by such constraints, this yield concept may not be seen as the most relevant objective. Relieving crops of all sorts of environmental stress, leaves light and temperature, together with varietal characteristics, as the only determinants of yield. That achieved, the same high yielding variety can be used all over a climatic zone, and target area for the variety is enormous. If stress can not be eliminated, however, adaptation to a usually site-specific environment becomes necessary. In that case, the target area for each variety will be small.

Do these different perspectives warrant different approaches to plant breeding? Conventional wisdom says no. In experiments where varieties are tested at various levels of inputs, the same high yielding variety usually comes out as top yielder at all levels. Therefore, plant breeders often claim that their varieties not only perform excellently under high-input conditions, but would also be better than traditional varieties in a low-input environment. However, when these trials are taken outside of experimental farms and the varieties are tested under local or farm conditions, researchers discover what is called 'crossover in performance'. At a certain level of stress there is a crossover point beyond which local varieties or landraces perform better than the high yielding varieties Farmers who experience such situations are not a tiny minority.

To some degree, most if not all farmers have to cope with local stress conditions. The stress-free environment is hard to achieve, and even harder to sustain. In many favourable areas, farmers abandon the high-input technology for economic reasons or because of ecological problems, thus increasing the need for locally-adapted germplasm. But modern plant breeding, whether public or private, can not supply adapted germplasm everywhere. Only a system of local seed selection can ensure that. And that means devolution of plant breeding.

Can such decentralized breeding be compatible with development needs in a changing world and meet economic aspirations in a poor society? If the answer to these questions is going to be 'yes', the decentralized breeding must be able to take advantage of the power of science as well as of the capacities of local communities. Three aspects should be considered and made mutually compatible: breeding technology, participatory research methods, and organization at community level.

History

Plant breeding is the purposeful manipulation of plant species in order to create desired genotypes and phenotypes for specific purposes. This manipulation involves either controlled pollination, genetic engineering, or both, followed by artificial selection of progeny. *Plant breeding* often, but not always, leads to plant domestication.

Plant breeding has been practiced for thousands of years, since near the beginning of human civilization. It is now practiced worldwide by government institutions and commercial enterprises. International development agencies believe that breeding new crops is important for ensuring food security and developing practices of sustainable agriculture through the development of crops suitable for their environment.

Domestication

Domestication of plants is an artificial selection process conducted by humans to produce plants that have fewer undesireable traits of wild plants, and which renders them dependent on artificial (usually enhanced) environments for their continued existence. The practice is estimated to date back 9,000-11,000 years. Many crops in present day cultivation are the result of domestication in ancient times, about 5,000 years ago in the Old World and 3,000 years ago in the New World. In the Neolithic period, domestication took a minimum of 1,000 years and a maximum of 7,000 years. Today, all of our principal food crops come from domesticated varieties.

A cultivated crop species that has evolved from wild populations due to selective pressures from traditional farmers is called a landrace. Landraces, which can be the result of natural forces or domestication, are plants (or animals) that are ideally suited to a particular region or environment. An example are the landraces of rice, *Oryza sativa* subspecies *indica,* which was developed in South Asia, and *Oryza sativa* subspecies *japonica,* which was developed in China.

Classical plant breeding uses deliberate interbreeding (crossing) of closely or distantly related species to produce new crops with desirable properties. Plants are crossed to introduce traits/genes from one species into a new genetic background.

For example, a mildew resistant pea may be crossed with a high-yielding but susceptible pea, the goal of the cross being to introduce mildew resistance without losing the high-yield characteristics. Progeny from the cross would then be crossed with the high-yielding parent to ensure that the progeny were most like the high-yielding parent, (backcrossing), the progeny from that cross would be tested for yield and mildew resistance and high-yielding resistant plants would be further developed. Plants may also be crossed with themselves to produce inbred varieties for breeding.

Classical breeding relies on homologous recombination of two genomes to generate genetic diversity. The Classical plant breeder may also makes use of a number of *in vitro* techniques such as protoplas fusion, embryo rescue or mutagenisis to generate diversity and produce plants that would not exist in nature.

Traits that breeders' have tried to incorporate into crop plants in the last 100 years include:

- Increased quality and yield of the crop
- Increased tolerance of environmental pressures (salinity, extreme temperature, drought)
- Resistance to viruses, fungi and bacteria
- Increased tolerance to insect pests
- Increased tolerance of herbicides

Before World War II

Intraspecific hybridization within a plant species was demonstrated by Charles Darwin and Gregor Mendel, and was further developed by geneticists and plant breeders. In the early 20th century, plant breeders realised that Mendel's findings on the non-random nature of inheritance could be applied to seedling populations produced through deliberate pollinations to predict the frequencies of different types. In 1908, George Harrison Shull described heterosis, also known as hybrid vigour. Heterosis describes the tendency of the progeny of a specific cross to outperform both parents. The detection of the usefulness of heterosis for plant breeding has lead to the development of inbred lines that reveal a heterotic yield advantage when they are crossed. Maize was the first species where heterosis was widely used to produce hybrids.

By the 1920s, statistical methods were developed to analyse gene action and distinguish heritable variation from variation caused by environment. In 1933, another important breeding technique, cytoplasmic male sterility (CMS), developed in maize, was described by Marcus Morton Rhoades. CMS is a maternally inherited trait that makes the plant produce sterile pollen, enabling the production of hybrids and removing the need for detasseling maize plants.

These early breeding techniques resulted in large yield increase in the United States in the early 20th century. Similar yield increases were not produced elsewhere until after World War II, the Green Revolution increased crop production in the developing world in the 1960s.

After World War II

Following World War II a number of techniques were developed that allowed plant breeders to hybridize distantly related species, and artificially induce genetic diversity. When distantly related species are crossed, plant breeders make use of a number of plant tissue culture techniques to produce progeny from other wise fruitless mating. Interspecific and intergeneric hybrids are produced from a cross of related species or genera that do not normally sexually reproduce with each other. These crosses are referred to as *Wide crosses*. The cereal triticale is a wheat and rye hybrid. The first generation created from the cross was sterile, so the cell division inhibitor colchicine was used to double the number of chromosomes in the cell. Cells with an uneven number of chromosomes are sterile.

Failure to produce a hybrid may be due to pre- or post-fertilization incompatibility. If fertilization is possible between two species or genera, the hybrid embryo may abort before maturation. If this does occur the embryo resulting from an interspecific or intergeneric cross can sometimes be rescued and cultured to produce a whole plant. Such a method is referred to as *Embryo Rescue*. This technique has been used to produce new rice for Africa, an interspecific cross of Asian rice *(Oryza sativa)* and African rice *(Oryza glaberrima)*. Hybrids may also be produced by a technique called protoplast fusion. In this case protoplasts are fused, usually in an electric field. Viable recombinants can be regenerated in culture.

Chemical mutagens like EMS and DMSO, radiation and transposons are used to generate mutants with desirable traits to be bred with other cultivars. Classical plant breeders also generate genetic diversity within a species by exploiting a process called somaclonal variation, which occurs in plants produced from tissue culture, particularly plants derived from callus. Induced polyploidy, and the addition or removal of chromosomes using a technique called chromosome engineering may also be used.

When a desirable trait has been bred into a species, a number of crosses to the favoured parent are made to make the new plant as similar as the parent as possible. Returning to the example of the mildew resistant pea being crossed with a high-yielding but susceptible pea, to make the mildew resistant progeny of the cross most like the high-yielding parent, the progeny will be crossed back to that parent for several generations. This process removes most of the genetic contribution of the mildew resistant parent. Classical breeding is therefore a cyclical process.

It should be noted that with classical breeding techniques, the breeder does not know exactly what genes have been introduced to the new cultivars. Some scientists therefore argue that plants produced by classical breeding methods should undergo the same safety testing regime as genetically modified plants. There have been instances where plants bred using classical techniques have been unsuitable for human consumption, for example the poison solanine was accidentally re-introduced into varieties of potato though plant breeding.

Exploiting heterogeneity and crop evolution in farmers' fields are outside the scope of most plant breeding research. One exceptional experiment, however, has shed some scientific light on the issue. It was started at the University of California (UC) in 1928. Composite cross populations of barley were produced, some of which were extremely diverse in origin of sources. These populations were exposed to continuous natural selection in current modern farming environments and became the subject of studies during the career span of several generations of UC professors.

It appears that after low yields in initial years, the composite cross populations gradually improved in performance and eventually became quite good yielders, with excellent yield stability and disease resistance. These results inspired Suneson to propose an evolutionary plant breeding method. After assessment of later generations of the same material, Soliman and Allard concluded that such evolutionary breeding "is unwarranted" if yield potential is the major goal. However, if disease resistance and yield stability are two major objectives, "the composite cross approach is an efficient method". This amounts to saying that a major part of world agriculture, many high-input systems included, could be well served by this approach.

In short, this means constructing a body of broadly diversified germplasm and exposing it to natural selection in areas of contemplated use. For those who are familiar with traditional farmers' breeding, this may sound like reinventing the wheel. In fact it is an improvement of the old wheel of plant breeding.

The first step, constructing a body of of broadly diversified germplasm, is not all that straightforward. Science has access to world collections and information sources that are unavailable to farmers. A research institute can chose relevant germplasm and make composite cross populations with an evolutionary potential, most probably far beyond that of locally available varieties.

The immediate outcome, the early generation composite cross population, will be unadapted everywhere and is likely to yield poorly. With time, however, recombinations and natural sorting will improve the adaptation, and, according to the Californian experience, narrow the gap with commercial varieties. The long term outcome could be populations that outperform commercial varieties in disease resistance and yield stability and that may be used as a source of artificial selection for high yield.

The disease resistance appears to have evolved through the building up of polygenic complexes. Therefore, it provides a durable resistance as opposed to the monogenic and, therefore, mostly non-durable resistance usually bred into commercial pure line varieties. The stability, at least to some degree, depends on the buffering effect of crop heterogeneity. The Californian experiment shows that when the population is propagated in isolation for a very long time, diversity will start declining, resulting eventually in reduced stability.

These experimental findings into the context of current development needs, a few conclusions can be drawn. We need breeding populations with a very high evolutionary potential, and these populations must be exposed to the stress conditions of, or similar to, current farm environments. Furthermore, a certain level of diversity within populations must be maintained in order to sustain evolutionary potential and yield stability.

An Age-old Tradition

Modern plant breeding is a sophisticated, high investment business, but its origins stretch back thousands of years to primitive farmers who selected the best plants in one year to provide seed for their next crop. This selective breeding was the first human refinement of natural plant evolution. Recent scientific and technological developments have allowed a greater rate of improvement. It was Gregor Mendel who, in the mid- 19th century, first provided a scientific explanation of genetic inheritance. His conclusions on the relationship between inherited characteristics in the offspring and the genetic makeup of the parents were the theoretical basis for classical plant breeding. Mendel's work went largely unrecognized in his own lifetime, and it was not until the early 20th century that it was rediscovered to become the basis of modern scientific plant breeding.

A Modern Industry

Until the early 1960s, plant breeding in Britain was largely confined to publicly funded research. This situation changed dramatically in the mid-1960s, with the passing into UK law of the 1964 Plant Varieties and Seeds Act. This legislation introduced a system of royalty payments on individual plant varieties, known as Plant Breeders' Rights, and triggered a rapid expansion of plant breeding as a commercial enterprise in its own right. Today, much of the basic research into crop science is still conducted by public sector research organisations, but the majority of commercial plant breeding takes place within the private sector. Some

60 plant breeding companies, based in the UK, are active across the entire spectrum of plant species from the major arable crops through to ornamental garden shrubs and flowers. In total, the plant breeding sector employs around 5,000 people, and supports a further 5,000 jobs in seed production and distribution. Plant breeding remains a vital industry to keep Britain competitive on world markets. The need for new varieties, adapted to our unique growing conditions, is never ending, driven by the challenges of new disease pressures, changing market requirements and shifts in agricultural and environmental policies.

- *Maize*: Maize crops used for grain and animal fodder are derived from wild races originating in Central America
- *Sugar beet*: Modern varieties of sugar beet have been developed from wild ancestors native to Central Europe.
- *Potatoes*: Wild ancestors of the modern potato still grow in parts of South America.
- *Wheat and Barley*: Small grain cereals, the mainstay of UK crop production, are derived from wild grasses of the Middle East.
- *Oilseed Rape*: Like many crops within the brassica family, oilseed rape has its origins in wild species native to China.

Participatory Plant Breeding

Certain trends have made the world ripe for adoption of participatory plant breeding methods.

- The shielding of crops from environmental stress in high-input systems is facing increasing economic and ecological problems.

Scientists are changing their attitudes, and a new paradigm is being formulated. Instead of modifying the environment to suit the requirement of high yielding varieties, the varieties need to be modified to suit the environment.

- The claim that modern varieties can be made broadly adapted and be superior across most farming environments within an ecogeographic region is being challenged.
- If relevant diversity exists in a locale, the combined action of natural and artificial selection within a local environment may be an efficient breeding method. Experiments show that this may work also in a fertilizer-intensive system.
- In recent years farmer groups working with local seeds have been organized all over the world. They are not primarily conservers of old seeds. They want their seeds to be improved, in an evolutionary, slow and steady way, and under their own control.
- Finally, participatory methods have been developed in order to facilitate the involvement of farmers together with scientists as active and equal partners in research to generate relevant farm technology. Such methods can be applied also to plant breeding.

Commercial seeds often diffuse into areas where the traditional seed supply system is still predominant. Farmers try them with an open mind and adopt or reject them according to their own criteria. If grown and multiplied in the villages, diversity will start to appear within them and local reselection will be possible. In that way commercial varieties eventually might become like landraces. It is also commonly observed that farmers change and exchange seeds. A traditional farming system rarely functions as an environment for the static preservation of old landraces.

Often practices range from neglect to very simple mass selection, but with a few scattered individuals who devote an exceptional amount of effort to the maintenance or improvement of seed quality. These exceptional persons, very often women, may be the source of good seeds for others in the community. Once such a community is organized for seed management and improvement, it becomes possible for it to establish links to scientific institutions.

Community resources for participatory approaches to seed management and breeding are not limited to indigenous culture and traditional practices. The educational status and experiences of modern farmers may also be turned into a resource for community action. In the Philippine group, a number of the members were high school graduates and a few had a university degree. And moreover, most of them had a couple of decades' experience with modern input-intensive farming. Seed activities opened their minds towards the traditional societies, towards themselves and towards the modern world. The traditional societies supplied them with their seeds, and through the seeds, they learned to appreciate the values and achievements of these societies. They discovered their own potential, and saw that the outside world could bring more than technology packets: it could bring knowledge and ideas to be exploited and further developed by themselves.

GENERAL BREEDING TECHNIQUES

EQUIPMENT

The equipment required for plant breeding is relatively inexpensive and easy to use.

Here are some items that you may find useful:

- Magnifying glass (10 or 15 power)
- Tweezers
- Small sharp-pointed scissors
- Camel-hair brush
- Small containers or vials
- Alcohol
- Rubber bands or soft wire
- Paper or cellophane bags
- Paper clips
- Tags
- Notebook

Prepollination steps generally should begin just before the flower opens. If you wait until after it opens, it may be pollinated by natural means, which would make the flower useless for experimentation. Since most plants bloom over a period of time, even if some of the flowers on a plant have bloomed before you have prepared them for breeding, there will probably be others that haven't opened yet and that you can still experiment with.

Extremely high temperatures or moist conditions are harmful to pollen. For best results you should pollinate plants on dry days during the cool hours of the morning or as soon as the anthers have split open.

SELECTING THE PARENTS

Some plants have natural barriers to cross- or self-pollination. It is advisable to check for this before you begin breeding, for although you frequently can overcome these barriers, some plants cannot be artificially pollinated. An example of a barrier that can be overcome is the natural cross-pollination prohibitor of snapdragons; snapdragon flowers are constructed so as to prevent entry of wind-borne pollen from other flowers. It is possible, however, to open the flower by hand. An example of a barrier that cannot be overcome is the self-pollination prohibitor of some orchids; the stigmas of certain orchids produce a substance which kills the pollen of flowers of the same plant. The mechanism that performs this cannot be removed without destroying the pistil.

The plants you select for breeding should be sturdy and healthy. It is usually easier to tell which ones are healthy after a few flowers on the plant have bloomed.

In choosing a pollen parent, select one that has a heavy yellow powder on the anther. This powder is the pollen. If you brush your fingernail against the anther, a trace of pollen should adhere to your nail. A fresh flower is more likely to have healthier pollen than one that has started to wilt or dry out.

In choosing a seed parent, examine the stigma. It should have either a glistening substance on it that is sticky to the touch or a "hairy" surface. It is this substance or surface that retains the pollen, thus making fertilization possible.

Once you have selected the pollen and seed parents, you are ready to begin pollination.

PREPOLLINATION STEPS

The first step is to mark those flowers that are to serve as pollen parents and those that are to serve as seed parents. This can be done with coloured thread, one colour for the male and another colour for the female. Or you can use paper labels, covered with varnish to protect them from the weather. Some plant breeders use bands designed for marking chickens.

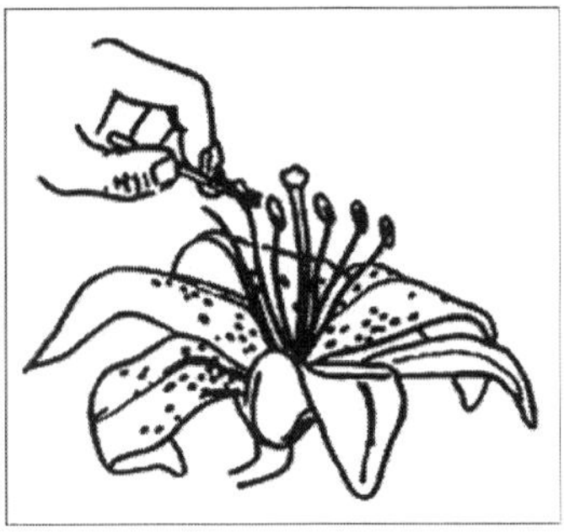

Fig. Emasculation is Necessary to Prevent Self-pollination. Remove all Anthers or Stamens.

Your next step is to protect the plant from unwanted pollen. If the plant is to be cross-pollinated, the stamens will have to be removed to prevent the possibility of selfing. The removal of the stamen is called emasculation. It should be performed before the anthers split open to release pollen. This may require opening the flower by hand before it is ready to bloom.

Emasculation can be accomplished by:

- Pinching off the stamens or anthers with tweezers (Fig.), or
- Snipping off the stamens or anthers with sharp-pointed scissors, or
- Removing the petals to which the stamens are sometimes attached.

A magnifying glass will be particularly useful in emasculation. Both the seed and pollen parents should be protected from contamination by foreign pollen. This can be done by one of the following methods:

Closing the Flower

In many flowers, such as morning-glories, petunias, and lilies, the petals can be closed around the floral organs with a piece of soft wire, string, or rubber band (Fig.) Care should be taken not to tear the petals.

Covering the Flower

Some flowers, such as composite flowers, cannot be closed. To protect them from unwanted pollen, you can cover the flower with a paper bag. Or, if you wish to observe the flower at all times, you can cover it with a cellophane bag. The bag should be held securely in place with a paper clip or string (Fig.).

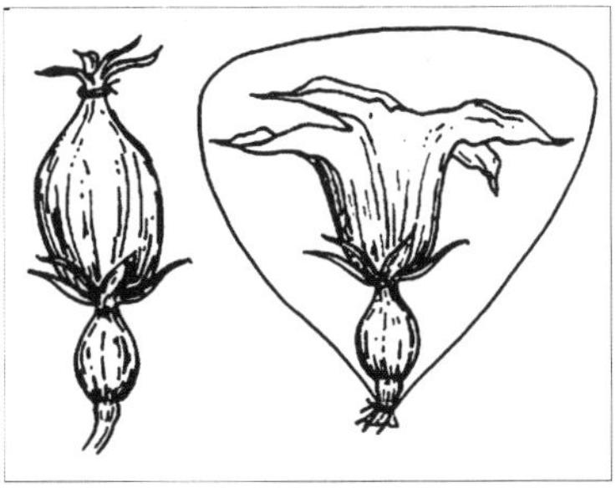

Fig. To Protect the Flower from Unwanted Pollen, Close it with a String or Soft Wire (left) or Cover it with a Bag (right).
A Cellophane or Plastic Bag will Permit you to Observe the Flower at All Times.

Flowers that are to be self-pollinated should likewise be protected from foreign pollen by either closing or covering the flower. If the plant is grown indoors there is little likelihood of contamination by foreign pollen and you do not have to cover or close the flower. However, indoor as well as outdoor plants require emasculation to avoid self- pollination.

POSTPOLLINATION STEPS

Immediately after pollination, close or cover the flower again. The next step is to label the seed parent.

The standard method of labeling is as follows:

Each pollinated flower should bear a label containing the name of the seed parent, the letter X (to signify a cross), the name of the pollen parent, and the date of the cross.

1. Write on the label in the following order:
   (a) The number that you have assigned or the variety name of the seed parent;
   (b) The letter X;
   (c) The number or variety name of the pollen parent; and
   (d) The date the cross was made (Fig.).
2. Attach the label to the stem just below the flower that has been pollinated.

Once the seed parent is labeled, your next step is to record the cross or self in a notebook. Keeping complete and accurate records of your breeding operations is very important. The information that you record should contain all essential facts regarding the cross or self so that you can refer to it at a later time and even do the job again from the beginning if necessary.

POLLINATION STEPS

Crossing. There are several methods that can be used for crosspollinating flowers.

Here are four of the most common methods:

1. Remove the stamens from the pollen parent with tweezers. Place the stamens in a small container. Remove the protector from the seed parent. Holding a stamen with the tweezers, gently brush the anther across the stigma.

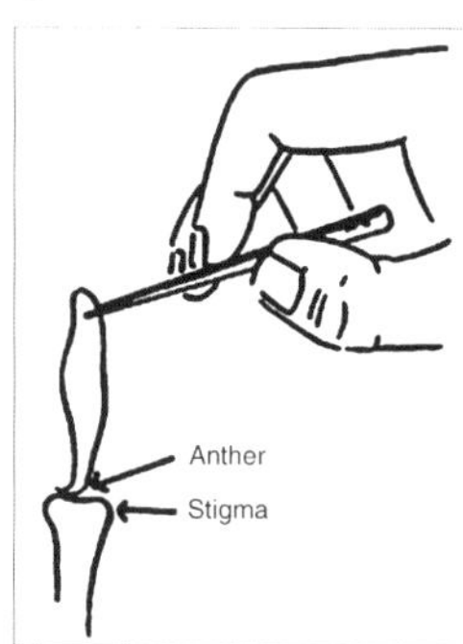

Fig. One way to cross-pollinate is to Remove the Stamens from the Pollen Parent and then, Holding a Stamen with Tweezers, Brush the Anther Across the Stigma of the Seed Parent.

2. Cut the flower that is serving as the pollen parent. Remove the protector from the seed parent. With tweezers remove a stamen from the pollen parent and brush an anther gently across the stigma of the seed parent.

3. Shake the bagged pollen parent so that the pollen is collected in the bag that is covering it for protection. Remove the bag from the pollen parent, being careful not to spill the pollen. Remove the protector from the seed parent and place the bag containing the pollen over the seed parent and shake the bag so that pollen falls on the stigmatic surfaces. This is usually done on corn.
4. With a camel-hair brush, transfer the pollen from the anthers of the pollen parent into a small container. Remove the protector from the seed parent and brush the pollen across the stigma (Fig.).

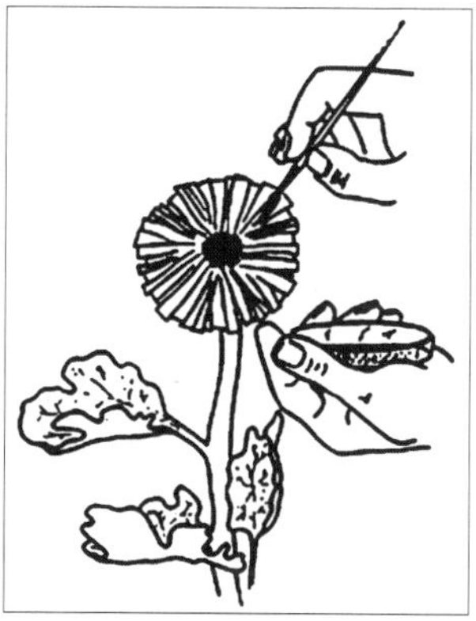

Fig. If the Stamens are Too Small or Too Difficult to Grasp, you can cross-pollinate by Transferring the Pollen from the Anthers with a Brush into a Container and then Brushing the Pollen onto the Stigma.

Each time you use different pollen, be sure to first wash with alcohol the camel-hair brush, tweezers, and any other item which might have touched some pollen. This step is very important to prevent pollination of the seed parent with unwanted pollen that has adhered to the equipment. After you wash the instruments be sure that they are dry before using them again.

Selfing

Procedures for self-pollinating flowers will depend on the type of flower. For perfect flowers, your job is done once you have closed or covered the flower, although you can sometimes help the pollen land on the stigma by shaking the flower once a day for several days after the pollen develops.

Only those composite flowers containing both disc and ray florets can be self-pollinated. Since they have both pistils and stamens, they can be selfed in the same way as perfect flowers.

With imperfect flowers, you will be able to self only those flowers that are on the same plant. In selfing imperfect flowers, the pollen from the staminate flower must be transferred to the stigma of the pistillate flower on the same plant. To do this you can use any of the methods given above for cross-pollination.

OBSERVING THE OUTCOME

Once pollination has been completed, a period must elapse during which fertilization and subsequent development of the fertilized ovule take place. When the seed is completely developed, it should be harvested. In plants grown for flowers, harvest the seed as soon as the seed pod is dry or when it just starts to split open. In fruits and vegetables the seed will be fully developed and ready for harvesting when the seed-bearing parts have reached maturity.

As you harvest the seed, place them in packets bearing the assigned offspring number. Be careful to keep seed from different crosses and selfs in separate packets. Air-dry the seed in a fairly warm spot for a week or so and then store them in a cool, dry place.

As soon as practicable, the seed should be planted and the results entered in your record book. The offspring can then be used for further study. Any seeds that are not planted should be saved in case of crop failure.

An important point to remember is that if both parents are pure, segregation usually will not become apparent until the F2 generation, and you may have to breed several generations before you can obtain a plant that will breed true.

TECHNIQUE OF CROSS-BREEDING

Here is hardly any plant-breeding project which does not involve cross-breeding. In most cases it is the object to produce variability, in order to obtain the material for selection to work upon. The quickest way to obtain variability is cross-breeding.

We should distinguish between the work of the plant breeder and that of the scientific worker. It is true that the great majority of plant-breeding problems are tackled at the larger institutes by genetically trained specialists, and yet the distinction should be made.

In scientific experiments only the most painstaking precautions are good enough. We must be absolutely certain about the paternity in our hybrid plants, and for this reason both the collection of the desired pollen and the elimination of the pollen from the mother-plant must be done with the minutest precautions.

If our object is a wholly practical one, we may often relax such precautions to a certain extent. For instance, in some plants castration—the elimination of stamens—is not only extremely difficult, but the manipulation brings a risk of seeing the flower wither and die. In such plants (soybeans, oats) castration can often be dispensed with, when we choose the parents in such a way that we can recognize the hybrids from inbred plants.

In every plant-family our first step must be to study the biology of the flower, in so far as this has not been done by others. We must know at what stage of growth of the flower- bud the anthers dehisce, so as to hit upon the safe time at which we must perform the operation of removing them. We must also know when the stigma is receptive, to find the moment at which to apply the pollen.

In most plants castration and the application of pollen can be done at the same time. This is the case in rice, in most of the cereals, in tobacco, Ipomoea, cocoa, peas, beans and a host of other plants. In some plants with large flowers the stamens are attached to the corolla, and it is possible to cut off the corolla with the stamens in one operation. This can some- times be done with the fingernail (apple, raspberry, straw- berry) or with a scalpel.

When a scalpel is used, it is sometimes advantageous to cover the blade almost up to the tip with adhesive tape, to avoid cutting down too deep. Where this is impossible, the stamens have to be extracted one by one. A set of fine pincers with differently curved tips should be in the possession of every worker. Very useful fine instruments can often be made to fit particular occasions. A very fine hook, made out of a bent pin, and mounted into a needle-holder can sometimes be used to great advantage, especially in castrating such minute flowers as in alfalfa or clovers.

For fine work a watch-maker's eyeglass, or even a binocular dissecting glass, may come in useful. On the whole, dispensing with castration in plants with minute flowers, and to use the method of straight cross-breeding, leaving the finding of hybrids till next season.

In some plants male and female flowers are separate. In such plants no castration is necessary; all we have to do is to protect the flowers from unwanted pollen. In cucurbits, melons, squashes, cucumbers, we found it quite easy to close unopened flower-buds, by means of small pieces of copper, or preferably lead-wire. Of course both male and female buds should be so protected. Lead-wire has the advantage of being very soft, and it can be used many times in one season.

Especially in wind-pollinated plants, care should be taken to protect the pollen used against contamination with unknown pollen. For this purpose the flowers from which pollen must be collected should be bagged. In some plants pollen is so abundant that we can use a separate anther for every flower to be pollinated, and we can hold this with fine forceps. Where the pollen is dry, it can be shaken out of the anthers into some small receptacle, for which a small gelatine capsule is just the thing. Such capsules can be closed, and the pollen contained in them can often be kept in very good condition for days, if the labelled capsules are kept cool in a well-closed bottle over calcium.

In maize both the male and female inflorescence should be bagged early. When the silks protrude they can be trimmed off short, and the pollen shaken over them from a tassel. If it is wished to obtain a cob full of kernels, it is of some advantage to insert a flowering tassel into the bag that covers the cob, and we can prolong its life and usefulness by standing this tassel in a small bottle of water hooked to the stalk.

In certain plants the operation of castration and fertilization can be performed on cut-off stalks that have been taken into the labouratory. In sugar-cane, in millet and sorghum, and even in wheat and barley, such panicles and ears, when stood in water, will ripen their seeds well, especially if only a few flowers are left on them for this purpose. Paper bags should be of ample size, and it is necessary to leave them on no longer than is absolutely essential for full protection during the few days in which the stigma is receptive. In certain cases (coffee, cocoa) a finely woven tartan may have advantages over paper; it is softer, and can be arranged around awkwardly placed flowers better than paper. Of course it is useful in certain cases to give the ripening fruit or seeds some protection against loss or damage (stone-fruit, potato, tobacco, (Enothera), and paper bags may be necessary at that stage.

For keeping small quantities of seeds, small paper or preferably cellophane envelopes can be used. Gelatin druggist capsules, however, are preferable. The labels should be en- closed with the seeds.

LABELLING, NUMBERING AND GADGETS

Every plant-breeder and every geneticist has his special methods of keeping his records. To be able to do this he must have suitable methods of numbering or labelling his plants or seeds.

Any system of labelling or marking must be legible and in- destructible by weather or insects. For merely marking, as an aid to tracing a special plant or part in a field, a coloured thread may be quite sufficient. If such a thread is used in addition to a number or a label, it helps to distinguish that label. Paper labels, even if made from very good parchment paper, tend to become discoloured.

To depend on such labels only, let us say in marking pea-pods, is courting trouble, for they closely resemble dried leaves or empty pods. The well-known Vilmorin red woollen thread is of very great value; it may even serve the double purpose of protecting the cross-fertilized ear of wheat or barley and of making it easy to find it again.

To mark seed capsules, fruits and such things, we found that lead-wire was much better than string. We can string a glass or china bead on a small piece of very soft lead-wire, and if we need a few dozen distinguishing marks, combinations of coloured beads will be excellent. We used this method in marking cross-fertilized squash blossoms and melon blossoms. If we make a loop of sufficient width, we can squeeze one end of this against the fruit-stalk, and it will accompany the fruit until harvesting time, or even beyond.

A very good label that will never lose its legibility is a poultry leg-band with a perforated number. These are made out of thin aluminium, and can be opened and adjusted before closing. Such labels are just the thing to mark plants that must be taken up and wintered in a cellar, like dahlias. Most seedsmen use stout painted stakes for carrying the name and the number of rows and individual plants, the name facing the row it marks. To obviate mistakes it is safer to have duplicate stakes, or to have the name and number of each lot written on a separate stake*

To save time and worry, good use can be made of poultry leg-bands or of the newer coloured plastic numbered wing- shields. These are numbered on both faces, and can be bought in consecutively numbered hundreds or thousands.

They can be hung from wire stakes with eyelets, and they accompany such plants as cereals or beans when they are bundled for harvesting. In some countries one has to guard against theft and mischief. A European blackbird pull out dozens of numbered yellow wooden labels and scatter them all over the place. A system of book-keeping which allows one to find the missing numbers from the book is indispensable if we want to avoid mistakes.

Numerous methods have been invented for protecting flowers and branches against insects or wind-borne pollen. Paper bags come first; they must be strong, rainproof and light. Some cotton-wool may come in very useful in making the connection of bag and stalk pollen-proof. Glassine bags and lengths of seamless sausage casing are being widely used in the United States. They allow one to see the inside of the bag, and so may save much time and error.

In some sympetalic plants flower-buds may be closed hermetically by the application of a length of very soft wire (lead- wire) wrapped round the bud, and used again after fertilization. We have used this method with squashes, melons and Ipomoea, The American plant-breeding journals and publications abound with methods for castrating and cross-breeding plants. Removing the pollen from the stigmas by washing it off or killing it by heat are methods but they sound excellent.

Gelatin capsules, made in two pieces, of which one fits over the other, are handy receptacles for keeping very small quantities of seeds or of pollen, for slightly larger quantities sachets for internal use may be used; they are cheap, and can be written upon, but if we close them we can never use them again. Paper bags must be used a good deal in storing seeds and similar objects (bulbils, bulbs, corms). They can be closed in different ways. The simplest system is the method of the French seed firms, in which each bag is closed by three folds, in such a way that it will never open spontaneously, even if submitted to considerable shaking.

The gadgets used in castration and cross-pollination are very numerous, and must be adapted anew to every new plant. Of course it is necessary to kill adherent pollen with alcohol, and extra scalpels, forceps or scissors are of great help. In working with very minute flowers a good binocular microscope is invaluable.

Avoid castrating minute flowers as much as possible. Everyone doing this sort of work occasionally regrets he was born with only two hands and without a prehensile tail. A ring to which has been soldered a minute cup that holds a small gelatin capsule with pollen, and that can be worn on the index-finger of one hand, is almost as useful. This is another American invention.

Hard seeds can be treated in different ways to make them grow. Some seeds, like those of canna or musa, can be scratched with a file, others may be rubbed on stone. Enothera seeds can be soaked in water; some seeds will only take up water rapidly when we exhaust the air above the water in which we keep them and then readmit the air.

Seeds of apricots, cherries and plums can be extracted after carefully cracking the stone. The extracted embryos (and the same is true for embryos of apples, pears and citrus) must be protected against bacteria and mould. Thymol solutions, even weak ones, will do this without injuring the seedlings.

SPEEDING-UP PLANT-BREEDING WORK

Plant-breeding work involving crosses generally takes a great number of generations before the superior new plant has been obtained. In the routine work of a large plant-breeding station so many breeding projects arc under way simultaneously that it matters little how much time elapses between the initial cross and the final purification of a promising commercial variety. Every year there are many projects being started, several are under way, and a few arc in their last stages.

It often pays to cut down the time required for work of this kind. One of the most promising means of doing this is to grow more than one generation of plants in one year.

This usually means that one of those generations of plants is grown in a season in which the plant is not ordinarily grown, and that for this reason a correct evaluation of the quality of the individual plants is out of the question.

In Europe or in America growing an extra generation on the spot means growing plants in the greenhouse. To grow cereals or beans in winter in greenhouse conditions not only requires heat; we must also have the means of regulating the length of day. The simplest method is to use large electric lights to furnish both the heat and the light, to lengthen the day by those means, and to let the temperature go down during the dark part of the night. Under those conditions cereals and beans will flower and set seed very well in the winter months,

The electricity required per square yard is roughly that used by lamps that require a kilowatt. It is clear that under such conditions the winter generation will not show the normal qualities of the plants in that group which are grown in summer-time in the field, but there are many cases in which this is not of much importance.

If we want to proceed by the method of making some hybrids and then growing a second and third inbred generation from this stock, we can do our crossing in the field, grow our hybrids in the winter, and then grow another crop of second-generation plants in normal field conditions. When we are using the method of mating back hybrid stock to some pure kind of plants we want to improve it matters but little if we cannot very well select the best plants during some of the generations involved.

It is evident that we must grow a generation under normal field conditions whenever we want to make our selections. In recent years a few seed-firms have hit upon the perfect scheme of growing two generations of annual plants. This does not involve any greenhouse generations. They work in two different localities with very similar climate, but so chosen that one of those localities is situated in the Southern Hemisphere.

There are regions of Chili where Californian firms can find conditions of plant-growth that are identical with conditions at home, with the exception that the growing season comes during the Californian winter. In those circumstances it is quite possible to grow two absolutely normal generations during one year. Our present greatly improved methods of communication make such schemes perfectly feasible.

With many tropical plants one generation yearly is grown of plants that need special conditions of either a wet or a dry season. Here it is easy enough to interpolate an additional generation, by looking for a suitable spot, or by providing the necessary irrigation.

Sometimes it takes a very long time to bring the plants to a condition where we can evaluate them. A good example is that of the fruit trees. It takes many years for an apple seedling or a seed-grown cherry to come into bearing. Here we can save a great deal of time by budding or top-grafting parts of the seedlings upon mature trees in full bearing. By this method we may save more than half the number of years normally required for bringing the seedlings to maturity.

In this connection we might also treat of those cases where certain seeds take a very long time to germinate. Storing the seed in suitable conditions of moisture, and especially of temperature, will speed-up germination in such plants as gooseberries and celery. In stone-fruits it has been found possible to crack open the stone, to extract the kernel and grow the seedlings under aseptic or even under antiseptic conditions (thymol solution).

SHOWS AND SHOWING

The exhibiting of plants and seed at the shows is a very curious thing. It pays extremely well for a seed-firm to employ some gardeners who know just exactly when and' how to get enough plants full of luscious beans or covered with flowers on the exact day for each show.

Florists and lovers of gardening see the quality of such plants at the shows, and many people who admire them there try in their turn to win the coveted prizes at their local shows the next year. This show game is certainly very nice and pleasant; it gives the gardeners something to strive after; it may even help to keep some of the lads in the country.

On the other hand, there is very, very little connection between the qualities judged at the shows and actual value of the material shown. The few selected mangolds, or the mammoth pumpkins, or the bundle of wheat-ears, or even of the maize-ear contests of the American agricultural shows.

All along the line, in horticulture as well as in agriculture, the selected samples shown give no indication of the value of the seed or plant stock the producers have for sale. Of course we can see that those few mangolds have the ideal shape and size, but the seed-firm must give us some guarantee about the percentage of mangolds that will grow to this size and shape from the seed they sell, just as the owner of a prize-winning bull must give some guarantee of this animal's ability to give profitable daughters, in addition to the blue ribbon awarded to the beautifully coloured and shaped prize-winner.

In agriculture this is beginning to be so well realised that the classes for beautiful corn cobs or for the largest mangolds are becoming very rare indeed, even if we still see stock-judging in dairy cows and similar farces. In the horticultural shows much interest is still shown in regard to flowers and vegetables.

It seems evident that when we strive for beauty, beauty contests for roses and begonias, for asparagus and delphiniums and dahlias are indicated.

The show is a shop-window. The seller must show if his competitors show, if only to keep his name before the public. To the public the show is only just a pleasant occasion for a half-day outing. To the buyer the show is much nicer to see than a description in a catalogue.

Long lists of novelties are bought at the shows by people with an experimental turn of mind. We see the representatives at the shows kept busy noting down long lists of orders. Showing certainly pays the seller.

A very much more sensible system of showing a great many varieties is that of the demonstration gardens and experimental stations, both stations in which the agricultural authorities try to compare promising new and good old varieties with the aid of objective standards, and such gardens as show-plots arranged by rose-lovers or by societies of growers of gladioli or dahlias. Here the varieties can be judged throughout the growing season in favourable conditions.

THE ANNUAL SELF-FERTILIZING CROPS

The number of plant species now being cultivated by man is really enormous, while only relatively few animal species are kept under domestication. It is possible for the author of a book on animal breeding to take those species almost one by one, and to give the specialist breeders a few hints over and above those which are contained in the general book.

This would be quite out of the question in a book on plant breeding. One way out of the difficulty would be to give one example of each of the three groups—self-fertilized plants, cross-fertilizers and plants that are usually propagated by vegetative means.

These plants in two sections— grains and legumes—giving some hints to the breeders of wheat, barley, oats, sorghum, rice, etc., and going into some details about the breeding of beans, soybeans, cow-peas, ground-nuts, peas, etc., separately. But, after all, the self-fertilizing annuals have so much in common from the point of view of plant-breeding methods.

Self-fertilization causes isolation. Each plant, as far as reproduction is concerned, is as self-contained and isolated in a crowded field as it would be if grown by itself in a greenhouse in a city. The inbreeding caused by self-fertilization has several results.

It makes all heritable variability disappear in every line, it makes every group consist of parallel lines, in which each plant descends from only one parent plant, so that the usual complicated network of ancestry is simplified into a system of isolated parallel lines. Where inherited variability exists, such as happens after a deliberate artificial cross or when accidental cross-fertilization occurs, it will automatically disappear.

That is to say, it will disappear in so far as each separate line is concerned, for Mendelian segregation will make the lines, descending from any hybrid heterozygous for a great many genes, diverse in their hereditary make-up.

The result is that a field of self-fertilizing annuals that has not been grown from just one pure line by a plant breeder always consists of a mixture of a number of genetically different lines. Even if accidental cross-fertilization or irregularities in crossing-over (mutation) are very rare, they do occur, and cause this state of things.

This genetic variability, however, is wholly different from what we find in other plants and animals. We are dealing with mixtures of pure lines, and almost every plant we take from any field is pure, homozygous for all its genes, and will, if its seeds are sown, give a very homogeneous descendance. From all this it follows that enormous progress can often be made in all self-fertilized plants by simply picking out a great number of promising-looking individual plants and comparing their descendance.

Many valuable widely grown wheats, tobaccos, peanuts and beans have been just found. They "just growed ", like Topsy.

The first step in plant-breeding in the self-fertilizing plants must always be to look for those ready-made pure lines. The next is to find whether a mixture of two of the very best pure lines can be discovered which will give us still higher production (10 per cent, higher production from a mixture of two strains as compared with that of the best one is quite common).

We have then to produce some more genetic variability by deliberate cross-breeding. This generally means that we try to combine the good qualities of two different strains in one new one.

Here there are two wholly different schools of thought, and two wholly different methods. One of them is to analyse second-generation plants, and to find those individuals that do combine, let us say, the disease resistance of line A with the large quantity of big seeds of line B. This is a method invented by the experimental geneticists who are trying to combine practical plant breeding with purely scientific gene- analysis. As a theoretical geneticist, interested in genes and their action.

It is great fun playing with a hundred different hybrid lots in one species, growing a few dozen second-generation plants in each lot and tabulating the results. But this is neither sound genetics nor sound plant breeding. We are dealing with so many unanalysed and unanalysable genes which all affect production and quality at the same time, that the chance of finding something really worth while in a few hundred plants is simply not good enough.

The other method is based upon the realization that hundreds of genes are commonly involved, and upon the fact that purification—reduction of the potential variability—is rapid and automatic in this material. It consists of growing the greatest possible number of seeds from first-generation hybrid plants. We can then proceed in two different ways.

One is an analytical method, which means that we are seeking for the very best-looking plants (for instance, by weighing their seeds), and progeny testing those plants by sowing a large number of seeds from each, in order to find the most profitable families. The other method is what Baur and Nilsson-Ehle called a "Ramsch" method.

This consists simply of growing the descendance of a hybrid in a vast mixture, refraining from any analysis, and continuing this hybrid lot as a miscellaneous collection for five or six generations, sowing as much of the mixed seed of each harvest as we have room for. If we do this, natural selection will weed out most of the unproductive and undesirable types, such as those plants t1 at ripen after harvesting time, all weaklings and all plants susceptible to diseases and pests. After five or six years that lot will consist of a mixture of pure lines, every line different from every other, with the successful lines represented by more individuals than the unsuccessful ones. After this we will treat the mixture just as if we were dealing with any other mixture, growing separate rows of beds each from one likely-looking plant.

This is a very sound method, which saves a great deal of un- necessary labour and time. It is extremely difficult to find just how much better this method is as compared with that of analysis in every generation, starting from the second generation. An analytical method could be made to work very well, provided we did all we could to make the number of plants of the second and third generations exceedingly large (in the cereals we can get almost as many seeds as in tobacco by splitting up the hybrid plant repeatedly before it starts to grow any stalks).

In actual practice there are only a few crosses that really produce anything worth while, and the general practice seems to be to make a great many attempts and grow a large number of first-generation hybrids. This renders it imperative to cut down on the number of plants per experiment. It is a vicious circle. This is all very wrong and wasteful. One cross every year and work with it, both analytically and by means of a "Ramsch ", doing the final selection after six years for one lot in every season, than waste my time castrating flowers and producing numerous hybrids that would have no really good chance of showing what they could do.

In all the self-fertilizing annuals we must be continually on the look-out for qualities that may be of special merit. Mostly of adaptations to special conditions. In wheaa on irrigated ground, resistance to drowning may be as important as disease resistance or quality of the grain. Conversely, rice could be grown much more extensively if we would take the trouble to do some plant-breeding work on dry-land rices, which abound in the mountain villages of all the tropical Asian islands.

In ground-nuts we should look for lines that are specially adapted to mechanical harvesting, as well as for lines adapted to the production of a first-quality product in the hands of tens of thousands of small cultivators. In all the plants of this group it is possible not only to improve the yield and the quality, but also to extend the area in which profitable production is possible. Tobacco, rice, soybeans and cow-peas are just as important in this respect as wheat, barley and oats.

Plant breeding with self-fertilizing plants is comparatively easy. It is even possible for the real amateur plant breeder to do some extremely interesting and useful work with some of the less well-known plants of this group—let us say with beans or cow-peas. Such work needs room, but very little special technical knowledge or talent for higher arithmetic. Frequently a few plants will stand out from an otherwise quite even field of barley, beans or sorghum.

Such "high birds"—climbing plants in bush beans, high, scmiawned plants in wheat—are often accidental hybrids. If we sow a few of their seeds, we can see immediately whether they are hybrids or admixtures, for hybrids will have very variable offspring. If then we sow all the seeds from such hybrids, we can start a series of experiments that may easily give us something worth while, even when the father of the hybrid is unknown.

PLANT BREEDING IN THE FUTURE

Success in any line of business depends on being able to anticipate and respond to changing market requirements. Such is the long-term nature of developing a new crop variety that plant breeders are in the unenviable position of forecasting their customers' demands at least five, and probably nearer ten years, in advance. Plant breeding objectives will continue to be shaped by changes in agricultural and environmental policy. There is no doubt that while improvements in crop yield and endues quality remain of paramount importance, the future for British agriculture lies in crop production systems which deliver both economic and environmental benefits.

Increasingly, plant breeders are targeting varieties suitable for low input and organic regimes. It is already estimated that genetic improvements in disease resistance alone save British farmers more than £100 million a year in reduced agrochemical use and improved yield. A range of other factors also determines a variety's suitability for low input, integrated farming systems. They include the crop growth cycle, straw strength and susceptibility to weed competition, as well as the variety's ability to thrive in a range of soil types, climatic conditions and rotational patterns.

Novel Crops

Breeders continue to adapt novel crops for UK growing conditions. New varieties of sunflowers and soya beans are already creating market opportunities in the food sector. Post-BSE, attention has focused on using more home-grown fodder in animal feed. For plant breeders, this means enhancing energy and protein levels in traditional forage crops, and developing the potential of new crop species, such as lupins.

Non-food Crops

The use of crop plants for industrial, nonfood applications is an area of increasing interest. Developments in plant breeding could underpin a transition from many crude, industrial processes towards renewable, field-based production of specialist chemicals, textiles and biofuels Quick-growing biomass crops such as willow coppice and miscanthus are already fuelling UK power stations, and there is huge potential to develop oilseed, starch and fibre crops for a wide range of product including lubricants, pharmaceuticals, dyes, plastics, flooring, paper and clothing.

New Traits

Pioneering research by British scientists is opening up new possibilities for crop improvement. Examples from current research include the use of genetic modification to overcome grey mould infestation in strawberries–not possible through conventional breeding because no resistance genes exist in the world strawberry germplasm. Similar techniques offer ways to control the ripening process in fruit and vegetables, to extend storage life, and improve flavour and texture for consumers. It will also increase the seasonal availability of homegrown produce, so reducing the distance food is transported. Modern biotechnology can be used to improve human nutrition. Recent research by British scientists includes the development of vegetables containing increased levels of cancer-fighting nutrients, and apples, which really do keep the dentist at, bay by protecting against tooth decay.

Patents

Recent developments in technology have also seen the emergence of patent applications in relation to plant breeding. This is particularly the case where technologies can be applied to several different varieties of the same crop, or across a range of different species. Such developments cannot be protected under the variety-specific system of Plant Breeders' Rights.

The use of patents where living organisms are concerned continues to arouse controversy, but future research will not take place unless some mechanism exists to reward and promote sharing of knowledge.

Issues and concerns

Modern plant breeding, whether classical or through genetic engineering, comes with issues of concern, particularly with regard to food crops. The question of whether breeding can have a negative effect on nutritional value is central in this respect. Although relatively little direct research in this area has been done, there are scientific indications that, by favouring certain aspects of a plant's development, other aspects may be retarded. Reductions in calcium, phosphorus, iron and ascorbic acid were also found. The study, conducted at the Biochemical Institute, University of Texas at Austin, concluded in summary: "*We suggest that any real declines are generally most easily explained by changes in cultivated varieties between 1950 and 1999, in which there may be trade-offs between yield and nutrient content*

The debate surrounding genetic modification of plants is huge, encompassing the ecological impact of genetically modified plants and the safety of genetically modified food.

8

Methods of Plant Breeding

CONVENTIONAL METHODS

Plant breeding is defined as identifying and selecting desirable traits in plants and combining these into one individual plant. Since 1900, Mendel's laws of genetics provided the scientific basis for plant breeding. As all traits of a plant are controlled by genes located on chromosomes, conventional plant breeding can be considered as the manipulation of the combination of chromosomes. In general, there are three main procedures to manipulate plant chromosome combination. First, plants of a given population which show desired traits can be selected and used for further breeding and cultivation, a process called (pure line-) selection. Second, desired traits found in different plant lines can be combined together to obtain plants which exhibit both traits simultaneously, a method termed hybridization. Heterosis, a phenomenon of increased vigour, is obtained by hybridization of inbred lines. Third, polyploidy (increased number of chromosome sets) can contribute to crop improvement. Finally, new genetic variability can be introduced through spontaneous or artificially induced mutations.

- Selection: Selection is the most ancient and basic procedure in plant breeding. It generally involves three distinct steps. First, a large number of selections are made from the genetically variable original population. Second, progeny rows are grown from the individual plant selections for observational purposes. After obvious elimination, the selections are grown over several years to permit observations of performance under different environmental conditions for making further eliminations. Finally, the selected and inbred lines are compared to existing commercial varieties in their yielding performance and other aspects of agronomic importance.y
- Hybridization: The most frequently employed plant breeding technique is hybridization. The aim of hybridization is to bring together desired traits found in different plant lines into one plant line via cross-pollination. The first step is to generate homozygous inbred lines. This is normally done by using self-pollinating plants

where pollen from male flowers pollinates female flowers from the same plants. Once a pure line is generated, it is outcrossed, *i.e.* combined with another inbred line. Then the resulting progeny is selected for combination of the desired traits. If a trait from a wild relative of a crop species, *e.g.* resistance against a disease, is to be brought into the genome of the crop, a large quantity of undesired traits (like low yield, bad taste, low nutritional value) are transferred to the crop as well. These unfavourable traits must be removed by time-consuming back-crossing, *i.e.* repeated crossing with the crop parent. There are two types of hybrid plants: interspecific and intergeneric hybrids. Beyond this biological boundary, hybridization cannot be accomplished due to sexual incompatibility, which limits the possibilities of introducing desired traits into crop Plants.

Heterosis is an effect which is achieved by crossing highly inbred lines of crop plants. Inbreeding of most crops leads to a strong reduction of vigour and size in the first generations. After six or seven generations, no further reduction in vigour or size is found. When such highly inbred plants are crossed with other inbred varieties, very vigorous, large sized, large-fruited plants may result. The term "heterosis" is used to describe the phenomenon of hybrid vigour. The most notable and successful hybrid plant ever produced is the hybrid maize. By 1919, the first commercial hybrid maize was available in the United States. Two decades later, nearly all maize was hybrid, as it is today, although the farmers must buy new hybrid seed every year, because the heterosis effect is lost in the first generation after hybridization of the inbred parental lines.

- *Polyploidy*: Most plants are diploid. Plants with three or more complete sets of chromosomes are common and are referred to as polyploids. The increase of chromosomes sets per cell can be artificially induced by applying the chemical colchicine, which leads to a doubling of the chromosome number. Generally, the main effect of polyploidy is increase in size and genetic variability. On the other hand, polyploid plants often have a lower fertility and grow more slowly.
- *Induced mutation*: Instead of relying only on the introduction of genetic variability from the wild species gene pool or from other cultivars, an alternative is the introduction of mutations induced by chemicals or radiation. The mutants obtained are tested and further selected for desired traits. The site of the mutation cannot be controlled when chemicals or radiation are used as agents of mutagenesis. Because the great majority of mutants carry undesirable traits, this method has not been widely used in breeding programmes.

Success in using conventional plant breeding principles and agricultural techniques reached its peak when high-yielding wheat and rice lines were cultivated in the 1960s. The doubling and tripling of productivity of these important crops in Asia signaled a agricultural revolution in the developing countries. This breakthrough in food production was termed the "Green Revolution" to describe the social, economic, and nutritional impact of the new high-yielding wheat and rice strains. Norman Borlaug was awarded the Noble Price in 1970 for his contribution in breeding new high-yielding strains of cereals. However, these strains were highly dependent on fertilizers, irrigation and agrochemicals and required energy-intensive investments. This led to degradation and loss of soils and other severe environmental problems all over the world. For these reasons, the "Green Revolution" has been both praised and damned.

Biotechnological Methods

Biotechnology is the discipline which deals with the use of living organisms or their products. In this wide sense, also traditional agriculture may be seen as a form of biotechnology. The European Federation of Biotechnology defines biotechnology as "the integrated use of biochemistry, microbiology and engineering sciences in order to achieve technological (industrial) application of the capability of microorganisms, cultured tissue cells and parts thereof". In recent years, biotechnology has developed rapidly as a practical means for accelerating success in plant breeding and improving economically important crops. The most important methods used to achieve these goals. The techniques of genetic engineering, which are a part of biotechnology.

- *In Vitro* Cultivation of Plant Cells and Regeneration of Plants from Cultured Cells Certain isolated somatic plant cells can be cultured in vitro (in the test tube) and are capable of proliferation and organization into tissues and eventually into complete plants. The process of regenerating whole plants out of plant cells is called in vitro regeneration. The three factors affecting plant regeneration are genotype, explant source, and culture conditions, including culture medium and environment. Different mixtures of plant hormones and other compounds in varying concentrations are used to achieve regeneration of plants from cultured cells and tissues. As the plant hormonal mechanisms are not yet understood completely, the development of in vitro cultivation and regeneration systems is still largely based on empirically testing variations of the three above mentioned factors.
- *In Vitro* Selection and Somaclonal Variation: Plants regenerated from cell cultures may exhibit phenotypes differing from their parent plants, sometimes at quite high frequencies. If these are heritable and affecting desirable agronomic traits, such "somaclonal variation" can be incorporated into breeding programmes. However, finding of specific valuable traits by this method is largely left to chance

and hence inefficient. Rather than relying on this undirected process, selection in vitro targets specific traits by subjecting large populations of cultured cells to the action of a selective agent in the Petri dish. For purposes of disease resistance, this selection can be provided by pathogens, or isolated pathotoxins that are known to have a role in pathogenesis. The selection will only allow those cells to survive and proliferate that are resistant to the challenge. Selection of cells also plays an important role in genetic engineering, where special marker genes are used to select for transgenic cells.

- Somatic Hybrid Plants: Somatic hybrid plants are plants derived from the fusion of somatic cells. Cell fusion was developed after the successful culture of a large number of plant cells stripped of their cell walls. The resulting cells without walls are referred to as protoplasts. Since also protoplasts from phylogenetically unrelated species can be fused, attempts have been made to overcome sexual incompatibility using protoplast fusion. In most cases, these attempts failed because growth and division of the fused cells did not take place when only distantly related cells were fused. Successful fusions between sexually incompatible petunia species and between potatoes and tomatoes did not lead to economically interesting products, but important contributions to the understanding of cell wall regeneration and other mechanisms were achieved.

SCOPE OF PLANT BREEDING

HYBRIDING PLANTS GROWN FROM SEEDS

The majority of named clones of fruit-trees, potatoes, strawberries, have been obtained by cross-breeding. This means that somebody has crossed two different kinds of plants and has found the new valuable clone among the hybrids. When we do this cross-breeding work we almost always obtain an enormously variable group, and this shows that the plants used in the cross were heterozygous for a very large number of genes. This is not surprising if we remember that those parents themselves have also been obtained by cross-breeding.

In work with plants grown from seeds, obtaining a hybrid is regarded as only the first step in our attempts to produce something new and very good. We know that we can never expect to obtain the full assortment of recombination's of the genes that went into the cross before we have seen the second generation, grown from the hybrids. Why is it that the breeders work in such a different way when they are out to improve their roses, apples or poplars?

In some cases this must be due to the fact that it frequently takes a long time for the hybrid plants to come to maturity. Another cause for this difference in method is certainly found in the great variability in the first hybrid generation, due to heterozygosity of the material.

Whatever the reason for the fact, it is easy to see that the plant breeders who are continually making new hybrids, and then look no farther than the first-generation hybrids, must very often discard hybrid plants that have no value in them- selves, but that would be potential parents of excellent new plants of the second generation. Even if many of the valuable qualities in this material are dominant, it stands to reason that some of the most valuable characters in every group are recessive; and we have not a very good chance of obtaining a full sample of the potential variability that results from a certain cross if we look no farther than the first hybrid generation.

We might even say, without grossly exaggerating, that most named clones in such plants are the result of a cross that has been made with a definite object which has not been attained by the breeder. When we cross two different plants with the object of combining some good quality of one of them with those of the second plant, our work is only half done when we have obtained the hybrids.

To a certain extent we might say that the qualities of the first-generation hybrids are not very important—we must simply regard them as the parents of the variable second generation. Now, in certain groups of plants self-incompatibility is the rule, so that it is impossible to obtain a self-fertilized second generation. In that case we need more than one first- generation hybrid to produce a second generation. But this self-sterility is the exception rather than the rule.

When in plants of this group we sow a great many seeds obtained by self-fertilization, or, by mating together hybrids from the same cross, it is quite possible that we actually realise the hope which set some other breeder working to obtain those hybrids.

Very often we find that some roses, ornamental shrubs or potatoes are setting seed spontaneously. It is frequently asked whether sowing this seed gives the breeder any chance of obtaining something worth while.

To answer this question we must remember that sowing seed with the object of obtaining a plant with good qualities—a valuable novelty—is always something of a lottery.

It is certainly not true that our chance of drawing a winner is made very much better when the seed sown is obtained by crossing two different sorts, unless again we follow through the cross by growing an extensive second generation.

As breeders generally are not in the habit of doing this, almost all the commercial clones have been obtained as hybrids, so that it will very often be worth our while to follow up their descendents.

If a plant breeder, working with vegetatively reproducing plants, has the habit of sowing only cross-bred seeds, he is certain to ascribe his success to this crossing. But we must realise that in so far as valuable new recessive qualities are concerned, we only find those among the hybrids because both parents were heterozygous. Groups in which cross-breeding has always been the rule— apples, lilac, roses, potatoes—sowing inbred seeds on an extensive scale is a method very much worth while.

In some plants, as in the potato, many valuable clones have been obtained from sowing "spontaneously" formed (and presumably inbred) seeds. The method can be especially recommended in those cases where, for some technical reason, such as minute size of the flowers, difficulties in castration, etc., it is extremely difficult to obtain cross-bred seeds. Sugar-cane is a case in point.

F1- Hybrids

In the autogamous plants where genetic purity, once obtained, is easily conserved, it is worth while concentrating our efforts to produce pure-breeding superior lines. But in other cases this may be so difficult that other working methods are preferable. One of those methods is to produce first-generation hybrids for direct use.

In such plants as maize or beets we often find that groups produced by crossing two distinct lines are superior in some way—for instance in yield—to the lines used in the cross. We must begin by giving an explanation of this phenomenon. If we could make a complete genetic analysis of a variable family of plants, we should see that in respect to a very large number of genes the group was pure.

All the individuals of such a group are alike in respect to a great number of genes, for which every individual is homozygous. On the other hand, there are other genes in regard to which such a line is impure, and when we would inbreed in this line, a number of offspring would be produced, that would differ in those genes from the majority. In a variable lot we should find that to a certain extent plants will differ in the set of genes for which they are either pure or impure. The same is true when we are com- paring separate lines.

When we make a cross between two different plants, and especially when we cross two plants of different lines, the result will be that the group of hybrids will be much more uniform than the descendants of either line. The explanation is very simple. The hybrids will inherit all the genes that are common to both parents. In addition to this, they will inherit all the genes for which only one of the parents was homozygous. If one parent had the composition AaBB and the other was AABb, the hybrids will all inherit both A and B.

The result will be that the hybrids are more likely to have valuable dominant qualities than inbred offspring of either line. Where dominance exists, however, and where there is no appreciable difference between homozygous and heterozygous plants (AA and Aa), individual selection will not quickly bring about purity. In other words, it is likely that good individual plants are still heterozygous for valuable genes, whose presence is desired for good quality.

We can understand why, at least in certain combinations of lines, first-generation hybrid lots of plants are superior in quality to and more uniformly good than, pure-bred lots from the same material. In other words, we could say that the homozygosity of one plant for certain genes will cover up the impurity of another plant in respect to those same genes.

We can now understand why hybrid lots are frequently better than inbred lots, and why we find this superiority of the hybrids especially in certain combinations of individuals and of strains. The simplest instances are those in which we combine two pure inbred lines, or even two clones, to produce the commercial seed. Some of the azalea specialists have found combinations of species that, when crossed, give hybrid seed from which beautiful large-flowered hybrid plants can be grown.

A second example is that of the Japanese hybrid egg-plants, and that of hybrid tomato seed. In Java a few tobacco estates have frequently used hybrid seed produced from a cross between quite different lines.

This practice of using first- generation hybrids is commercially possible only if producing such seed is not excessively difficult or expensive. In the cases enumerated, each fertilization gives us a very large number of seeds.

The best-known example of hybrid seed for the commercial production of first-generation hybrid plants is that of maize. Here it will suffice to say that it was found with maize that very often the first-generation hybrid plants produced by crossing different breeds gave a very high yield. The same was found true for hybrids between different inbred lines in one variety. Starting from small-scale genetical experiments, a very important hybrid-corn seed industry has grown up, first in the U.S.A. and later in other maize-growing countries.

In some cultivated plants (but not in others) inbreeding will have an adverse effect on vigour; and, conversely, crossing between different lines of plants will often give us remarkably vigourous hybrid offspring. Both from a purely scientific standpoint and from a practical one, this subject is of the greatest interest. The facts have been well known for a long time. Darwin made numerous inbreeding experiments, and found that some plants were much more subject to degeneration as a result of it than others.

Many different theories have been proposed to explain both the degeneration due to inbreeding and the great vigour due to crossing. The pre-Mendelian explanations do not concern us much. One of the chief difficulties in the way of a simple genetic explanation has certainly been the old conception of genes as determinants, each calling forth a special discrete unit character.

So long as most geneticists had this conception of genes and of their action, the mere sampling effect of Mendelian segregation of genes and the addition of genes by cross- breeding were thought to be insufficient to account either for the degeneration caused by inbreeding or for the very great special vigour of cross-bred individuals.

It is necessary to go into the different theories that have been proposed and have then been abandoned. The hypothesis that, to my mind, best explains all the facts.

If we look for evidence of loss of vigour by inbreeding, we are first struck by the fact that certain classes of plants are apparently exceptions to the general rule. In the habitually self- fertilized plants, such as tobacco, wheat, barley, and garden beans, the strictest kind of inbreeding—self-fertilization—does not lower the vigour or the yield.

On the other hand, in all the allogamous plants that have been tried—in beets, in maize, in rye—inbreeding brings along a rapid decline in almost all the desirable qualities; and in the same plants, cross-breeding such deteriorated stocks immediately restores full vigour. Now, if we start cross-breeding in the habitually self-fertilizing plants, we often find that the first-generation hybrids are extremely vigourous and productive, and if this happens we notice a very large general reduction in vigour in subsequent inbred generations.

This shows that we must not look for a mysterious special immunity to degeneration in some plants, but that resistance to inbreeding degeneration simply spells homozygosity.

In full accord with this is the circumstance that in animals the facts run wholly parallel to what we find in plants. Those groups in which cross-breeding has been the rule show a decline in quality after inbreeding, and those groups that have been regularly inbred—such as fancy pigeons—are inbreeding- resistant. Again, just as in plants, it has been possible both with chickens and with ducks to produce great purity by a system of inbreeding, while we retain full fertility and vigour as a result of strict selection for those very points.

Of course, the examples of the very fertile and vigourous highly inbred wheats and fowls show that the vigour we see in hybrid stock is not in any way due to the fact that the hybrids are heterozygous in respect to a great many genes, but only to the fact that they possess a favourable combination of genes. In theory it might be possible, by careful work, to make a group of plants (or animals) that would be homozygous in respect to that ideal combination of co-operating genes.

As a matter of fact, that the circum- stance that we do not attain this desirable end is most often due to faulty principles of selection. Essential mistakes are frequently made that render it almost impossible to attain the much-desired results. To understand the nature of inbreeding-degeneration we must remember that dominant good qualities, as far as the effects of selection are concerned, are in a class apart from the recessive virtues. If in a variable group some individuals have some unwanted dominant quality, it is quite simple to eliminate that dominant fault from the material.

If we prefer white maize to yellow, it is quite simple to produce a pure-breeding, white- seeded maize. But the reverse—to produce a pure-breeding, yellow-seeded maize from the same material—is much less simple, as some of the yellow-seeded plants chosen may again be heterozygous.

What is true for one monofactorial difference is true for all of them. If we subject a group of plants to individual selection, or even to some sort of selection by means of progeny tests (*e.g.*, ear to row in maize), and we do not have recourse to inbreeding in order to purify the genotype, the result will be that the superior strains of beets or maize we have built up in this way will be chockful of individuals heterozygous for very important genetic factors.

As soon as we start inbreeding, we reduce the potential variability of such material. In every succeeding generation from self-fertilization the number of genes for which the material was still heterozygous is halved. In every generation the plants become purer and purer.

Let us take the case of just one gene, A. If we inbreed a heterozygote, Aa, 25 per cent, of its descendants will be AA9 25 per cent, will be aa and 50 per cent, will again be heterozygous, Aa. Now, if A is an important gene, the AA and Aa individuals will be somewhat better than the aa's.

If we were concerned with this gene only, we could, by merely eliminating the aa plants, work towards better plants, and the proportion of pure-breeding AA plants would improve. But now we must remember that we are concerned with a multitude of genes, for all of which our original material may have been heterozygous.

If we were dealing with only three separate genes, A, B and C, we might retain only the 75 per cent. A plants (AA + Aa) and of these only the three-fourth B plants (AB -f Bb), and of these only the plants with C (again three-fourths). But we must remember that the fraction three-fourth power n becomes rapidly smaller and smaller with an increase of n (the number of desirable genes for which the stock is impure).

For this reason even strict selection for vigour while inbreeding will not prevent a loss of desirable genes from the material in an inbred series. And, of course, blind inbreeding, without any selection, will very rapidly bring down the proportion of plants with dominant desirable qualities, when the material was heterozygous for many genes that helped to build up those desirable dominants.

The mutual independence of genes scattered over the chromosomes of our plants will make it a matter of chance for which desirable genes each inbred line will become homozygous, and which ones will be dropped from the line. It is certain that if we start a number of parallel inbred lines—let us say in beets or in maize—by starting every generation from just one single plant, those lines will become homozygous each for its own genotype.

But the genotype of every line will become different from that of every other line. If we are lucky, and are running a good many lines, we may in the end still possess all the valuable genes that are responsible for the good qualities of superior plants. This does not mean that in one line we may have fixed the plants in respect to one desirable dominant virtue, and that in another we may find another separate good quality.

For we must realise that such good qualities do not depend each on a separate gene, but rather on the correct combination of genes. To a certain extent it is possible that we shall find good qualities in some of our lines—disease resistance in one, good quality of the product in another one. But if the foundation stock was highly cross-bred we shall seldom find highly inbred lines of sufficient all-round quality to make it worth while growing them commercially.

What happens when we cross-breed such inbred lines? As a general rule the vigour and all-round quality of hybrids between two such lines are very much better than those of the lines themselves. Sometimes, but not always, such hybrids are as vigourous and as good as the original material before we started our inbreeding.

The genetic explanation of this can only be that this is a result of the genetic difference between those lines. The loss of quality by an actual loss of genes partly responsible for desirable dominant qualities. Especially in blind inbreeding, chance determines which of such genes will be eliminated in a given line, and which will be consolidated in the line (for which the line would become pure).

If we think of the good quality and good vigour of the original strain as partly caused by the right combination of a great many genes, and of the inferior quality of inbred lines as resulting from incomplete sets of such genes, we can see that the hybrids, which inherit a combination of the genes remaining in inbred strain X with those for which strain Y happens to be homozygous, will be superior to both those strains.

In this superior quality of the hybrid generation between inbred lines we are concerned with the genotype, with the combination of useful genes present in the plants, and our present-day explanation is that it is this good combination of genes only which confers good quality upon those plants, and that the fact that they are heterozygous for many of them does not help. It would be very desirable if we could devise means for making a strain of plants that would be homozygous for all those important genes.

Hybrids of Maize and Beets

An Example of Utilizing "Hybrid Vigour"

From a small beginning in the experimental plots of the geneticists (G. Shull, East and Jones), the work with hybrid vigour in maize has grown to enormous proportions. The greater part of the corn harvest in the U.S. A. is now produced from hybrid seed, and the yield per acre has been much improved thereby. In the older well-known maize varieties that formerly were always grown from open fertilized plants the potential variability was very great. Many plants are heterozygous for very many different genes, and the best vigour and the best yield are found in those plants which happen to possess the most complete combination of those genes.

If we start inbreeding in series of lines, so that in every line we take one plant and force it to pollinate itself, we quickly reduce the potential variability in each of those lines. The plants will be homozygous for many genes for which the original mother-plant was still impure, and, on the other hand, we shall lose some of the useful dominant qualities.

The general quality and vigour of the inbred lines will decline, especially so if we use blind inbreeding without selection in continuing the lines. If, after some generations of inbreeding, we now cross those inbred lines, we shall find that, in some combinations more than in others, the resulting hybrids will be of very good quality. We have explained this by the hypothesis that those hybrids again have the full set of valuable genes. Now, the question may be asked why it should be advantageous to go to all this trouble of purifying those separate lines in order to reconstitute the original set of genes by cross-breeding them. Why should the hybrids between a good combination of those lines be more valuable than plants from the original non-inbred strain? There are at least two reasons for this. In the original lot—in the strain, before it was inbred—the potential variability was very high.

A large proportion of the plants—probably all of them—were heterozygous for several of the genes that help to make for quality. The germ-cells produced in the field were of the most diverse kinds, and, as a result, a certain percentage of the plants of the next generation always had undesirable recessive traits. Maize and sugar- beets are full of the most diverse undesirable recessive characters, that crop up in every generation.

The inbred strains may not have a full complement of the desirable genes, but in any case they are homozygous for those they do carry. If, for this reason, we are able to find a set of two inbred lines that "click" because together they happen to have a full complement of the desirable genes, the result will be that the hybrids grown from this combination will all carry the whole set of those genes. For this reason the hybrid lot will be much less variable than the original open-fertilized variety. In other words, the method of establishing many different inbred lines is a method of cutting down the heterozygosity in the material. For practical reasons, this reduction of the variability is not attempted in one single series.

The breeders know that it would be impracticable with our present methods to produce just one inbred strain, homozygous for all the desirable genes, so they do the work in parallel series, fixing the genotype of several lines by inbreeding, and hoping that after several generations of this they will discover a few inbred lines whose combination will give them the desired hybrid seed that will produce vigourous high-yielding plants.

It is obvious that the quality of such first-generation hybrids depends on the combination of genes present, but that the plants themselves are again heterozygous for all those genes. Those hybrid plants are meant to be harvested for their pro- duce—seed from such hybrid lots would give as variable a supply of plants as the original strain, or almost so. From the point of view of the seed firms this circumstance is rather happy; the users of the seed are satisfied, but they have to come back next year for a fresh lot of seed.

The fact that most inbred strains of maize and beets are of very low vigour and production makes the production of such hybrid seed expensive. In many maize-strains the cobs are miserable nubbins, the seeds are small and irregular in shape and size, and poorly adapted to the use of sowing by machinery. There are several ways out of this difficulty. Obviously the best way is to produce at least one line that will be a fairly good producer of seed.

The methods of blind inbreeding of the first years have been improved, and it is usual nowadays to combine inbreeding with selection. Sometimes this means straight selection in the inbred line, sometimes the selection is used to fit the inbred strain for its ultimate use—cross-breeding, We can do this by testing the pollen of several plants in cross- breeding (top-crossing) tests, and so choose the sub-line which is retaining most of the valuable genes.

Another method which is often used is that of the triple cross, in which the hybrid plants that furnish a lot of good- looking and very even seed are fertilized by pollen of yet a third line, that has been carefully chosen for this very purpose. If the ultimate product must be very uniform—such as in sweet-corn that is grown for canning, and that must be pre- pared by machinery—the use of one single cross between two inbred lines is indicated. The use of hybrid plants as parents of the commercial seed gives too variable a crop. In the ordinary use of maize some variability can be permitted, provided the average yield is good. In certain circumstances inbred strains can be produced that are valuable for "top-crossing ".

Such strains must be pure for a great many important genes, and they must be good pollen producers. The usual practice in producing hybrid seed for sale is to grow the mother-plants in a field, and grow the pollen parents throughout that field in every third or fourth row. The mother-plants are then carefully detasselled, and the field produces hybrid seed from these detasselled plants and inbred seed from the pollen-bearers.

Nowadays much of the commercial maize seed is grown by the system of "double-crosses ". In this system four different inbreds are used, and to grow the commercial seed, hybrids A × B are crossed with hybrids of two other strains, C × D. This, of course, makes the production of commercial seed much cheaper. It is doubtful if from the standpoint of the seed-user this system is as good as the triple cross.

We must realise that the hybrid plants of the first cross are heterozygous in respect to all the genes in which the parents differ. A uniform descendance from such plants can be expected only when the pollen used is from plants homozygous for almost all those genes; and we cannot expect that the pollen of another hybrid is in that condition.

It is certain that even commercial double-cross seed, when well-chosen inbreds are used, gives a better harvest than ordinary mass-selected stock of the same variety. But the results are never as good (and could not be expected to be as good) as those obtained from single-crossed seed. A certain extent smart advertising plays a role in this subject of the hybrid-corn seed production.

In the catalogues of the seed firms we often see pictures of "typical" ears of inbred strains, of the first hybrids, and of the "double-cross ". The enormous improvement in the first hybrids over the inbred strains, but they always show the largest and finest cob as the result of the double cross. Such pictures must be taken with a grain of salt!

If we look at the production of seed maize from the commercial angle, we must realise that the process of inbreeding those sub-strains costs money. During the first generations many plants have to be enclosed in paper bags, and to be pollinated by hand.

But, on the other hand, when the difficult and painstaking labour of purifying and testing the inbreds is done, those inbred strains can be further multiplied, without any manipulation, in fields, sufficiently far removed from other fields of corn. In fact, once we have found the two suitable inbred strains that "click", or the three necessary for a triple cross, we can reduce our selection to the very minimum, for once such strains are homozygous for the important genes, they remain in that condition, and all we have to do is to make the final cross to produce the seed-crop for sale.

The farmers, especially in the U.S.A., have become "hybrid- corn conscious ". They have experienced on their own farm how much more vigourous are the crops that can be obtained from the hybrid seed. The idea of the superiority of cross-bred seed has taken firm root, the name "double-cross" has more advertising value than the facts warrant. The only yield figures ever seen in which "double- cross" plants gave a (slightly) better harvest than their single-cross parents were taken from a cross-breeding experiment in which four different varieties were used.

With beets, especially sugar-beet, the commercial houses use a system of seed production that has great similarities to that of the hybrid maize. They generally use a system of mild inbreeding in isolated lines. Inbreeding is much less easy here, and whole plants must be enclosed in very well-constructed and ventilated cages. Next they try combinations of their selected strains, and finally they grow their commercial seed from a field in which two strains are planted in a mixture.

Until recently there has been no unfailing method of producing 100 per cent, hybrid seed, as a certain percentage of the seed must come from self-fertilization or from the inter- crossing between different plants of the same line. In practice, however, we find that the vast majority of the beets in a commercial field are first-generation hybrids.

If we deliberately cross two strains of beets, in which we can easily recognize the hybrids—by their leaf-colour or by their shape—we find that, with very few exceptions, the field contains only hybrid plants. The cause of this must be sought in the great vigour of the first-generation hybrids. In sowing beets many more seeds are sown than will ultimately grow up into mature beets.

Soon after the seedlings are up, the field is thinned, the supernumerary plantlets are hoed out. As this is done by hand, the man who does the thinning always favours the strongest, most vigourous plantlets. Even if we did not consciously select the favourable combinations between lines to be combined for seed-growing, the percentage of mature beets that have been grown from cross-fertilized seed is always very high.

Further Improvement in Maize and Beets

We saw that inbreeding will reduce the potential variability of the families subjected to it, that those lines will become homozygous for some of the valuable genes and that they will lose some of the others. One question that is of the greatest theoretical, but especially practical interest, is what we could do to influence this process of purification in such a way that each line would become pure for the maximum number of useful genes.

For discussing this, we must again look at this process of automatic reduction of the variability during inbreeding. We start with a plant of maize or of sugar-beet that is heterozygous for a great number of valuable genes—genes that we would like to have in every plant. The ordinary selection work ensures that such genes remain present in the stock, even if ordinary mass selection does not quickly ensure purity in respect of those genes.

It does not matter whether we know or whether we do not know all such genes separately in their action upon the end result. But we do know that a plant heterozygous for a gene A (Aa)y on being self-fertilized, will give us 25 per cent, homozygous A A plants, 50 per cent Aa's and 25 per cent aa's (which last will tend to be discarded).

The chance that we will hit upon a plant homozygous for one of those genes is one in four; if we are dealing with five genes, we have only one chance in eight hundred of finding a descendant homozygous for all of them—in fact, the chance is one in 4" (n being the number of genes involved) that we find a plant homozygous for all the genes concerned. In a limited number of inbred plants all we can hope is that we conserve some of those useful genes in the heterozygotes.

In the work with hybrids between lines we help ourselves by continuing a number of parallel inbred lines, hoping that in the end we shall have two such lines which between them carry the whole series of important genes in homozygous condition. It is evident that (theoretically) we should be able to produce inbred lines of maize or sugar-beets that were as productive as some of the best strains we now possess, and that would be homozygous and breed true.

In fact, if the number of inbred plants from a very good parent plant were only large enough, we should be able to find such a plant homozygous for all those genes for which the mother was still impure, in the very fast generation. In practice we can never know whether we would have to sow two thousand or twenty thousand plants to find that ideal one. But one thing is certain: that if we sow a few dozen plants from self-fertilized seed we shall undoubtedly have no chance whatever!

The simplest genetic calculations show that in this selection work in such plants as maize and beets the usual number of individual plants grown per generation is far and away too small! Not only should we combine selection with inbreeding, we should see to it that the methods used gave us an adequate chance of finding what we want to select.

Of course a great difficulty lies in the fact that it is practically impossible to appraise the genetic quality of one plant by inspecting its individual quality. This difficulty is, of course, greater in beets than in maize. Nevertheless, it would pay well to start inbreeding series on a very wide front, sowing all the seeds from an excellent plant, and testing a number of very excellent daughter-plants by sowing their inbred progeny.

The circumstance that inbred series almost always deteriorate rapidly in vigour and yield does not prove a thing, so long as the numbers we work with are so inadequate. A technical difficulty, of course, is that in maize it is not possible to pick out the individually best plants before they have ripened their seeds. When we must pollinate them we do not know their quality. Beets give much less trouble in this respect.

A very great difficulty with which plant breeders have to contend in the allogamous plants, lies in the large number of genes in respect to which our material is variable. Inbreeding will decrease this potential variability, but, as we have already seen, it is extremely difficult to direct the purification in such a way that the inbred lines will be of good commercial value. There is one way out of this difficulty. We know that the inbred lines are highly homozygous.

If they are not good enough as they are, we know nevertheless that such lines are homozygous for many valuable genes. It is well worth while trying to improve such pure inbred lines, and this is being widely tried in America to-day. We can do this in two ways: by crossing to another line, followed by another series of inbreeding, or by the use of repeated back-crosses. A cross with another suitable line will give us the genes lacking in our inbred lines. But the hybrids will be highly impure, and we must try to reduce this impurity while profiting by the homozygosity of the inbred line. This we can best do by a combination of repeated back-crosses to that line, and selection for vigour and yield.

In the existence of good inbred lines in maize, that can be propagated separately, we have an excellent and sure method of producing first-class commercial seed for sale. In the beets we are far from this. But in the beet there are certain possibilities that should be worth trying out. In the first place, there is male sterility. Certain families of beets have been found that have no functioning pollen.

This femaleness is inherited in a very peculiar way; it follows purely maternal lines, and it seems to be due to a peculiarity of the cytoplasm (protoplasm outside the nucleus). By cross-breeding it can be brought into any strain of beets. This gives us a possibility of producing on those plants seed that is guaranteed cross-bred, by having such plants flower in a field with other plants.

The other possibility in beets is the use of vegetative reproduction. Let us suppose that we possess two plants of sugar- beet that happen to fit together so well that the hybrid seeds produced from their combination give us beets of high purity and sugar content and with an excellent yield. Now our difficulty would be to produce two different groups of beets that would have the same constitution as those two individual beets.

This would be extremely difficult and tedious work. On the other hand, if we had a very good method of rapid vegetative reproduction, all we would have to do would be to grow tens of thousands of beets of clone X and as many of clone T, and then to make them flower in a mixed field in the ordinary way. There exist a few methods of vegetative reproduction in beets. Some of these methods, involving making the beets produce shoots and reproducing new beets from cuttings from those shoots, work fairly well.

If the methods can be perfected, they could lead to an avoidance of the very costly methods of selection now in use.

The Choice of Plant-breeding Material

The advantage of having at one's disposal a large collection of plant varieties, both plants of commercial importance and imported plants of no immediate practical importance, is enormous. A difficult question for the plant breeder is always what crosses to make, what plants to combine with the greatest hope of ultimate success.

In the breeding of such plants as potatoes, raspberries and roses, certain clones have the reputation of being profitable parents of hybrid lots. The Early Rose potato, the Lloyd George raspberry, the Chilean strawberry.

In some special instances the choice is obvious. When our present assortment of flaxes or potatoes or roses happens to be deficient in some such quality as disease resistance or winter- hardiness, we are on the look-out for plants that will show those desirable qualities to the greatest extent. In other words, we shall, when choosing the parents of our future hybrids, try to find material that promises the possibility of a combination of the very good points of both.

This is obvious. Often crosses that seemed very little promising have given excellent results. It would seem that here no definite rule could be indicated. It is always worth while trying crosses with new material, with newly imported species—in fact with material which has not been used previously. This is especially so when, as in horti- culture, everybody is on the look-out for something new.

Another rule that could be given would be that when we make very wide crosses we cannot hope to find commercially valuable strains in the second hybrid generation, unless we grow extremely large numbers of plants in that generation. When we make a very wide cross to incorporate some special quality into the existent stocks, the best thing we can do is to use the hybrids in back-cross series.

We can then strive to retain just one special quality of the new species and incorporate it into the general set of genes with which we are familiar in our older stock.

In several instances very wide crosses have been successfully used, and sometimes plants have been used in cross-breeding that in themselves showed very little promise. The inedible cattle gourd that Orton crossed with his water- melons, of some of the potato crosses with Andean wild species (Salaman) and of the back-crosses to hybrids with wild cane from which all our modern resistant varieties of sugar-cane have been derived.

It is here—in the choosing of plant material for breeding work—that cytological studies and chromosomal counts are of the greatest practical value.

Methods of Evaluation and Comparison

The aim of every plant breeder is to produce a group that will be worth a place in the assortment of cultivated plants. In some cases the new plant will have only ornamental value, in others it will, for any of a dozen reasons, be better adapted to agriculture than other similar ones. It is easy enough in all classes of plants to produce novelties, to produce variability. The experimental grounds of the plant breeder would soon be cluttered up with novelties if he did not keep on weeding out the comparatively worthless. After producing our novel plants, we must judge them.

Some qualities are much easier to judge than others. It is comparatively easy to weed out a lot of seedling dahlias and to pick out a few that may be worth keeping on for further testing. In the selection of our agricultural plants, in cereals, in beets, in maize, the assortment of valuable kinds is already so great that only relatively few of the newer selections are worth keeping. In the vegetatively propagated plants this work of testing novelties is comparatively easy.

In the first place, we are not bothered by considerations of purity—once we have produced a very good specimen, we can just multiply it with all its properties intact. In the second place, we can easily, for purposes of testing and comparison, produce as many individual plants of the novelty as we require.

When the differences are relatively small—as in cereals and beets, where the novel plant produced must compete with excellent old stock that has been made by a process of selection and comparison during centuries—a correct system of evaluation becomes of very great importance. What we really want to test is the inherited composition of the material.

This may be difficult, as the final quality of the plants to a large extent depends on favourable or unfavourable conditions under which they grew up. It will perhaps be as well to begin with the simplest problems. In the production of novelties in such flowers as roses or dahlias the value of a new variety depends entirely on the appeal it makes to the gardening public.

The man who is proud to have found a novel dahlia seedling or a new and strikingly beautiful rose sends it to a show and hopes that it will take the judge's eye and win him the much- "prized ribbon or certificate. But of course the insider knows that many highly commended novelties will never be heard of again, or that they will drop out of cultivation in a few years.

There is a difference between the fact that the grower can bring half a dozen perfect blooms to the show on time, and the question whether that same rose or chrysanthemum will ever be popular in the gardens, or even be worth while growing for the commercial producer of cut flowers. A dahlia that wins the medal may be a very weak grower, or it may be very susceptible to rain.

The strikingly beautiful rose may be very susceptible to mildew, or its flowering season may be extremely short. If we visit the growers of greenhouse roses, or the gardens of the man who produces cut flowers for the market, we find that the number of varieties that have real commercial value—varieties "with bread in them ", as our gardeners express it—is extremely small. We shall probably find only three or four of the relatively older dahlias worth growing for cut flowers, and it is seldom that a novelty is added to that list.

In testing out agricultural plants, both during the process of selection and after they have finally been introduced, many methods are in use. The direct, simple examination of a number of small plots or rows will show us the qualities of those lots, but this method will only serve to find the groups with definitely inferior quality.

It is not good enough to find exceptional merit, as circumstances may have favoured those qualities we appreciate. The only way to avoid reaching the wrong conclusions is to see to it that environmental conditions do not vitiate our results. The best way to do this is to com- pare the quality of a number of plants with that of other plants that were grown in wholly comparable conditions.

One of the best ways to do this is to grow the plots of our plant between similar plots of an older variety, that can be used as a standard of comparison. The choice of this standard is of great importance, and we will revert to this subject when treating of the breeding of special plant groups. As an example we can here take the evaluation of a newly produced (or of an imported) potato clone, or pure line of wheat or barley, In a case like this the standard may be some line or clone that is of commercial importance, against which the tested plant will be called upon to compete. If there is sufficient material, it is wise to repeat the comparison—that is to say, we should grow the test plot between standard plots in different parts of the field, or preferably in different fields. To equalize accidental conditions in small parts of the field, oblong plots are generally better than square ones. If we are testing a number of experimental lots, we can grow them side by side, and interpolate standard plots to be used for purposes of comparison. The experimental garden will need an efficient system of book-keeping. We must note down anything of value about the experimental and the standard plots, date of maturity, disease resistance, quality, yield.

We can calculate average yield for the sum of the plots of one kind, and then compare those figures with the average computed yield of the standard plots or with the average yield of the whole field; and a comparison of the deviation of the yields in the different plots with calculated mean errors will give us valuable data about the question in how far the superiority or inferiority of each separate number is significant as an indication of merit.

The practice of plant breeding in different parts of the world. A multitude of miserably small checkerboard fields being weighed at harvest time to produce the figures for an office staff to juggle with. A successful establishments where the field-work was under the supervision of men who were thoroughly conversant with the material, and where excellent results were obtained with a minimum of calculating machine work.

A method of annotating the results obtained in experimental plots which gives the plant breeder an excellent insight into the merits of his new material is that of graphical registration. This method can be adapted to the most diverse projects; it is very simple, and gives us our results at a glance. The method consists of registering the quality (yield, height, resistance) of each plot, both standard plots and testing plots, on ruled paper.

If we do this, the line that can be drawn through the points of the standard plots will probably rise in one part of the field, and descend in some other part of it, and by drawing a line that joins those points we obtain a basis upon which we can found our judgement of the results obtained in the experimental plots.

If we want to test our experimental material in two or three different qualities, we must construct separate graphs for each of those qualities, or sometimes we can use the same graph, employing different colours for separate qualities (weight, sugar percentage, purity in beets, etc.).

Reading those graphs will now show us the quality of our plant plots, in comparison with the standards, and we can use this relative value as a basis for our selection work. If we compare half a dozen groups in one experiment, and we find that one or two of them will in every duplication remain well below the standard in quality, this may be a very good reason for discarding them.

On the other hand, if we find one or two numbers that all through the experiment are well above the standard line in quality, we have a very good reason for preferring them.

It may be that in the case of one or two numbers the results are very variable, so that in one part of the field the number is above and in other parts it is below the standard in quality. Then we may well retain this number for another trial in the next season.

The influence of one row of plants on those growing close to it may be considerable, and we must guard against this. If we can grow three rows and use only the middle one for our records, we take a step in the right direction. If we have enough seed to plant six rows, and we examine only the two middle ones, discarding two rows on each side, this is still better.

The greatest difficulty arises in those cases where we are dealing with very few plants of each number; or, worse, where, as in the second generation of a cross, every plant is probably different from every other. It then becomes very important to equalize conditions for each plant as far as this is possible. A very good plan is to grow the plants in rows, and alternate those experimental rows with rows of some pure standard variety.

Each plant may then have two diverse neighbours in its own row, but it will touch four other plants that will be of the same kind as four out of six neighbour plants in the case of every other plant in the experiment. This means that to save time we must make a preliminary choice out of the many numbers in those early generations, and it is very important that we shall make the cultural conditions as nearly equal as possible for every plant. This is the place to warn against using apparent correlative qualities in judging the value of our experimental plants. In the breeding of animals even to-day we often see that almost as much attention is paid to correlative qualities as to real merit.

Some of this is certainly due to the fact that in some material real merit can only be judged in one of the two sexes. There is no good reason to give any attention to correlative qualities. In so far as a real correlation exists, selection for real quality will automatically bring along the correlative character, and in as far as we are mistaken (and this will nearly always be so) we are simply led astray by paying attention to apparent correlations. In plant-breeding selection according to qualities believed to be indirect indications of good quality used to be very frequently attempted.

This has gone out of fashion, and we only meet traces of it in very special fields (sugar-cane, apples). In our selection work we should always consider actual practical merit, and we should care- fully study in how far the qualities we can examine, measure and weigh are indications of this real merit.

The Importance of Collections

Most plant-breeding institutes own a collection of plants, sometimes consisting of imported and old varieties, sometimes completed with a collection of wild species. Such collections are of the utmost importance to plant breeding. We must remember that by selection we cannot create new genes, we cannot even remove existing genes that are present in all the plants.

For this reason the possibilities for plant- breeding work are restricted to recombination of the material present in our collection, The importance of a collection of old varieties is sometimes seen in the historical interest, and occasionally old breeds are retained for sentimental reasons. This is wrong, for at any time somebody responsible for the collection may have other ideas; he may be less of a sentimentalist than his predecessors, or he may think the value of all those old varieties is over- rated now that we have so many superior new ones. The truth is that the collections furnish material for serious future plant- breeding work. Every new gene added to the complement of genes in our present material doubles the possibilities of recombinations. Every novel (relative) gene-absence does the same. It is some- times possible to use very important qualities of intrinsically worthless old kinds of plant.

A garden pea, called "Acacia". In this pea each leaf consists of an uninterrupted series of pairs of leaflets like a Robinia leaf, tendrils being wholly absent. The plants therefore cannot attach themselves to supports or neighbouring plants. One of the difficulties in growing field peas in wet summers in some countries lies in the circumstance that the young plants intertwine themselves into veritable mats before the time for hoeing the weeds is past. To give the plants another cultivation in May would mean tearing apart the tendrils; in most years this would be quite out of the question.

For this reason the "Acacia" variety with a good, green, round field pea, and by selecting among the plants of the third and fourth generation succeeded in remaking a green field pea with a good yield and a high resistance to mildew, but without tendrils, a plant which could be given a last hoeing fully three weeks after the ordinary peas in neighbouring fields had grown together inextricably.

When we want to extend the region in which we can grow some group of plants it is often possible to find plant varieties in an extensive collection which carry qualities that are the very thing we want. It is uneconomical to duplicate and triplicate such collections; it is much better to have one huge central "ecole" of cultivated plants than to have a number of incomplete collections in the hands of private plant breeders and plant-breeding firms.

Moreover, keeping up such a collection costs money, and for this reason the collection, or some of its less- important-looking species, always runs the risk of being sup- pressed some day. It is evident that the Government should take over the task of keeping up such plant collections. The American Government has acted upon this idea, and its Plant Importation Bureau has done an enormous amount of very valuable work in this connection.

The keeping of a collection and the importation of foreign plants should go hand in hand. Especially in backward countries a wealth of plants is often present, and we sometimes find a great many curious plants well adapted to special conditions of season and climate. Plants as the Hopi maize, which comes up from almost two feet underground, the pea that buries its pods like a peanut, and the cauliflower-like sugar-cane of some of the islands around Java.

Hull-less oats and hull-less barleys may be out of fashion in some countries just now, nevertheless they carry extremely valuable characteristics, well worth preserving. Think of the extremely hardy peaches and apricots of the higher Himalayan valleys, of blight-free Chinese chestnuts, of the enormous seedless tangerines from Java. Keeping a collection is no work for a geneticist; it needs its own specialists with a very special turn of mind—a zest for systematic work and for collecting. Perhaps philatelists might make excellent keepers for such collections! The chief source of plants in such collections must be importations from newly explored countries. Sometimes, however, it might be worth while to add some of the curious plants that turn up in the descendants of hybrids—plants carrying absolutely novel qualities, new for the group.

Having an important collection of cultivated plants on hand makes it unnecessary to start hunting for suitable material whenever we want to take the amelioration of some plant group in hand. Most plant-breeding firms and plant-breeding institutes keep some sort of a collection, mostly of plants that have or that have had some economic significance. It is of the utmost importance to centralize this work in every country.

MONOPLOIDS IN PLANT BREEDING

The number of chromosomes in a basic set is called the monoploid number (x). Organisms with multiples of the monoploid number of chromosomes are called euploid. Haploids and diploids, then, are both cases of normal euploidy. Euploid types that have more than two sets of chromosomes are called polyploid. The polyploid types are named triploid ($3x$), tetraploid ($4x$), pentaploid ($5x$), hexaploid ($6x$), and so forth. Polyploids arise naturally as spontaneous chromosomal mutations and, as such, they must be considered aberrations because they differ from the previous norm.

However, many species of plants and animals have clearly arisen through polyploidy, so evidently evolution can take advantage of polyploidy when it arises. It is worth noting that organisms with one chromosome set sometimes arise as variants of diploids; such variants are called monoploid ($1x$). In some species, monoploid stages are part of the regular life cycle, but other monoploids are spontaneous aberrations. The haploid number (n), which we have already used extensively, refers strictly to the number of chromosomes in gametes. In most plants with which we are familiar, the haploid number and monoploid number are the same.

Hence, n or x (or $2n$ or $2x$) can be used interchangeably. However, in certain plants, such as modern wheat, n and x are different. Wheat has 42 chromosomes, but careful study reveals that it is hexaploid, with six rather similar but not identical sets of seven chromosomes. Hence, $6x = 42$ and $x = 7$. However, the gametes of wheat contain 21 chromosomes, so $n = 21$ and $2n = 42$.

Monoploids

Male bees, wasps, and ants are monoploid. In the normal life cycles of these insects, males develop parthenogenetically—that is, they develop from unfertilized eggs. However, in most species, monoploid individuals are abnormal, arising in natural populations as rare aberrations. The germ cells of a monoploid cannot proceed through meiosis normally, because the chromosomes have no pairing partners. Thus, monoploids are characteristically sterile.

(Male bees, wasps, and ants bypass meiosis in forming gametes; here, mitosis produces the gametes.) If a monoploid cell does undergo meiosis, the single chromosomes segregate randomly, and the probability of all chromosomes going to one pole is $(1/2)^{x-1}$ where x is the number of chromosomes. This formula estimates the frequency of viable (whole-set) gametes, which is a small number if x is large.

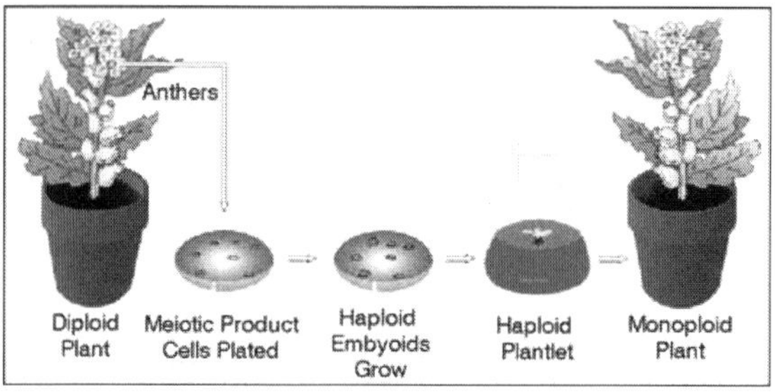

Fig. Generating Monoploid Plants by Tissue Culture. Pollen Grains (Haploid) are Treated so that they will Grow and are Placed on Agar Plates Containing Certain plant Hormones.

Monoploids play an important role in modern approaches to plant breeding. Diploidy is an inherent nuisance when breeders want to induce and select new gene mutations that are favourable and to find new combinations of favourable alleles at different loci. New recessive mutations must be made homozygous before they can be expressed, and favourable allelic combinations in heterozygotes are broken up by meiosis. Monoploids provide a way around some of these problems. In some plant species, monoploids can be artificially derived from the products of meiosis in a plant's anthers. A cell destined to become a pollen grain can instead be induced by cold treatment to grow into an embryoid, a small dividing mass of cells. The embryoid can be grown on agar to form a monoploid plantlet, which can then be potted in soil and allowed to mature.

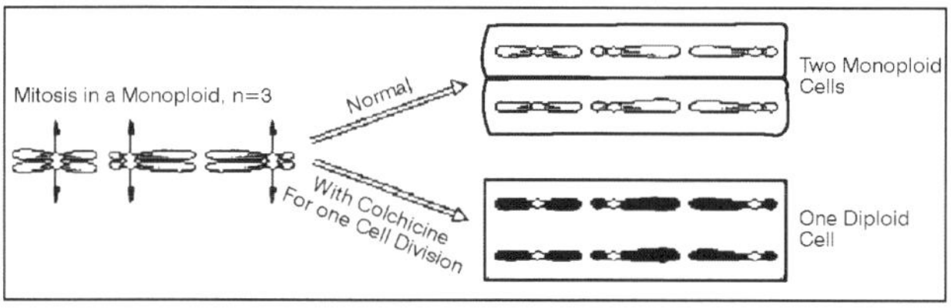

Fig.The use of Colchicine to Generate a Diploid from a Monoploid. Colchicine added to Mitotic cells

Plant monoploids can be exploited in several ways. In one, they are first examined for favourable traits or allelic combinations, which may arise from heterozygosity already present in the parent or induced in the parent by mutagens. The monoploid can then be subjected to chromosome doubling to achieve a completely homozygous diploid with a normal meiosis, capable of providing seed. How is this achieved? Quite simply, by the application of a compound called colchicine to meristematic tissue. Colchicine—an alkaloid drug extracted from the autumn crocus—inhibits the formation of the mitotic spindle, so cells with two chromosome sets are produced. These cells may proliferate to form a sector of diploid tissue that can be identified cytologically.

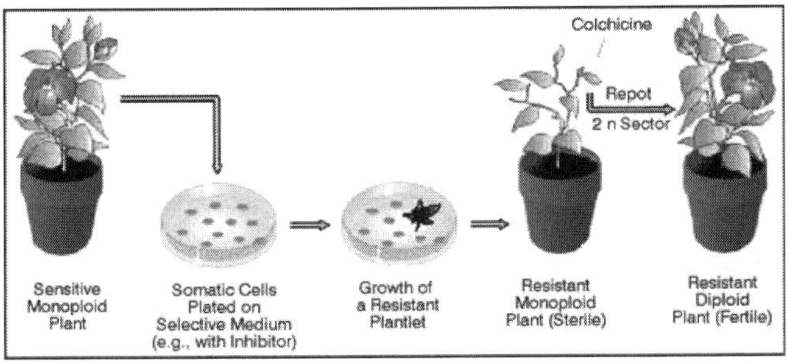

Fig. Using Microbial Techniques in plant DEngineering. The cell Walls of Haploid cells are Removed Enzymatically.

Another way in which the monoploid may be used is to treat its cells basically like a population of haploid organisms in a mutagenesis-and-selection procedure. A population of cells is isolated, their walls are removed by enzymatic treatment, and they are treated with mutagen. They are then plated on a medium that selects for some desirable phenotype. This approach has been used to select for resistance to toxic compounds produced by one of the plant's parasites and to select for resistance to herbicides being used by farmers to kill weeds. Resistant plantlets eventually grow into haploid plants, which can then be doubled (with the use of colchicine) into a pure-breeding, diploid, resistant type. These powerful techniques can circumvent the normally slow process of meiosis-based plant breeding. The techniques have been successfully applied to several important crop plants, such as soybeans and tobacco. The anther technique for producing monoploids does not work in all organisms or in all genotypes of an organism. Another useful technique has been developed in barley, an important crop plant. Diploid barley, *Hordeum vulgare,* can be fertilized by pollen from a diploid wild relative called *Hordeum bulbosum.*

This fertilization results in zygotes with one chromosome set from each parental species. In the ensuing somatic cell divisions, however, the chromosomes of *H. bulbosum* are eliminated from the zygote, whereas all the chromosomes of *H. vulgare* are retained, resulting in a haploid embryo. (The haploidization appears to be caused by a genetic incompatibility between the chromosomes of the different species.) The chromosomes of the resulting haploids can be doubled with colchicine. This approach has led to the rapid production and widespread planting of several new barley varieties, and it is being used successfully in other species too.

Polyploids

In the realm of polyploids, we must distinguish between autopolyploids, which are composed of multiple sets from within one species, and allopolyploids, which are composed of sets from different species. Allopolyploids form only between closely related species; however, the different chromosome sets are homeologous (only partly homologous)—not fully homologous, as they are in autopolyploids.

Triploids

Triploids are usually autopolyploids. They arise spontaneously in nature or are constructed by geneticists from the cross of a $4x$ (tetraploid) and a $2x$ (diploid). The $2x$ and the x gametes unite to form a $3x$ triploid.

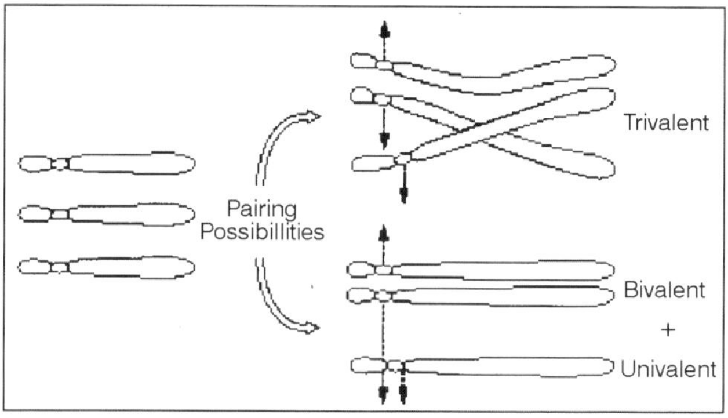

Fig. Two Possibilities for the Pairing of three Homologous Chromosomes before the first Meiotic Division in a Triploid.

Triploids are characteristically sterile. The problem, like that of monoploids, lies in pairing at meiosis. Synapsis, or true pairing, can take place only between two chromosomes, but one chromosome can pair with one partner along part of its length and with another along the remainder, which gives rise to an association of three chromosomes. Paired chromosomes of the type found in diploids are called bivalents. Associations of three chromosomes are called trivalents, and unpaired chromosomes are called univalents. Hence in triploids there are two pairing possibilities, resulting either in a trivalent or in a bivalent plus a univalent.

Paired centromeres segregate to opposite poles, but unpaired centromeres pass to either pole randomly. If all the single chromosomes pass to the *same* pole and simultaneously the other two chromosomes pass to the opposite pole, then the gametes formed will be haploid and diploid.

The probability of this type of meiosis will be $(1/2)^{x-1}$, and this proportion is likely to be low. All other possibilities will give gametes with chromosome numbers intermediate between the haploid and diploid number; such genomes are aneuploid—"not euploid." It is likely that these aneuploid gametes will not lead to viable progeny; in fact, this category is responsible for the almost complete lack of fertility of triploids.

The problem is one of genome imbalance. For most organisms, the euploid chromosome set is a finely tuned set of genes in relative proportions that seem to be functionally significant. Multiples of this set are tolerated because there is no change in the relative proportions of genes. However, the addition of one or more extra chromosomes is nearly always deleterious because the proportions of genes in those extra chromosomes are altered. Although the action of some genes can be regulated to compensate for extra gene "dosage," the overall effect of the extra genetic material seems too great to be overcome by gene regulation. The deleterious effect can be expressed at the level of gametes, making them nonfunctional, or at the level of the zygote, resulting in lethality, sterility, or lowered fitness. In triploids, it is possible that some haploid or diploid gametes will form, and some may unite to form a euploid zygote, but the likelihood of this possibility is inherently low.

Consider bananas. The bananas that are widely available commercially are triploids with 11 chromosomes in each set ($3x = 33$). The probability of a meiosis in which all univalents pass to the same pole is $(1/2)^{x-1}$, or $(1/2)^{10} = 1/1024$, so bananas are effectively sterile. The most obvious expression of the sterility of bananas is that there are no seeds in the fruit that we eat. Another example of the commercial exploitation of triploidy in plants is the production of triploid watermelons. For the same reasons that bananas are seedless, triploid watermelons are seedless, a phenotype favoured by some for its convenience.

AutotetrAploids

Autotetraploids arise naturally by the spontaneous accidental doubling of a $2x$ genome to a $4x$ genome, and autotetraploidy can be induced artificially through the use of colchicine. Autotetraploid plants are advantageous as commercial crops because, in plants, the larger number of chromosome sets often leads to increased size. Cell size, fruit size, flower size, stomata size, and so forth, can be larger in the polyploid.

Epidermal leaf cells of tobacco plants, showing an increase in cell size, particularly evident in stoma size, with an increase in autopolyploidy. (a) Diploid, (b) tetraploid, (c) octoploid. Because 4 is an even number, autotetraploids can have a regular meiosis, although this is by no means always the case. The crucial factor is how the four homologous chromosomes, one from each of the four sets, pair and segregate.

There are several possibilities. Pairings between four chromosomes are called quadrivalents. In tetraploids, the two-bivalent and the quadrivalent pairing modes tend to be most regular in segregation, but even here there is no guarantee of a 2:2 segregation.

If all chromosome sets segregate 2:2 as they do in some species, then the gametes will be functional and a formal genetic analysis can be developed for such autotetraploids.

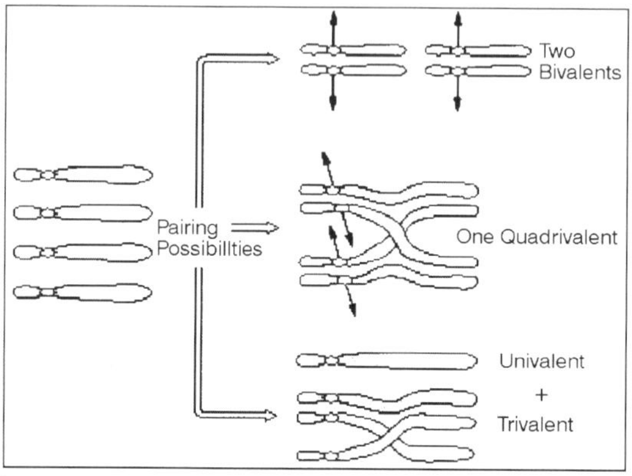

Fig. Meiotic Pairing Possibilities in Tetraploids.

Let's consider the genetics of a fertile tetraploid. We can consider an experiment in which colchicine is used to double the chromosomes of an *A/a* plant to form an *A/A/a/a* auto-tetraploid, which we will assume shows 2:2 segregation. We now have a further concern because polyploids such as tetraploids give different phenotypic ratios in their progeny, depending on whether the locus in question is tightly linked to the centromere. First, we consider a centromere-linked gene. The three possible pairing and segregation patterns are presented in Figure; these patterns occur by chance and with equal frequency. As shows, the 2*x* gametes produced are *A/a, A/A*, or *a/a,* in a ratio of 8:2:2, or 4:1:1. If such a plant is selfed, the probability of an *a/a/a/a* phenotype in the offspring is 1/6 × 1/6 =1/36. In other words, a 35:1 phenotypic ratio of *A/_/_/_:a/a/a/a* will be observed if *A* is fully dominant over three *a* alleles.

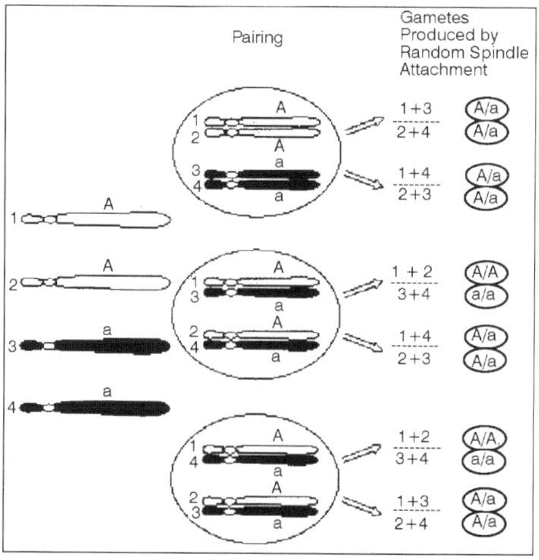

Fig. Gene Segregation in a Tetraploid Showing Orderly Pairing by Bivalents.

If, in the same kind of plant, a genetic locus having the alleles *B* and *b* is very far removed from the centromere, crossing-over must be considered. This example forces us to think in terms of chromatids instead of chromosomes; there are four *B* chromatids and four *b* chromatids. Because the number of crossovers in such a long region will be large, the genes will become effectively unlinked from their original centromeres. The packaging of genes two at a time into gametes is very much like grabbing two balls,

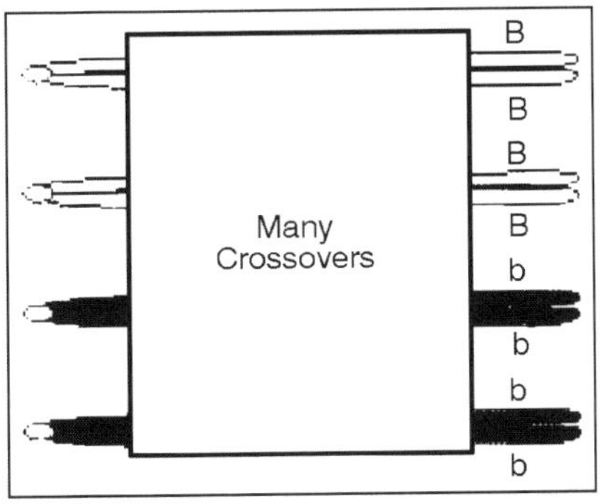

Fig. Highly diagrammatic Representation of a Tetraploid Meiosis in which a Heterozygous locus is Distant from the Centromere.

at random from a bag of eight balls: four of one kind and four of another. The probability of picking two *b* genes is then

$$\frac{4}{8}(\text{the first one}) \times \frac{3}{7}(\text{the second one}) = \frac{12}{56} = \frac{3}{14}$$

So, in a selfing, the probability of a *b/b/b/b* phenotype is 2/4 × 3/14 = 9/196, which is approximately 1/22. Hence, there will be a 21:1 phenotypic ratio of *B/_/_/_:b/b/b/b*. For genetic loci of intermediate position, intermediate ratios will result.

Allopolyploids

G. Karpechenko synthesized the "classic" allopolyploid in 1928. He wanted to make a fertile hybrid that would have the leaves of the cabbage *(Brassica)* and the roots of the radish *(Raphanus)*. Each of these species has 18 chromosomes, and they are related closely enough to allow intercrossing.

A viable hybrid progeny individual was produced from seed. However, this hybrid was functionally sterile because the nine chromosomes from the cabbage parent were different enough from the radish chromosomes that pairs did not synapse and disjoin normally.

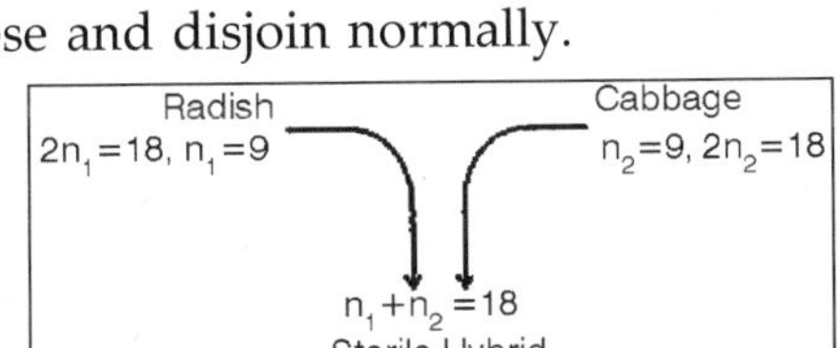

However, one day a few seeds were in fact produced by this (almost) sterile hybrid. On planting, these seeds produced fertile individuals with 36 chromosomes. All these individuals were allopolyploids. They had apparently been derived from spontaneous, accidental chromosome doubling to $2n_1 + 2n_2$ in the sterile hybrid, presumably in tissue that eventually became germinal and underwent meiosis. Thus, in $2n_1 + 2n_2$ tissue, there is a pairing partner for each chromosome and balanced gametes of the type $n_1 + n_2$ are produced.

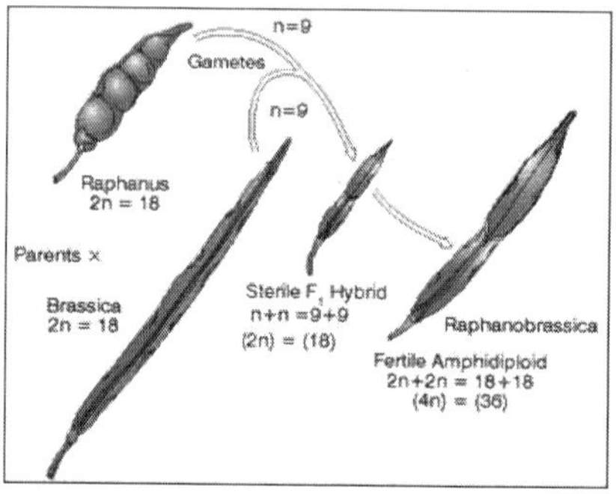

Fig. The Origin of the Amphidiploid *(Raphanobrassica)* formed from Cabbage *(Brassica)* and Radish *(Raphanus)*. The Fertile Amphidiploid Arose in this case from Spontaneous Doubling in the $2n$ = 18 Sterile Hybrid. Colchicine can be used to Promote Doubling.

These gametes fuse to give $2n_1 + 2n_2$ allopolyploid progeny, which also are fertile. This kind of allopolyploid is sometimes called an amphidiploid, which means "doubled diploid". (Unfortunately for Karpechenko, his amphidiploid had the roots of a cabbage and the leaves of a radish.) When the allopolyploid was crossed with either parental species, sterile offspring resulted.

The offspring of the cross with radish were $2n_1 + n_2$ constituted from an $n_1 + n_2$ gamete from the allopolyploid and an n_1 gamete from the radish. The n_2 chromosomes had no pairing partners, so sterility resulted. Consequently, Karpechenko had effectively created a new species, with no possibility of gene exchange with its parents. He called his new species Raphanobrassica.

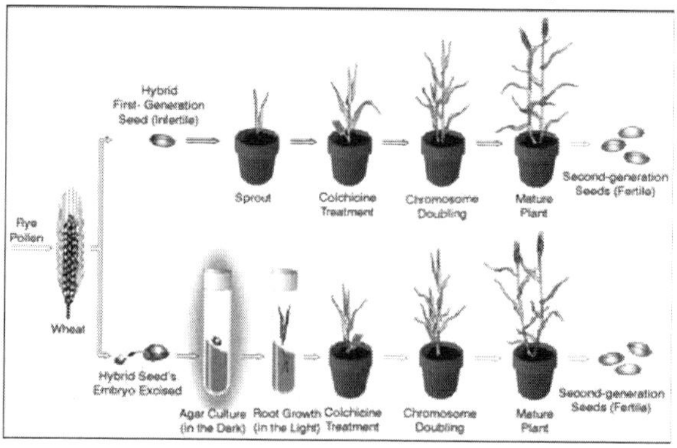

Fig. Techniques for the Production of the Amphidiploid *Triticale*.

Today, allopolyploids are routinely synthesized in plant breeding. Instead of waiting for spontaneous doubling to occur in the sterile hybrid, the plant breeder adds colchicine to induce doubling. The goal of the breeder is to combine some of the useful features of both parental species into one type. This kind of endeavor is very unpredictable, as Karpechenko learned. In fact, only one synthetic amphidiploid has ever been widely used. This amphidiploid is *Triticale,* an amphidiploid between wheat.

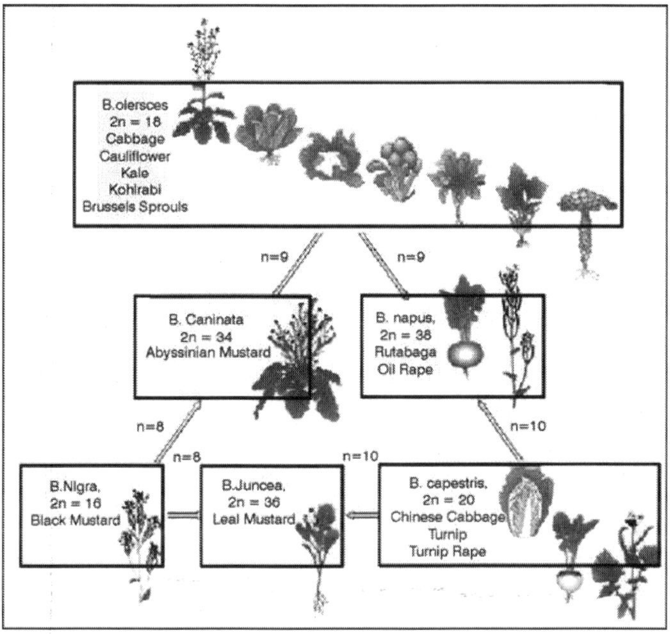

Fig. A Species Triangle.

In nature, allopolyploidy seems to have been a major force in speciation of plants. There are many different examples. One particularly satisfying one is shown by the genus *Brassica,* as illustrated in. Here three different parent species have hybridized in all possible pair combinations to form new amphidiploid species.

A particularly interesting natural allopolyploid is bread wheat, *Triticum aestivum*($2n = 6x = 42$). By studying various wild relatives, geneticists have reconstructed a probable evolutionary history of bread wheat. In a bread wheat meiosis, there are always 21 pairs of chromosomes. Furthermore, it has been possible to establish that any given chromosome has only one specific pairing partner (homologous pairing)—not five other potential partners (homeologous pairing). The suppression of such homeologous pairing (which would make the species more unstable) is maintained by an allele, *Ph,* on the long arm of chromosome 5 of the B set. Thus, *Ph* ensures a diploidlike meiotic behaviour for this hexaploid species. Without *Ph,* bread wheat could probably never have arisen. It is interesting to speculate about whether

Western civilization could have begun or progressed without this species—in other words, without the *Ph* mutation.

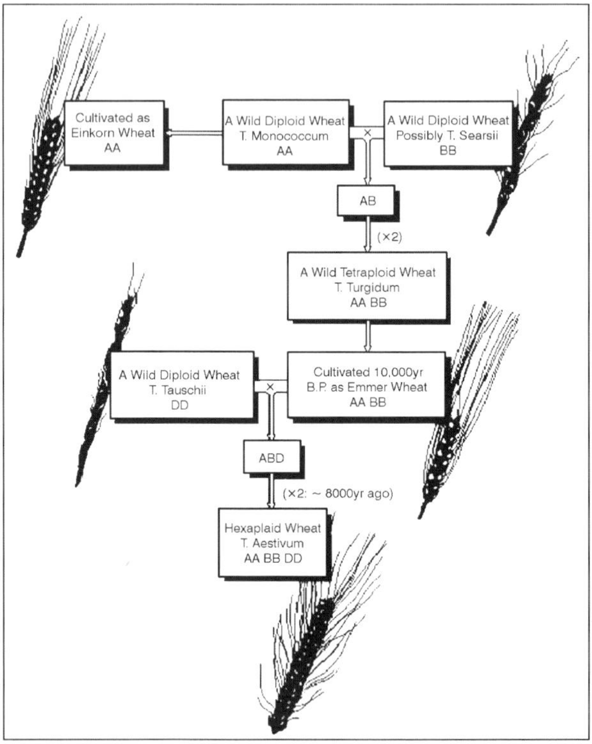

Fig. Diagram of the Proposed Evolution of Modern Hexaploid

INBREEDING DEPRESSION AND HETEROSIS

Inbreeding depression and heterosis are phenomena which may occur at positive and negative values of the inbreeding coefficient (F) of the considered plant material, respectively. These phenomena may occur if F deviates from 0. Their size appears from the difference between the expected genotypic value (EG) at the value for F in force and the expected genotypic value of the same plant material at F = 0(EGRM). For self-fertilizing crops the latter is for $p = q = 1$

2 equal to EGF2; for cross-fertilizing crops it is equal to the expected genotypic value of the population with the Hardy–Weinberg genotypic composition corresponding to the actual gene frequencies. The inbreeding depression or heterosis amounts thus to:

$$\mathrm{E}\underline{\mathcal{G}} - \mathrm{E}\underline{\mathcal{G}}_{\mathrm{RM}}$$

$$\left[m + \sum_{i=1}^{K}(p_i - q_i)a_i + 2(1-\mathscr{F})\sum_{i=1}^{K} p_i q_i d_i\right] - \left[m + \sum_{i=1}^{K}(p_i - q_i)a_i + 2\sum_{i=1}^{K} p_i q_i d_i\right]$$

$$= -2\mathscr{F}\sum_{i=1}^{K} p_i q_i d_i$$

If EG – EGRM = 0 at F _= 0 there is a strong indication of absence of dominance at the relevant loci. If EG – EGRM _= 0 at $F > 0$, inbreeding depression occurs, whereas EG – EGRM _= 0 at $F < 0$ implies the presence of heterosis. At F _= 0 the frequency of heterozygous plants is $2pq(1–F)$, at $F = 0$ it is $2pq$. The difference is $–2Fpq$, *i.e.* there is a deficit of heterozygous plants at $F > 0$ and an excess at $F < 0$. Considered in this way inbreeding depression and heterosis are due to a deficit or an excess of heterozygous plants, measured in comparison with the Hardy–Weinberg frequency.

It has been observed that continued selfing is very often associated with a decreasing average phenotypic value. This applies especially to cross-fertilizing crops. Thus there is a general tendency for *Ópiqidi* to be positive, implying that $d > 0$ for most loci or for many of the most important loci. This unidirectional dominance of the alleles giving, in homozygous genotypes, rise to higher genotypic values has already been mentioned in Section.

There is an obvious reason to measure both inbreeding depression and heterosis in comparison to the performance of the corresponding population with the Hardy–Weinberg genotypic composition. In a cross-fertilizing crop, such as maize, heterosis is relevant if the outbred plant material performs better than conventional open-pollinating varieties. (Likewise, heterosis of selffertilizing crops is measured by comparing the performance of F1 hybrids to the performance of conventional pure line varieties.)

Measuring heterosis in a cross-fertilizing crop in comparison to the performance of pure lines would not be of practical interest. Superiority of an F1 hybrid over its homozygous parents is called hybrid vigour. In self-fertilizing crops hybrid vigour is less conspicuous than in cross-fertilizing crops and is hardly exploited.

The F2 and later generations may show transgression. This means that the segregating population contains plants with a genotypic value outside the range of the genotypic values expressed by the homozygous parents. If transgression does not occur one may conclude that the population did either not comprise enough plants in relation to the number of segregating loci to give rise to such genotypes, or that the involved parents represented already the genotypes with the extreme genotypic values.

Equation shows that among the segregating loci only loci with *di* _=0 contribute to inbreeding depression or heterosis. Thus only such loci get attention in Section. Furthermore, the equation also shows that these two phenomena are linearly related to F and that they are affected by

- The allele frequencies of the relevant loci
- The number of relevant loci.

The Effect of the Allele Frequencies

For $p = q = 1/2$, which applies to plant material derived from an F1, Equation simplifies to

$$E\underline{G} - E\underline{G}_{RM} = -\frac{1}{2} f \sum_{i=1}^{k} d_i$$

The Effect of the Number of Loci

For a smaller number of segregating loci, *i.e.* a smaller value for parameter *K* in Equation, the inbreeding depression or heterosis will be smaller than for a higher number of segregating loci. It is, indeed, not a good idea to develop a hybrid variety from related pure lines. In self-fertilizing crops fixation of alleles giving rise to homozygous genotypes with high genotypic values is pursued. Thus, for such crops inbreeding depression and heterosis are understandably smaller than for cross-fertilizing crops.

This may also explain why the recently started selection from cross-fertilizing crops for inbred lines that perform well has been rather successful. Due to this, seed representing single-cross hybrids of maize can economically be produced.

With regard to the extreme values for inbreeding depression or heterosis, one should also take into consideration *K*, the number of relevant loci. Equation indicates that the probability that a plant is completely homozygous is _1+*Ft*/ 2 _*K*. This probability is smaller as *K* is larger. In the process of inbreeding it will amount to 0.99 or more, sooner when *K* is small than when *K* is large.

Thus at low values for *K* the maximum inbreeding depression is reached relatively quickly. According to Allard (1960), Jones established the maximum inbreeding depression for plant height in maize as early as in the S5 population; for yield, in contrast, it had not yet occurred by S20. According to Equation EG–EGRM depends linearly on *F*. Crow and Kimura derived that EG – EGRM is a quadratic function of *F* in the occurrence of epistasis. A non-linear relation between the observed inbreeding depression and *F* may thus be due to epistasis.

In Section it was shown that selfing in autotetraploid crops leads to a slow decrease in the frequency of heterozygous plants. Yet a single round of reproduction by means of selfing of a natural cross-fertilizing autotetraploid population yields strong inbreeding depression. Allard reported for alfalfa that the S1 yielded 32% less than the original variety. Busbice and Wilsie attributed the strong inbreeding depression to the strong reduction of the frequency of plants with a tri- or tetra-allelic heterozygous genotype, *i.e.* *BB*β*b* or *BB*β*b*.

In artificially made autotetraploid plant material, *e.g.* rye, the inbreeding depression is less than in natural autotetraploid material. The difference is attributed to the lower frequency of plants with a tri- or tetra-allelic heterozygous genotype in artificial autotetraploid populations, but it might equally be due to the expression of deleterious recessive genes.

Both inbreeding depression and heterosis are due to unidirectional dominance of *B*-alleles, *i.e.* incomplete dominance, complete dominance, or even overdominance. Jinks (1981) concluded that the failure to find examples of 'true' overdominance is general. Thus, if epistatic effects are absent or of minor importance, inbreeding depression and heterosis will mainly occur in the case of dispersion of alleles with (in)complete dominance. This implies that it should be possible to develop pure lines performing as well as F1 hybrids. The phenomenon of pseudo-overdominance may give rise to erroneous conclusions about the genetic control of the considered trait.

Heterosis is exploited by developing varieties containing an excess of heterozygous plants in comparison to their frequency at the Hardy–Weinberg equilibrium. Such excess occurs after bulk crossing. The heterosis of the plant material obtained by the bulk cross is:

$$\frac{1}{2}\sum_{i=1}^{k}(Pl_i - Pl_i)^2 d_i$$

$$\frac{1}{2}(Pl_i - Pl_i)^2$$

Represents the excess of plants with genotype *Bibi* if the difference in the frequency of allele *Bi* between the two parental populations amounts to *p*1*i* – *p*2*i*. Equation implies that heterosis will be large:

- If (*p*1*i*-*p*2*i*)2 is large. A bulk cross involving contrasting pure lines, *i.e.* lines with genotypes *bibi* and *BiBi*, yields the maximum value for (*p*1*i* - *p*2*i*)2, *viz.* 1. The resulting plant material is then heterozygous (and genetically uniform).
- If *K* is large, *i.e.* if the parental populations, preferably pure lines, have a different homozygous single-locus genotype for a high number of loci.
- If the parental populations, preferably pure lines with a different homozygous single-locus genotype for many loci, have homozygous genotypes for alleles differing in such a way that *di* is at its maximum. This should be pursued by trial and error.

According to Note the above conditions describe, in quantitative genetic terms, the requirements for a high specific combining ability.

Heterosis with regard to a complex trait, *i.e.* a trait of which the genetic variation is the result of the variation of a number of component traits, may tentatively be explained on the basis of additive inheritance (absence of dominance) of the components.

The explanation is clarified by considering yield (*Y*) data of some crop, where yield is determined by number of fruits and (average) single fruit weight. When observing each candidate plant with regard to the following traits:

- Number of fruits
- Number of harvested grammes of product, *i.e.* yield (thus: *B* = *Y*

One may, in the following way, calculate phenotypic values of the yield components *X*1 and *X*2:

$X1 = A$: number of fruits per plant of the considered candidate
$X2 = B/A$: single fruit weight

$$Y = A \times \frac{B}{A} = B$$

A specific case which pointed to the importance of components of complex characters, was the unexpected superiority of hybrids between African and Asian oil-palms. The latter were also of African origin but had undergone several generations of selection under totally different climatic conditions.

Under African conditions, the local palms produced a high number of small bunches, whereas the imported Asian palms produced a few very large bunches. The hybrid was intermediate for both number and average weight of the bunches. This resulted in an overall yield far exceeding the mid-parent value.

It has often observed that parents having mutually complementing phenotypic values with regard to yield components, produce a single-cross hybrid with heterosis for yield or other complex characters. Example illustrates this phenomenon for a self-fertilizing and a cross-fertilizing crop. It has become known as recombinative heterosis.

One may speculate with regard to this phenomenon as follows. The yield of a plant may be assumed to be at its maximum if all organs and functions are mutually tuned.

This may occur if the plant has an intermediate phenotypic value for each of a number of yield components, *e.g.* number of stems, number of flowers per stem, number of seeds per flower and seed size. If the intermediate phenotypic values for the components are due to heterozygous single-locus genotypes, it is understandable that plants with a heterozygous complex genotype have a superior value for the complex character.

The idea that a complex trait, *e.g.* grain yield, should be indirectly improved via improvement of its components may lead to an interest in the physiological processes underlying the complex trait. Thus, in addition to plant architectural features, *e.g.* ear size, crop physiological parameters may be used to describe the features of the ideal genotype, the so-called ideotype.

The ideotype for rice is, for instance, characterized by erect leaves, compact and large panicles on a short and firm culm, a vigorous root system and absence of unproductive tillers.

An ideotype may be designed on the basis of estimates of the crop physiological parameters that are relevant to the crop growth model used. These estimates are usually obtained from evaluation of a limited set of genotypes. After having designed an ideotype, crop physiologists simply advise breeders to create it.

In practice there are, however, complications: the majority of the traits that are to be assessed with this approach are hard to measure with the required accuracy. The assessment, for example, of the rate of reallocation of dry matter from stems and leaves to seeds is not feasible in a segregating population with many genotypes, each of which is represented by a single plant or by, at most, a small number of plants. Selection for such traits is thus mostly beyond the breeder's capability.

Furthermore it is assumed when designing an ideotype that parameter values can be combined at will in a single genotype. The possible existence of constraints, *e.g.* lack of genetic variation, and correlations among the parameters, especially correlations due to pleiotropic loci, is ignored. Sparnaaij and Bos (1993) and Bos and Sparnaaij (1993) considered the analysis of complex characters as well as the phenomenon of recombinative heterosis and its prediction.

Equation shows that inbreeding depression is due to a deficit of heterozygous plants in comparison with their Hardy-Weinberg frequency. Random variation of allele frequencies also leads to a decrease in the frequency of heterozygous plants. If *Pnf,*0 designates the probability that fixation with regard to locus *Bi-bi* has not yet occurred in the initial population, *Pnf,t* is expected to be ψ*Pnf,*0, where ψ represents the remaining part of *Pnf,*0.

The initial contribution of locus *Bi-bi* to EG is *(pi–qi)ai+2piqidi*. At fixation of genotype *BiBi,* which occurs with probability *pi,* the contribution is *ai*; at fixation of genotype *bibi,* which occurs with probability *qi,* it is *–ai*. Thus, at fixation, the expected contribution of this locus is *(pi–qi)ai*. Consequently, at fixation due to random variation of allele frequencies its expected contribution to 'inbreeding' depression amounts to *–2piqidi*. The expected depression, due to fixation, is thus equal to the depression occurring in the case of continued inbreeding.

HYBRID VARIETIES

Comparison of a number of the annual Dutch lists of varieties shows both an increase in the total number of varieties for grain and silage maize, and a gradual shift in the most frequently included type of variety. The increase in the total number of varieties reflects the increase in acreage since 1970.

Apparently breeders responded by offering more and more varieties. The main type of variety offered changed simultaneously: from open-pollinating varieties via double-cross hybrids (DC-hybrid) and threeway-cross hybrids (TC-hybrids) to single-cross hybrids (SC-hybrid).

The table shows that, in the past, DC-hybrids were more popular than SChybrids. Because DC-hybrid seed is produced by a vigorous SC-hybrid, it was much cheaper than SC-hybrid seed. (The latter is produced by an inbred line suffering from inbreeding depression). At present, however, relatively high yielding pure lines are available as maternal parent of a SC-hybrid.

Already in 1980 about 80% of the acreage of maize grown in the Corn Belt of the USA consisted of SC-hybrids.

Table: The Number of Varietes of Grain and Silage Maize included in Dutch lists of Recommended Varieties and their Distribution Across Open-Pollinating Varieties (OP), Double-cross(DC), Threeway-cross (TC) and single-cross (SC) Hybrid Varieties.

	Type of variety				
Year	OP	DC	TC	SC	Total
1967	4	4	0	0	8
1977	0	3	6	0	9
1980	0	2	8	0	10
1984	0	1	12	0	13
1988	0	2	14	0	16
1990	0	2	19	0	21
1992	0	2	19	3	24
1994	0	1	26	16	43
1996	0	0	19	17	36
1998	0	0	19	19	38

Two reasons for the present popularity of SC-hybrids are

1. Farmers prefer their greater uniformity

2. Breeders prefer to evaluate the lower number of all conceivable SC-hybrids instead of all conceivable TC- or DC-hybrids

Numbers of conceivable SC-, TC- and DC-hybrids

When having available *N* promising inbred lines, one might produce and test The costs of producing 1 tonne of SC-hybrid maize seeds are not necessarily higher than those required to produce 1 tonne of TC- or DC-hybrid seed, the reasons being:

- Because of mutual isolation of maize fields, grown for maintenance of inbreds or their crossing, the production of TC- or DC-hybrid seed is more demanding than the production of SC-hybrid seed: to produce DC-hybrid seed at least seven isolated fields are required, instead of three when producing SC-hybrid seed.
- For a given successful SC-hybrid the alleles may be reshuffled to produce a new maternal and a new paternal inbred line, such that the new maternal line has a higher seed yield.

Of course, growers will be interested in the performance of G1, *i.e.* the plant material obtained by open pollination in the hybrid variety. If the performance of G1 would be satisfactory, they might decide to grow G1-, G2-, *etc.* material.

In the case of the absence of epistasis a single round of panmictic reproduction will yield plant material (G1) with an expected genotypic value equal to that of any later generation obtained by panmixis, *i.e.* equal to EGRM. Then the reduction in the performance, occurring when growing G1,G2, *etc.* instead of the hybrid, is EGhybrid – EGRM, which is equal to the heterosis as defined by Equation. Example illustrates the reduction occurring when growing plant material obtained by panmictic reproduction of a hybrid. In addition to the reduction in performance, the plant material will show a reduced uniformity.

If the number of SC-hybrid plants is insufficient to produce the desired amount of DC-hybrid seed, one may apply open pollination within both of the SC-hybrids underlying the DC-hybrid. Next the two G1s are crossed. This procedure yields plant material with (approximately) the same genotypic composition as expected when crossing the two SC-hybrids.

The explanation for this is as follows. The population resulting from open pollination of a SChybrid is identical to the population resulting from self-fertilization of the SC-hybrid.

When applying selfing, the haplotype frequencies with regard to unlinked loci do not change. (In the case of linkage the change is insignificant). Thus a single round of panmictic reproduction of each of the two SC-hybrids hardly affects the genotypic composition of the DC-hybrid to be produced.

INCREASING ADOPTION OF PLANT BREEDING

The adoption of molecular plant breeding approaches has occurred at different rates among crop species and institutions engaged in crop improvement, due to the combined influence of scientific, economic, and sociological factors. Important early scientific barriers included the recalcitrance of cereal crop species to Agrobacterium-mediated transformation and lack of knowledge about genetic control of traits already defined as important breeding targets. Continued research and technology development has largely overcome obstacles for plant transformation of nearly all important crop and horticultural species. Similarly, information gained from genomics research in plant species and other organisms has generated a wealth of information about gene structure and function, as well as large numbers of molecular markers for use in plant breeding.

Despite these resources, genetic specificity of robust QTLs remains elusive, unless breeding programs and associated information management systems are restructured to fully integrate knowledge of pedigrees, phenotypes, and marker genotypes that can be leveraged to optimize response to selection. Even with such integration, modifying regulatory functions remains a scientific challenge for molecular breeding because it is difficult to determine the sequence basis for regulatory changes and to predict their phenotypic effects. Once enabling technologies in biotechnology and genomics become available, economic factors often dictate the degree to which these innovations are integrated into existing plant breeding programs.

The expense of gaining governmental regulatory approval for commercial release of transgenic varieties is a significant economic barrier. The costs associated with the development, establishment, and operation of molecular plant breeding are greater than conventional plant breeding practices, requiring significant investments in new research infrastructure and intellectual capacity.

Such resources initially existed only in private agribusiness firms and a handful of larger public institutions, further accelerating an ongoing trend for increased industrialization of plant breeding programs among major crops such as corn, soybeans, cotton (Gossypium hirsutum), and wheat. Where there has been adoption in companies, the balance may favour biotechnology over QTLs for improving complex traits, despite greater product development costs, because transgenics can be designed to produce stronger, even dramatic, phenotypic effects, and can often be more rapidly deployed across a broader range of germplasm, resulting in new solutions sooner.

Though molecular breeding is now considered an essential component of current crop improvement efforts for major crops by large companies, the broad applicability of modern molecular approaches to conventional plant breeding remains a source of debate among some practicing plant breeders in the public sector, particularly for minor crops. In addition to the valid scientific and economic factors that have delayed or prevented adoption of molecular approaches in meeting some plant breeding objectives, there are at least three additional reasons contributing to this view. First, molecular plant breeding requires training and expertise in both molecular biology and plant breeding. Educational efforts that delivered such interdisciplinary training were initially established in the early 1990s, but still remain limited to a relatively small group of academic institutions with historic strengths in plant breeding.

A second reason for reduced enthusiasm to embrace biotechnology among some plant breeders is the problems with acceptance of transgenic crops among certain governments and groups of consumers, as exemplified by the shelving of wheat varieties with transgenes for resistance to glyphosate herbicides.

Finally, excitement about the potential of molecular plant breeding also stimulated shifts in funding at public institutions to enhance intellectual capacity and infrastructure for molecular genetics and genomics research, which ironically often occurred at the expense of conventional plant breeding. This emphasis may have been temporarily necessary to establish the foundations for 21st century plant biology, but there is currently a growing recognition that increased investment in plant breeding capacity and translational research linking molecular methods with breeding objectives is necessary to fully realise the potential of recent advances in biotechnology and genomics.

BREEDING SCHEMES AND THE GENETIC GAIN CONCEPT

Conceptually, plant breeding is simple: cross the best parents, and identify and recover progeny that outperform the parents. In practice, plant breeding is a three step process, wherein populations or germplasm collections with useful genetic variation are created or assembled, individuals with superior phenotypes are identified, and improved cultivars are developed from selected individuals. A wide diversity of approaches, tailored to the crop species and breeding objectives, have been developed for improving cultivars. These breeding methods feature different types of populations, selection procedures, and outcomes. Figure summarizes the three breeding methods that are commonly employed in crop improvement programs.

As mentioned previously, when the goal is to upgrade an established elite genotype with trait(s) controlled by one or a few loci, backcrossing is used either to introgress a single gene or to pyramid a few genes. For genetically complex traits, germplasm improvement instead requires reshuffling of the genome to produce new favourable gene combinations in the progeny. The pedigree breeding method produces such novelty via crossing and recombination among superior, yet complementary, parents and selection among segregating progeny for improved performance. Recurrent selection aims to simultaneously increase the frequencies of favourable alleles at multiple loci in breeding populations through intermating of selected individuals. For hybridized crops such as maize, recurrent selection may be extended to improve the performance of distinct complementary populations (*e.g.* heterotic groups) that are used as parents to form superior hybrid combinations. This practice is referred to as reciprocal recurrent selection.

Quantitative genetic principles have been particularly powerful as the theoretical basis for both population improvement and methods of selecting and stabilizing desirable genotypes. An important concept in quantitative genetics and plant breeding is genetic gain (ΔG), which is the predicted change in the mean value of a trait within a population that occurs with selection. Regardless of species, the trait of interest, or the breeding methods employed, ?G serves as a simple universal expression for expected genetic improvement.

Figure shows the genetic gain equation and an expansion of its terms to fundamental parameters of quantitative genetics. Though clearly an oversimplification of the advanced quantitative genetic principles employed in plant breeding, the genetic gain equation effectively relates the four core factors that influence breeding progress: the degree of phenotypic variation present in the population (represented by its SD, sP), the probability that a trait phenotype will be transmitted from parent to offspring (heritability, h2), the proportion of the population selected as parents for the next generation (selection intensity, i, expressed in units of SD from the mean), and the length of time necessary to complete a cycle of selection (L). L is not only a function of how many generations are required to complete a selection cycle, but also how quickly the generations can be completed and how many generations can be completed per year.

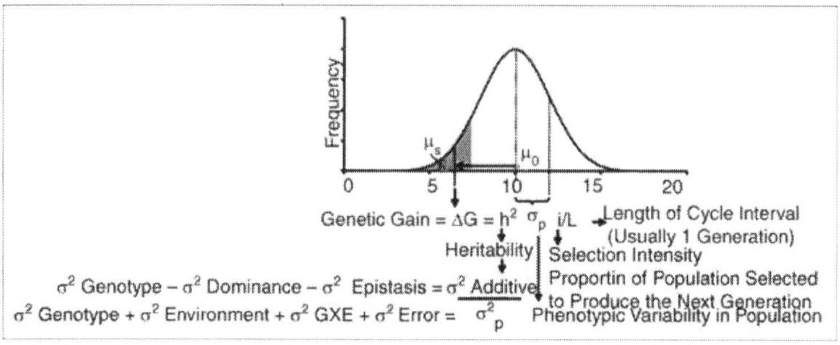

Fig. The Genetic Gain Equation and its Component Variables.

GENETIC DIVERSITY FOR CROP IMPROVEMENT

The maximum potential for genetic gain is proportional to the phenotypic variation (σ_P) present in the original source population and maintained in subsequent cycles of selection. Phenotypic variation is positively associated with genetic diversity, yet also depends on environmental factors and the interactions between genotype and environment. Genetic diversity may be derived from breeding populations (either naturally occurring or synthetic), segregating progeny from a cross of selected parental lines, exotic materials that are not adapted to the target environment, wide interspecific crosses, naturally occurring or induced mutations, the introduction of transgenic events, or combinations of these sources.

However, not all phenotypic variation is equal. For example, the use of exotic germplasm has been extremely successful for improving many crop species, but difficulties may be encountered through the introduction of undesirable alleles associated with lack of adaptation. The need for genetic diversity must be balanced by elite performance, because choosing the best parents is key to maximizing the probability for successful improvement. In contrast, the expected increase in linkage disequilibrium among elite populations derived from intense prior selection may also limit the creation of new genetic combinations for future gain. Intermating source populations for genetic recombination may overcome this problem, but delays cultivar development.

Molecular markers and more recently, high-throughput genome sequencing efforts, have dramatically increased knowledge of and ability to characterize genetic diversity in the germplasm pool for essentially any crop species.

Using maize as one example, surveys of molecular marker alleles and nucleotide sequence variation have provided basic information about genetic diversity before and after domestication from its wild ancestor teosinte, among geographically distributed landraces, and within historically elite germplasm. This information enriches investigations of plant evolution and comparative genomics, contributes to our understanding of population structure, provides empirical measures of genetic responses to selection, and also serves to identify and maintain reservoirs of genetic variability for future mining of beneficial alleles. In addition, knowledge of genetic relationships among germplasm sources may guide choice of parents for production of hybrids or improved populations.

While molecular markers and other genomic applications have been highly successful in characterizing existing genetic variation within species, plant biotechnology generates new genetic diversity that often extends beyond species boundaries. Biotechnology enables access to genes heretofore not available through crossing and creates an essentially infinite pool of novel genetic variation. Genes may be acquired from existing genomes spanning all kingdoms of life, or designed and assembled de novo in the laboratory. Both subtle and extreme examples of the power of transgenes to introduce novel phenotypic variation can be found in the three different transgenes developed for resistance to glyphosate herbicides in maize and other crops.

The first glyphosate-tolerant maize hybrids used a modified version of the endogenous maize gene encoding 5-enol-pyruvylshikimate-3-P synthase, which was followed later by events produced with a 5-enol-pyruvylshikimate-3-P synthase gene isolated from Agrobacterium. More recently, a synthetic gene with enhanced glyphosate acetyltransferase activity was created via gene shuffling and selection in a microbial system. Each of these glyphosate-tolerant maize events also illustrates another benefit of biotechnology, where new combinations of regulatory sequences (*e.g.* the cauliflower mosaic virus 35S and rice actin1promoters) may be used to achieve optimal trait expression with respect to overall activity and tissue distribution relative to what might be possible with endogenous genes.

FAVOURABLE GENE ACTION

Quantitative genetics uses the theoretical concept of heritability to quantify the proportion of phenotypic variation that is controlled by genotype. In practice, heritability is greatly influenced by the genetic architecture of the trait of interest, which is described by the number of genes, the magnitude of their effects, and the type of gene action associated with phenotypes. Better knowledge of genetic architecture and favourable gene action (that which is more amenable to selection) often has the greatest impact on improving genetic gain. For the genetic gain formula, heritability (h^2) is used in its narrow sense, representing the proportion of phenotypic variation due to additive genetic effects (those that reflect changes in allele dosage or allelic substitutions). Additive genetic effects are also referred to as the breeding value because they are predictably transmitted to progeny. Deviations from additive effects are significant for many traits, and are partitioned into either dominance effects that reflect the interactions between different alleles at the same locus or epistatic effects resulting from interactions among different loci. Gene action and breeding values are characterized by progeny testing, where the phenotypes of individuals in a population are compared to their parents and siblings produced from either self-pollination or outcrossing.

Previous efforts to develop large numbers of molecular markers, high density genetic maps, and appropriately structured mapping populations have now made routine for many crop species the ability to simultaneously define gene action and breeding value at hundreds and often thousands of loci distributed relatively uniformly across entire genomes. The results from such mapping studies provide greatly improved estimates for the number of loci, allelic effects, and gene action controlling traits of interest. More importantly, genomic segments can be readily identified that show statistically significant associations with quantitative traits (quantitative trait loci [QTLs]).

In addition to genetic mapping in families derived from biparental crosses, new advances in association genetics with candidate genes and approaches that combine linkage disequilibrium analysis in families and populations further enhance power for QTL discovery.

Information about QTLs can be used in a number of ways to increase heritability and favourable gene action. For traits exhibiting low to moderate heritability, such as grain yield, QTLs, and their associated molecular markers often account for a greater proportion of the additive genetic effects than the phenotype alone. Furthermore, knowledge of genetic architecture can be exploited to add or delete specific alleles that contribute to breeding value. When either genetic linkage or epistasis among loci with antagonistic effects on a trait limits genetic gain, QTL information can be used to break these undesirable allelic relationships.

Success in using information about QTLs to increase genetic gain depends greatly on the magnitude of QTL effects, precise estimation of QTL positions, stability of QTL effects across multiple environments, and whether QTLs are robust across relevant breeding germplasm. Prediction of QTL positions is enhanced by further fine mapping, which facilitates testing QTL effects and breeding values in additional populations. When the density of observed recombinations approaches the resolution of single genes, the causal genetic change for a QTL can be determined (for review, see Salvi and Tuberosa, 2005; Yu and Buckler, 2006; Beló *et al.*, 2008; Harjes *et al.*, 2008). Molecular isolation of QTLs permits the development of perfect or functional molecular markers at the potential resolution of the fundamental unit of inheritance, the nucleotide, and dramatically increases the specificity and precision by which genetic effects are estimated and manipulated in breeding programs.

The use of transgenes can further simplify the genetic architecture for desirable traits, in ways that may be superior to or not possible even when perfect markers are available for robust QTLs of large effect. Transgenes typically condition strong genetic effects at operationally single loci, which also exhibit dominant gene action where only one copy of the event is needed for maximal trait expression in a hybrid cultivar. These features of transgenes can reduce complex quantitative improvement to a straightforward, often dramatic, solution. Excellent examples are provided by the expression in transgenic corn hybrids of insecticidal toxin proteins from Bacillus thuringiensis (Bt) to reduce feeding damage by larvae of the European corn borer (Ostrinia nubilalis) or the corn rootworm beetle. Partial resistance in maize germplasm to these insect pests had been previously characterized as quantitatively inherited traits with low heritability, but the Bt transgenic events offer a simply inherited alternative that is efficiently manipulated in breeding programs.

By simplifying genetic architecture, transgenes may also permit disruption of allelic interactions between factors controlling the trait of interest and other important performance characteristics. For example, employing a transgenic source of insect resistance (*e.g.* a single locus Bt transgene) may facilitate selection for favourable alleles for yield improvement that are tightly linked in repulsion with endogenous genes for resistance to the same class of insect pests. In addition, transgenic events may be engineered to uncouple negative pleiotropic effects from beneficial phenotypes conditioned by recessive mutations. This application is illustrated by the use of RNA interference to specifically down-regulate zein seed storage protein gene expression. This strategy mimics the effects of the opaque2 mutation on improving the amino acid profile of maize grain for animal feed, while circumventing the softer endosperm texture and susceptibility to fungal pathogens typically associated with opaque2.

Transgenic events can also be designed to intervene at key regulatory steps for entire metabolic or developmental pathways, such that gene action for the corresponding traits are largely inherited as single dominant factors that are less sensitive to environmental effects. Examples include the expression of a transcription factor that increases drought tolerance, and altering the balance between levels of the GLOSSY15 transcription factor relative to its repressor, microRNA172, to delay flowering time in maize hybrids.

Biotechnology also facilitates the molecular stacking of transgenes that control a trait or suite of traits into a single locus haplotype defined by a transgenic event. Examples of such an approach include the initial Golden Rice, recently released YieldGuard VT Triple transgenic maize hybrids where herbicide tolerance and multiple insect resistance traits are integrated as one genomic locus, or the combination of transgenes that simultaneously increase synthesis and decrease catabolism of Lys in maize seeds.

Recent reports of improvements in gene targeting technology and the construction of meiotically transmissible plant minichromosomes pave the way for introducing more traits with increasing complexity. With such advances, biotechnology is now poised to assemble useful genetic diversity from essentially any source into constructs that concentrate favourable gene action and maximize heritability for a greatly expanded set of traits.

The molecular cloning of QTLs has yielded novel insights about the biology of quantitative traits that were not likely to be discovered from the analysis of gene knockouts or overexpression strategies, in particular the impacts of regulatory variation on phenotypic variation and evolution. Furthermore, molecular markers, genomics, and biotechnology are now applied in an iterative network to exploit genetic diversity for crop improvement.

EFFICIENCY OF SELECTION

Conventional plant breeding that relies only on phenotypic selection has been historically effective. However, for some traits, phenotypic selection has made little progress due to challenges in measuring phenotypes or identifying individuals with the highest breeding value. The effects of environment, genotype by environment interaction, and measurement errors also contribute to observed differences. Evaluation of genotypes in multiple environments with replicated designs allows better estimation of breeding values, but requires additional time and expense. For some traits, it may be necessary to sacrifice the individual to measure phenotypes, or trait expression may depend on variable environmental conditions (*e.g.* disease pressure) and the stage of development (*e.g.* grain quality can only be assessed after flowering). Furthermore, plant breeders typically must simultaneously improve a suite of commercially valuable traits, which may limit gains from selection. Just as molecular plant breeding helps to expand genetic diversity, characterize genetic architecture, and modify gene action, its methods can also be applied to increasing the efficiency of selection.

An extensive body of literature has considered the utility of molecular marker-assisted selection and its fit with different breeding methods, with the reader being referred to a number of recent excellent reviews on this topic. Molecular marker genotypes that are either within genes or tightly linked to QTL influencing traits under selection can be employed as a supplement to phenotypic observations in a selection index. In cases where genetic correlations are high, further efficiencies can be gained by substituting genotypic for phenotypic selection during some selection cycles, which can reduce phenotyping efforts and cycle times by permitting the use of off-season nurseries.

Johnson summarized an early example of combining phenotypic data and molecular marker scores to increase selection gains for maize grain yield and resistance to European corn borer. An effective strategy to simultaneously modify multiple traits is the use of selection indices that consider multiple factors in choosing the final improved genotype. Eathington *et al.* recently reported on results obtained from the use of multiple trait indices and marker-assisted selection for nearly 250 unique corn breeding populations. Use of molecular markers increased breeding efficiency approximately 2-fold relative to phenotypic selection alone, with similar gains also observed in soybean (Glycine max) and sunflower (Helianthus annuus) populations.

Marker-assisted selection can also significantly enhance genetic gain for traits where the phenotype is difficult to evaluate because of its expense or its dependence on specific environmental conditions. Molecular markers may be used to increase the probability of identifying truly superior genotypes, by focusing testing resources on genotypes with the greatest potential (*i.e.* early elimination of inferior genotypes), by decreasing the number of progeny that must be screened to recover a given level of gain, and by enabling simultaneous improvement for traits that are negatively correlated. Successful examples include resistance to soybean cyst nematode, resistance to cereal diseases and drought tolerance in maize.

The efficiency of phenotypic selection for some complex traits can be enhanced by including physiological or biochemical phenotypes as secondary traits, if these exhibit strong genetic correlations with the target trait and possess high heritability.

Recent advances in functional genomics permit the population-scale profiling of RNA abundance, protein levels and activities, and metabolites that are associated with important traits. In addition to molecular markers that tag DNA sequence variation, such genetical genomics approaches may provide additional secondary phenotypes as selection targets, particularly for traits defined by responses to environmental, developmental, or physiological cues. Marker-assisted selection also accelerates the deployment of transgenes in commercial cultivars.

Typically, this has been achieved through marker-assisted backcrossing. However, for future biotechnology improvements such as tolerance to drought or nutrient limitation, forward breeding may be required to cooptimize transgene expression and genetic background because endogenous genes and environmental factors may have the potential to influence the phenotypes resulting from transgenic modifications. Of course, use of molecular markers could aid forward breeding efforts as well. Alternatively, discovery efforts for additional genes or QTLs that are necessary for dependable trait performance may suggest design of new transgene constructs that stack primary transgenes with known genetic modifiers into second-generation transgenic events.

Index